WILDLIFE MANAGEMENT AND LANDSCAPES

Wildlife Management and Conservation

Paul R. Krausman, Series Editor

Wildlife Management and Landscapes

Principles and Applications

EDITED BY

William F. Porter
Chad J. Parent
Rosemary A. Stewart
David M. Williams

Published in Association with *THE WILDLIFE SOCIETY*

 JOHNS HOPKINS UNIVERSITY PRESS | BALTIMORE

In memory of William "Bill" F. Porter, 1951–2020
For his dedication and love for his family, friends, students, and the wildlife profession.
A true giant in wildlife management and conservation.

———————————

© 2021 Johns Hopkins University Press
All rights reserved. Published 2021
Printed in the United States of America on acid-free paper
9 8 7 6 5 4 3 2 1

Johns Hopkins University Press
2715 North Charles Street
Baltimore, Maryland 21218-4363
www.press.jhu.edu

Library of Congress Cataloging-in-Publication Data

Names: Porter, William F., editor. | Parent, Chad J., editor. |
    Stewart, Rosemary A., 1976– editor. | Williams, David M.,
    1974– editor.
Title: Wildlife management and landscapes : principles and
    applications / edited by William F. Porter, Chad J. Parent,
    Rosemary A. Stewart, and David M. Williams ; published
    in association with The Wildlife Society.
Description: Baltimore : Johns Hopkins University Press,
    2021. | Series: Wildlife management and conservation |
    Includes bibliographical references and index.
Identifiers: LCCN 2020018589 | ISBN 9781421440194
    (hardcover) | ISBN 9781421440200 (ebook)
Subjects: LCSH: Landscape ecology. | Wildlife management.
    | Habitat conservation.
Classification: LCC QH541.15.L35 W55 2021 | DDC 577.5/
    5—dc23
LC record available at https://lccn.loc.gov/2020018589

A catalog record for this book is available from the British
    Library.

Special discounts are available for bulk purchases of this book.
For more information, please contact Special Sales at special
sales@press.jhu.edu.

Johns Hopkins University Press uses environmentally
friendly book materials, including recycled text paper that
is composed of at least 30 percent post-consumer waste,
whenever possible.

Contents

Contributors

Mohammed A. Al-Saffar
Upper Mississippi / Great Lakes Joint
    Venture
US Fish and Wildlife Service
East Lansing, Michigan, USA

Jocelyn L. Aycrigg
Department of Fish and Wildlife
    Sciences
University of Idaho
Moscow, Idaho, USA

Guillaume Bastille-Rousseau
School of Biological Sciences
Southern Illinois University
Carbondale, Illinois, USA

Jon P. Beckmann
Rocky Mountain West Program
Wildlife Conservation Society
Bozeman, Montana, USA

Joseph R. Bennett
Department of Biology
Carleton University
Ottawa, Ontario, Canada

William M. Block
Rocky Mountain Research Station
US Forest Service
Flagstaff, Arizona, USA

Todd R. Bogenschutz
Wildlife Bureau
Iowa Department of Natural Resources
Boone, Iowa, USA

Teresa C. Cohn
Department of Natural Resources and
    Society
University of Idaho, McCall Field
    Campus
McCall, Idaho, USA

John W. Connelly
Department of Fish and Wildlife
    Resources
University of Idaho
Moscow, Idaho, USA

Courtney J. Conway
Idaho Cooperative Fish and Wildlife
    Research Unit
US Geological Survey
University of Idaho
Moscow, Idaho, USA

Bridgett E. Costanzo
Natural Resources Conservation
    Service
US Department of Agriculture
Richmond, Virginia, USA

David D. Diamond
Missouri Resource Assessment
    Partnership
University of Missouri
Columbia, Missouri, USA

Karl A. Didier
Wildlife Conservation Society
Bronx, New York, USA

Lee F. Elliott
Missouri Resource Assessment
    Partnership
University of Missouri
Columbia, Missouri, USA

Michael E. Estey
Habitat and Population Evaluation
    Team
US Fish and Wildlife Service
Fergus Falls, Minnesota, USA

Lenore Fahrig
Department of Biology
Carleton University
Ottawa, Ontario, Canada

Cameron J. Fiss
Department of Biology
Indiana University of Pennsylvania
Indiana, Pennsylvania, USA

Jacqueline L. Frair
College of Environmental Science and
    Forestry
State University of New York
Syracuse, New York, USA

Elsa M. Haubold
Landscape Conservation Cooperative
    Network
US Fish and Wildlife Service
Falls Church, Virginia, USA

Fidel Hernández
Department of Rangeland and Wildlife
    Sciences
Caesar Kleberg Wildlife Research
    Institute
Texas A&M University–Kingsville
Kingsville, Texas, USA

Jodi A. Hilty
Yellowstone to Yukon Conservation
    Initiative
Canmore, Alberta, Canada

Joseph D. Holbrook
Haub School of Environment and
    Natural Resources
University of Wyoming
Laramie, Wyoming, USA

Cynthia A. Jacobson
Innovative Outcomes
Carbondale, Colorado, USA

Kevin M. Johnson
Landscape Conservation Cooperative
    Network
US Fish and Wildlife Service
Lakewood, Colorado, USA

Jeffrey K. Keller
Habitat by Design, LLC
Pipersville, Pennsylvania, USA

Jeffery L. Larkin
Department of Biology
Indiana University of Pennsylvania
Indiana, Pennsylvania, USA

Kimberly A. Lisgo
BEACONs Project
University of Alberta / Yukon College
Whitehorse, Yukon, Canada

Casey A. Lott
Department of Biology
Indiana University of Pennsylvania
Indiana, Pennsylvania, USA

Amanda E. Martin
Environment and Climate Change
    Canada
Ottawa, Ontario, Canada

James A. Martin
Warnell School of Forestry and Natural
    Resources
University of Georgia
Athens, Georgia, USA

Darin J. McNeil
Department of Natural Resources
Cornell University
Ithaca, New York, USA

Michael L. Morrison
Department of Rangeland, Wildlife and
    Fisheries Management
Texas A&M University
College Station, Texas, USA

Betsy E. Neely
The Nature Conservancy (retired)
Boulder, Colorado, USA

Neal D. Niemuth
Habitat and Population Evaluation
    Team
US Fish and Wildlife Service
Bismarck, North Dakota, USA

Chad J. Parent
Wildlife Division
North Dakota Game and Fish
Bismarck, North Dakota, USA

Humberto L. Perotto-Baldivieso
Department of Rangeland and Wildlife
    Sciences
Caesar Kleberg Wildlife Research
    Institute
Texas A&M University–Kingsville
Kingsville, Texas, USA

Ronald D. Pritchert
Habitat and Population Evaluation
    Team
US Fish and Wildlife Service
Bismarck, North Dakota, USA

Fiona K. A. Schmiegelow
University of Alberta / Yukon College
Whitehorse, Yukon, Canada

Amanda L. Sesser
21sustainability
New Orleans, Louisiana, USA

Gregory J. Soulliere
Upper Mississippi / Great Lakes Joint
    Venture
US Fish and Wildlife Service
East Lansing, Michigan, USA

Leona K. Svancara
Idaho Department of Fish and Game
Moscow, Idaho, USA

Stephen C. Torbit
US Fish and Wildlife Service (retired)
Golden, Colorado, USA

Joseph A. Veech
Department of Biology
Texas State University
San Marcos, Texas, USA

Kerri T. Vierling
Department of Fish and Wildlife
    Sciences
University of Idaho
Moscow, Idaho, USA

Greg Wathen
Tennessee Wildlife Resources Agency
Nashville, Tennessee, USA

David M. Williams
Department of Fisheries and Wildlife
Boone and Crockett Quantitative
    Wildlife Center
Michigan State University
East Lansing, Michigan, USA

Mark J. Witecha
Bureau of Wildlife Management
Wisconsin Department of Natural
    Resources
Madison, Wisconsin, USA

John M. Yeiser
Warnell School of Forestry and Natural
    Resources
University of Georgia
Athens, Georgia, USA

Foreword

As a graduate student, I distinctly remember a faded bumper sticker slapped on a university freezer that read, "Habitat Is the Key to Wildlife Conservation." The idea of habitat as a key component of wildlife management was a central theme in my university training, and that sticker made sense; each species and population requires certain habitats to be sustained. My academic upbringing also coincided with important habitat conservation milestones. The Northwest Forest Plan responded to the value of old-growth forests and the viability of northern spotted owl (*Strix occidentalis caurina*) populations. Important advancements were made in what we learned about causes and implications of spatial patterns in landscapes on wildlife populations.

While many early habitat studies and management strategies focused on local phenomena, the field of landscape ecology emerged as a dominant field of study. It became apparent that landscape-level patterns and processes were vitally important in understanding wildlife–habitat relationships and sustaining wildlife populations. Despite what I was reading and hearing on these topics decades ago, I sensed a slight disconnect between landscape theory and wildlife management. Fast-forward 25 years, and we have come a long way with integrating landscape ecology and wildlife management. Although we have had successes, there is a continued need to connect landscape ecologists and wildlife managers so that they speak the same language and understand one another's perspectives, methods, and limitations. This book fulfills an important niche by promoting communication between wildlife managers and landscape ecologists in practical terms that will help both think big and embrace landscape-level management strategies.

Habitat is a central component of landscape ecology and wildlife management; it is the bedrock on which both fields are based. Aldo Leopold's *A Sand County Almanac* expresses the virtues of thinking about landscapes, and his book *Game Management* describes composition and interspersion of resources on the land as important in affecting animal abundance. Although habitat is a common theme, there are several impediments that limit a full merger of landscape ecology and wildlife management. Part I of this book outlines these issues while making the case that a landscape perspective is necessary to address management questions. Part I also provides important definitions of habitat, associated terms, and misuse of terms surrounding habitat and niche. Within these pages it becomes evident that landscape ecologists need land managers to implement their ideas if they hope to positively influence conservation. Conversely, land managers need landscape ecologists to develop relevant concepts and tools that help them influence factors they care about, such as species distribution, abundance, harvest, or diversity. Authors in Part I expand on these ideas and provide an important historical basis, and they review many of these overlapping concepts and applications such as island biogeography theory, habitat fragmentation, disturbance ecology, and niche theory. Readers will appreciate the historical context in addition to the up-to-date thoughts on important concepts.

In a 1939 issue of the *Journal of Wildlife Management*, Leopold wrote (3:158) that "The basic skill of the wildlife manager is to diagnose the landscape, to discern and predict trends in biotic community, and to modify them where necessary in the interest in conservation." He goes on to add that knowing the interrelationships of the component parts is necessary. Through tools and fundamental concepts, landscape ecologists help fill this role in diagnosing the landscape. Developments in methodology and technology—such as geographic information systems, satellites, computing advances, and analytical tools—have improved our capacity to assess temporal-spatial events at appropriate scales. As we learn in Part II of this book, these advances help diagnose the landscape and develop solutions for habitat management, yet studying important habitat-wildlife relationships relevant to management is still inherently difficult. Part II goes on to help us identify these challenges and diagnose solutions. The dynamic and heterogeneous structure of natural systems complicates the effectiveness of management strategies, and factors such as soil, hydrology, fire, grazing, and timber harvest, which are only sometimes controllable, lead to uncertainty in management decisions. The discussion of uncertainty in this book is important, as we never completely observe or measure true states, environmental stochasticity results in normal variation in the environment, there is structural uncertainty in models that inhibit our ability to determine how a system will respond to a management action, and we only have partial controllability of management actions.

Part III of this book should be read by all landscape ecologists who have an interest in applying their work to on-the-ground activities. And all wildlife managers will benefit from reminders about how they can be effective collaborators and use decision-support tools. In Part III, there are important discussions of matching scale between population processes and management, limitations to management across jurisdictional boundaries, the sometimes competing management objectives of private landowners and management agencies, and land management mandates. Tangible and applicable examples ranging from ideas for disseminating research to encouraging personal interactions seek to ensure that communication from early stages are addressed. Such communication will lead to co-produced science and management recommendations that are most likely to be supported, applied, and successful. Part III also identifies the use of decision-support tools, how they are implemented, and their limitations. Because private lands are an important provider of wildlife habitat and the role of private lands in conservation will only increase, I was pleased to see an overview of the Farm Bill and other incentive and partnership programs.

The chapters in Part IV translate concepts in landscape ecology to management and highlight several current approaches to landscape management; namely, the joint ventures, Landscape Conservation Cooperatives, private conservation initiatives, and work by various nongovernmental organizations. I can think of no better demonstration of applying the principles of landscape ecology in wildlife management and planning than these examples. Such programs are built around the notion that effective conservation for many species requires an integrative approach. Part IV is where the rubber meets the road because important principles in conservation design and explicit consideration of humans as part of the landscape are included. Success in such endeavors requires that multiple entities—such as private landowners, researchers, and agency personnel—trust one another and are willing to coordinate actions. Effective strategies and important principles are discussed that encourage integration of social science information and an adaptive co-governance framework.

Most forewords comment on the future, so I feel obligated to do so in a general way. The proliferation of technology will continue to have an important role in the future of landscape ecology and wildlife management. Such trends are predictable, and we will witness an ever increasing suite of analytical options with data at much finer spatiotemporal scales. While I readily embrace these trends, we should not become overly reliant on technology and work only in virtual worlds. Landscape ecologists and wildlife managers need to assess which tools are best suited to address questions, hopefully through addressing well-defined hypotheses. We need to consider how to use these techniques most effectively rather than assuming technology will solve our problems and automatically lead us to greater insight. To be successful in large-landscape programs, we will need to find win-win solutions that consider wildlife along with economics, jobs, and other factors that society values. Landscape ecologists and wildlife managers will increasingly need one another to address the multifaceted challenges we face, such as climate change. They will need to co-produce science and management recommendations. Collaboration across agency, cultural, and political lines will be necessary for large-landscape management. Last, other methods of communication will be necessary. The field of human-centered design could help us ensure we are delivering information to the end user in the most effective way, particularly when communicating with private landowners and other constituents. This book sets the stage for these future endeavors.

The blending of theory and management in this book helps create a common language across disciplines while breaking down unnecessary barriers and promoting effective diagnosis and stewardship of landscapes and wildlife populations. This book will help unleash and leverage the important theory, technology, and tools at our disposal. Habitat will continue to be a key component of conservation in the future, and this book will serve as a key for unlocking solutions to the most challenging problems of our time.

Joshua J. Millspaugh
Boone and Crockett Professor of Wildlife
Conservation
University of Montana

Preface

The question at the heart of this book is, What does landscape ecology have to offer wildlife management? For nearly a century, wildlife management has been profoundly successful by focusing on direct manipulation of habitat at the scale of the small landowner or public landholding. The geographic extent of these lands is managed as hundreds to thousands of hectares. As wildlife management has become more common on national parks and industrial forestlands, the extent over which decisions affect the landscape has increased to millions of hectares.

As most practicing wildlife managers will attest, the principles of management learned in the 20th century fit a large portion of their day-to-day planning and decision making. Yet 21st-century wildlife conservation presents challenges that extend beyond those principles. Adapting to these challenges requires a shift in the paradigm of management. The habitat characteristics that managers use as the focus of management on geographic scales of millions of hectares and the tools they have to affect habitat on these landscapes are different. The habitat characteristics of habitat suitability models such as stem densities, distance to edge, and mast production are too detailed for decisions at the large landscape scale. Increasingly, managers must make decisions with information such as configuration of course-grain habitat classes and corridors for seasonal migration, and land ownership patterns that

influence the coupling of human-ecological communities.

We argue that the tenets and tools of landscape ecology will be instrumental in bridging the gap to that new paradigm. Landscape ecologists would benefit from understanding the constraints that managers face and therefore how to translate their work into more practical applications. Wildlife managers would benefit from understanding the vocabulary and conceptual processes of landscape ecologists and gain comfort with landscape perspectives.

Thus this book addresses a vital need for an authoritative resource that bridges a gap between wildlife managers and landscape ecologists. This gap is predicated on differences in the long-held principles of habitat and population management that are core to wildlife agencies, and rapidly evolving concepts of landscape ecology that are central to many nongovernmental wildlife conservation organizations and university ecology programs. Today, perhaps more than ever before, it is important to bridge wildlife management and landscape ecology. Significant shifts in the spatial scale of extractive, agricultural, ranching, and urban land uses are upon us, and these shifts are having profound effects on the habitats that support wildlife.

Our objective in creating this book is to provide an applied resource that explains the concepts of landscape ecology and wildlife conservation and demonstrates their application. Part I lays a foundation for the innate similarity between wildlife management and landscape ecology, noting that habitat represents a common thread that is woven throughout both of these disciplines. Part II addresses the wildlife manager, building a case for embracing landscape ecology as a means to effectively manage the resource. Part III addresses the landscape ecologist, shining a light on the challenges that must be overcome to make their work meaningful to wildlife managers. Finally, Part IV offers concrete examples of the application of landscape ecology to management in terms that specifically promote clear communication and mutual understanding. We developed the book for students studying the important relationships between wildlife management and landscape ecology, and for practicing wildlife biologists, managers, and landscape ecologists that must work together to maintain and enhance the habitats of wildlife.

Acknowledgments

Our deepest appreciation to the many colleagues, friends, and family whose patience and encouragement helped get us through the editing process. We are especially grateful to the Boone and Crockett Program at Michigan State University and the Department of Fisheries and Wildlife at Michigan State University for their in-kind and financial support of this endeavor.

We owe a significant debt of gratitude to Paul Krausman, emeritus professor at the University of Arizona and editor of The Wildlife Society's Wildlife Conservation and Management book series. Paul supported our vision for this project from the beginning and was instrumental throughout the editing process. Vincent J. Burke, executive editor, and Tiffany Gasbarrini, senior science editor, at Johns Hopkins University Press (JHUP) provided valuable guidance while we assembled contributors for this volume. We also thank them for their counsel. Esther Rodriguez, editorial assistant at JHUP, coordinated manuscript submission and fielded many questions from us during the finalization process. Julie McCarthy and Andre Barnett of JHUP coordinated manuscript editing. We are grateful to the JHUP design team for optimizing figures in several chapters. Our copy editor, Ashleigh McKown, greatly improved the quality of this book with her meticulous attention to detail. Kathy Patterson provided indexing for the book.

We supplied the lead author for each chapter with comments from one or more anonymous reviewers. This book improved substantially through comments made by these reviewers, and we are grateful for their time. We also appreciate the comments from two anonymous reviewers of the book in its entirety. We list each chapter reviewer below.

CHAPTER REVIEWERS

Steve Backs, Indiana Department of Natural Resources

Leonard Brennan, Texas A&M University–Kingsville

Mary Jo Casalena, Pennsylvania Game Commission

Joanne Crawford, Michigan State University (current: Minnesota Department of Natural Resources)

Steve DeMaso, US Fish and Wildlife Service

John Edwards, Texas A&M University–Kingsville (current: Cross Timbers Consulting, LLC)

Kristine Evans, Mississippi State University

T. J. Fontaine, University of Nebraska–Lincoln

Timothy Fulbright, Texas A&M University–Kingsville

Jim Giocomo, Oaks and Prairies Joint Venture

Michale Glennon, Wildlife Conservation Society (current: Adirondack Watershed Institute at Paul Smith's College)

Brett Goodwin, University of North Dakota (current: Fleming College–Frost Campus)

Fidel Hernández, Texas A&M University–Kingsville

Christopher Hoving, Michigan Department of
Natural Resources

Jason Isabelle, Missouri Department of
Conservation

Chuck Kowaleski, Texas Parks and Wildlife
Department (deceased)

Thomas Loveland, US Geological Survey

Stacy McNulty, State University of New York,
College of Environmental Science and Forestry

Michael Mitchell, Ducks Unlimited

Christine Ribic, University of Wisconsin–Madison

Brent Rudolph, Michigan Department of Natural
Resources (current: Ruffed Grouse Society)

Matthew Schnupp, King Ranch, Inc. (current:
Pennsylvania Game Commission)

Terry Sohl, US Geological Survey

Bronson Strickland, Mississippi State University

Wayne Thogmartin, US Geological Survey

CHAPTER ACKNOWLEDGMENTS

Chapter 2: M. McConnell and J. Maerz provided
insightful comments on the concepts in this
chapter.

Chapter 3: K. McKelvey, D. Naugle, and an
anonymous referee provided insightful
comments on earlier drafts of this chapter.

Chapter 11: We sincerely thank T. Lorenz (US
Forest Service) for her many insights; our
research and management colleagues (R.
Niemeyer, A. Bentley Brymer, A. Suazo, J. D.
Wulfhorst, M. McGee, L. Okenson, and J. Suhr
Pierce); and the many stakeholders that were
part of the public participatory geographic
information systems (PPGIS) process. We
also thank the managers who provided input
on the species distribution models in the
Western Governors' Association Crucial Habitat
Assessment Tool process. The views, statements,
findings, conclusions, recommendations, and
data in this chapter are solely the work of the
authors and do not necessarily represent the
policies and positions of the US government.
This chapter and the ideas presented here
would not have been hatched and developed
without these collaborations. The PPGIS
process was funded by the National Science
Foundation IGERT program (Award 0903479),
Bureau of Land Management (Award
L14AC00060), and Idaho Governor's Office
of Species Conservation (Award BOSEIA14).
We thank the US Geological Survey Gap
Analysis Program for support under the
research work order #G12AC20244 to the
University of Idaho.

Chapter 12: We thank R. E. Reynolds, S. P. Fields,
A. A. Bishop, C. R. Loesch, B. Wangler, T. L.
Shaffer, A. J. Ryba, and D. H. Johnson for their
ideas, support, and discussion; S. P. Fields,
D. R. Hertel, C. R. Loesch, and T. L. Shaffer
for providing data used in examples; and
C. J. Parent and an anonymous reviewer for
reviews of earlier drafts of this manuscript. We
also thank the Refuges and Migratory Birds
Programs of the US Fish and Wildlife Service
for their support of our office. The findings
and conclusions in this chapter are those of the
authors and do not necessarily represent the
views of the US Fish and Wildlife Service.

Chapter 16: R. Pierce assisted with development
of figures and tables. J. Giocomo provided
helpful review comments on an early draft of the
chapter.

Chapter 17: We thank B. Matheson, formerly of
the Northwest Boreal Landscape Conservation
Cooperative, for his assistance in editing the
chapter. In addition, we want to thank former
members of the Landscape Conservation
Cooperatives Network for their commitment
to working together toward achieving a
common conservation vision, and we especially
recognize the amazing, enthusiastic staff who
helped coordinate this innovative conservation
approach. The findings and conclusions in this
chapter are those of the authors and do not
necessarily represent the views of the US Fish
and Wildlife Service.

Chapter 18: The following individuals contributed golden-winged warbler occurrence data that were used to delineate Priority Areas for Conservation boundaries: J. Larkin, K. Johnson, D. J. McNeil, and E. Bellush (Indiana University of Pennsylvania); S. Harding (Virginia Department of Game and Inland Fisheries); S. Barker and R. Rohrbaugh (Cornell Laboratory of Ornithology); K. Aldinger (National Wild Turkey Federation); P. Wood (US Geological Survey Cooperative Fish and Wildlife Research Unit at West Virginia University); L. Bulluck (Virginia Commonwealth University); C. Smalling (Audubon North Carolina); and R. Bailey (West Virginia Breeding Bird Atlas). Thousands of individual birders and citizen scientists contributed observations to the eBird database that we downloaded for these analyses. We are grateful to these individuals and to the eBird enterprise at Cornell University for providing this data set for public use. This analysis was supported by funding from the Natural Resources Conservation Service's Conservation Effects Assessment Program.

PART I • Understanding Habitat on Landscapes

1

CHAD J. PARENT AND
FIDEL HERNÁNDEZ

The Landscape Perspective in Wildlife and Habitat Management

Introduction

The landscape perspective is becoming an increasingly common framework for ecologists and managers who engage with contemporary wildlife and habitat management issues. It is easy to see that landcover and land use practices that shaped the landscape in the 20th century have changed and no longer resemble what is happening on the landscape in the 21st century. The literature is replete with evidence of these shifts, and there are increasing warning signs of the impending effect on wildlife populations. For countless wildlife species, a shift in landcover and land use over time is synonymous with catchphrases like "broad-scale habitat loss" and "range-wide population declines." These changes to the landscape have had major effects on population size, distribution, and diversity across a variety of taxa (Wilcove et al. 1998, Brennan and Kuvlesky 2005, Schipper et al. 2008, Gibbs et al. 2009). Habitat loss is only part of the dilemma, however. Not only have we reduced the amount of habitat on the landscape (composition), but we've also altered the arrangement of habitats on the landscape (configuration). Therefore a landscape ecological argument needs to be made, and, accordingly, the goal of this book is to advocate a landscape perspective of wildlife and habitat management.

In this chapter we establish the underpinnings of what is meant by landscape perspective. Advocating for a landscape perspective is not a new idea, and much has been written on this topic before. For example, the concept of land use planning emerged from the timber management literature in the late 1970s and called for the explicit integration of wildlife ecology into the management of the country's vast timber resources (Thomas 1979). Later, in the early 1990s, the concept of scale was advanced as a new paradigm through which to study patterns and variability in ecological processes across all levels of organization (Levin 1992). Admittedly, we are reticent to pit our conceptualization of a landscape perspective against the ideas that came from prolific minds like J. W. Thomas and S. Levin. Yet here we are in the 21st century, wrestling with the same issues and clearly not lacking knowledge about the issue. Instead, we suggest this knowledge may not be transferred efficiently between those developing it and those applying it. Much of the past literature arguing for landscape perspectives was written by landscape ecologists for landscape ecologists. When we as authors—including a natural resource agency ecologist and a university professor who has been a principal investigator on numerous wildlife and habitat management-oriented projects throughout his career—struggle with some of the content in these books, we wonder how well students who are encountering ideas about landscape ecology for the first time process this information. In contrast, many of the chapters in this book are written by natural resource managers for natural resource managers, and the wildlife or landscape ecologists who wrote these chapters are entrenched in the management arena. Other texts from the landscape ecology literature may be more suitable if a reader's goal is to understand the theoretical foundations of landscape ecology. In contrast, this book is meant to be an introduction to the landscape ecological ideas that are needed for natural resource managers to have productive conversations with landscape ecologists. The overarching goal of the book is to present a focused view of how natural resource managers can take ideas from landscape ecology and apply them to managing wildlife and habitats on landscapes. The wildlife profession is fortunate to have so many smart men and women developing theoretical and quantitative approaches to studying wildlife on landscapes, but unless these ideas can be expressed in accessible, on-the-ground terms, they are of little value to natural resource managers. Our objectives for this chapter are to highlight the need for collaboration between landscape ecologists and natural resource managers, present a case for the landscape perspective, and explain why landscape ecology is a good framework for accomplishing this.

The Need for Collaboration between Ecologists and Managers

Adopting a landscape perspective will require professionals from multiple disciplines, including landscape ecology, wildlife ecology, and wildlife management. Therefore it is important to discuss each expert's role. In general, we assume landscape ecologists and wildlife ecologists (or ecologists when we're referring to both) are the subject matter experts; their role is to design research projects, perform quantitative assessments of data, and publish the results in each of their respective areas of expertise. Natural resource managers (hereafter referred to as managers) are the individuals who are conducting the on-the-ground management of habitats, or the species biologists in charge of managing harvests within natural resource agencies. Their role is to interpret the results of research and apply management on the ground.

Success in management on landscapes necessarily depends on landscape ecologists and managers working collaboratively. Without collaboration between landscape ecologists and managers, the old hammer-and-nail adage applies: if the only tool you have is a hammer, then every problem looks like a nail. In other words, if you give landscape ecologists a data set, they'll surely come up with sophisticated ways to model those data. Alternatively, if you give managers a drip torch, they'll have no shortage of ideas of how to use prescribed fire to improve the landscape. Both the landscape ecologist and manager are doing something, but it's not clear whether their efforts result in synergistic benefits to wildlife on the landscape; the ecologist may not be evaluating the right hypotheses, and the manager may not be improving the landscape in the right places. When ecologists and managers work on goals independently, it leads to ineffective research and management outcomes, and ultimately disconnect (Krausman 2012, Merkle et al. 2019).

Before ecologists and managers can work on resolving disconnect, there must be an explicit recognition that there is disconnect (Krausman 2012). Part of the challenge is that landscape ecologists, wildlife ecologists, and managers do not have enough opportunities to interact. They do not attend the same meetings, they do not publish in the same journals, and they are interested in fundamentally different questions. Ecologists studying landscape ecological questions primarily attend conferences coordinated by the International Association of Landscape Ecologists or the Ecological Society of America, and they publish in journals dealing with more theoretical content like *Landscape Ecology* and *Ecology*. They may not attend conferences that are wildlife oriented, such as those coordinated by The Wildlife Society, or publish in wildlife management-oriented journals such as the *Journal of Wildlife Management*. Managers working for state or federal natural resource agencies receive varied support to attend meetings and usually prioritize limited travel resources to attend working group meetings or technical committee meetings where business on topics more directly related to their job responsibilities is conducted. Accordingly, the disconnect is understandable, and a basic review of each's role in managing habitat on landscapes is necessary.

Landscape ecologists need managers. Without managers on the ground to collect data and convey field observations about wildlife and habitats, landscape ecologists would have little insight on which research hypotheses are relevant, or the most appropriate modeling approaches to evaluate research hypotheses. This is important because modern ecological models are developed with a specific species and system in mind (e.g., integrated population models, hierarchical models; Royle and Dorazio 2008). Such models come with strict assumptions that can only be answered by managers with institutional knowledge of wildlife, habitat, and the landscape. Moreover, if landscape ecologists want managers to embrace the results of their research, they need to work closely with managers to tailor research results to management. When it comes to translating research results to on-the-ground management actions, managers will have direct knowledge about which tools will be most effective at producing the desired conditions specified by research.

Similarly, managers need landscape ecologists. Without landscape ecologists, the data collected by managers and the knowledge and insights gained from time in the field would not get translated into a formal statistical framework to draw infer-

ences about the data collected, elucidate patterns, and make predictions for how wildlife populations respond to management input. It is essential that landscape ecologists and managers work together because the backbone of linking pattern and process to management are collaboratively developed hypotheses about the habitat-based mechanisms responsible for population demographics, paired with spatially explicit models to inform management.

Meaningful collaboration, therefore, entails not only meaningful communication but also a shared set of concepts and vocabulary, and a shared cause. Thomas (1979:preface) suggested the concept of habitat could accomplish such goals: "Habitat is something the manager can relate to, understand, and control. Most important, it is an entity for which a manager can be held accountable. The maintenance of appropriate habitat is the foundation of all wildlife management. Habitat, therefore, is the key to organizing knowledge about wildlife so it can be used in forest management."

Thomas's words continue to be prescient, and we also suggest meaningful collaboration begins with habitat. Both landscape ecologists and managers are conversant in the concept of habitat. Landscape ecologists have experience quantifying habitat and modeling its relationships with wildlife populations, and managers have experience applying management interventions to habitat to produce beneficial effects for wildlife. Such experience provides a convenient common denominator to communicate landscape ecology to managers and management to landscape ecologists. Accordingly, the framework we outline below relies on habitat-based concepts from the wildlife and landscape ecology literatures as the basis for linking pattern and process to management.

The Need for a Landscape Ecological Perspective

There are many excellent case studies from the wildlife literature that could be used to reinforce the need for a landscape perspective. Northern bobwhite (*Co-*

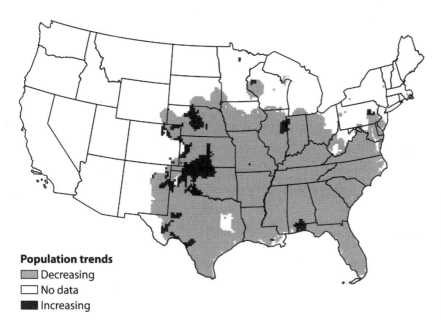

Figure 1.1. Distribution and population trend of northern bobwhite (*Colinus virginianus*) based on Breeding Bird Survey data between 1966 and 2015. Reproduced from Sauer et al. (2017).

Population trends
- ▨ Decreasing
- ☐ No data
- ■ Increasing

linus virginianus) are dear to our hearts because it is where our intellectual curiosity in landscape ecology originated (Parent et al. 2016). More importantly, there is comprehensive literature on bobwhite ecology and management (Guthery 2002). Their ecology is well understood, and this legacy has resulted in a comprehensive theory for their management (Guthery 1997, 1999). So, despite our biases, bobwhite provide an objectively good model for discussing the importance of a landscape perspective.

Bobwhite were likely not numerous on the landscape prior to European settlement of North America. European settlers introduced major continent-wide disturbances that benefited bobwhite, consisting of rudimentary practices associated with farming, forestry, and ranching. These disturbances scattered across private lands throughout the bobwhite's geographic range created important features that permitted bobwhite to exist at greater capacities and in more well-connected populations (Fig. 1.1). Leopold (1932:11) poetically described the landscape and its effect on the status of bobwhite populations: "Came now the settler, bringing axe, plow, cow, rail fence, hedges, weeds, and grain. The axe converted shady woods into brush stumplots. The plow flanked

them with weedy stubbles, bearing bumper crops of strange but nourishing seeds with a regularity hitherto unknown to quaildom. Plow furrows further out on the prairie checked the sweep of fires, and promptly the border shrubs romped outward up every draw and coulee, with bobwhite at their heels. Moreover the upland settlers planted thousands of miles of hedge around their new-broken grainfields, converting vast reaches of the hitherto forbidden prairie into quail-heaven."

The "quail-heaven" that Leopold described did not last forever. Over the course of the 20th century, farming, forestry, and ranching practices evolved (Brennan 1991). Rudimentary farming equipment and practices were replaced by highly efficient machines, hedges and fence rows were removed, and field sizes were increased to accommodate larger farming implements. Forest clearings were allowed to mature or were converted to silvicultural systems. Livestock were allowed to graze at densities that diminished important landcovers for grassland birds. Human population growth and urbanization occurred simultaneously with changes in land use and landcover. Taken collectively, the trajectory of bobwhite populations began to trend downward and was documented quantitatively on the basis of Breeding Bird Survey data (Fig. 1.1; Sauer et al. 2017). The decline likely began even earlier, given evidence of declines starting in the late 1800s (Stoddard 1931, Errington and Hammerstrom 1936).

When Brennan (1991) put the bobwhite population decline into perspective with his editorial in the early 1990s, it sparked a whole new set of research hypotheses aimed at understanding the ecology of bobwhite and patterns associated with their population dynamics (Hernández et al. 2013). For example, considerable research attention was given to predation, fire ants (*Solenopsis invicta*), and disease, factors that were probably only proximately associated with population declines (Hernández et al. 2013). Much of this research took place at local spatial scales on individual populations within a site (Williams et al. 2004). As a result, a lot of science contributes to our understanding of quail ecology at a local level. Where applicable, these results were translated into management actions to make positive individual effects on local bobwhite populations. When measured at the large spatial scales meaningful to managers (e.g., wildlife management areas, counties, states, ecoregions), however, bobwhite populations continued to decline across their range.

By the start of the 21st century, it was clear that quail ecologists needed to think more about the factors that affected bobwhite across much larger extents. It was generally accepted that habitat loss was a primary influence in bobwhite population declines across their range (Hernández et al. 2013). Because habitat loss was occurring at spatial scales that exceeded the extents of traditional research and management investigations, there was a need to expand the scale of research and management to the landscape (Williams et al. 2004, Brennan 2012).

Historically, quail ecologists were making simple ecological arguments about bobwhite populations and their resource needs. For example, when resources necessary for survival and productivity are limited, fewer bobwhites can be supported, and the logical result will be a decline in abundance. This example is easily conceptualized as a proportional relationship between bobwhite density (*y*-axis) and resource quantity (*x*-axis). But we suggest this is only part of the dilemma. Loss of habitat on the landscape can be manifested in a variety of different ways (Fig. 1.2), and the distribution of habitat losses is extremely important (Fahrig 1997). If reductions in resources (i.e., habitat) automatically lead to population declines in wildlife, then hypotheses like Leopold's Law of Interspersion (Leopold 1933) would have been dismissed long ago (the Law of Interspersion is a hypothesis that states a species density is proportional to the amount of edge habitat). Yet the Law of Interspersion still holds up today as a means of planning habitat management for low-mobility species (but see Guthery and Bingham 1992). Of course, we get to make such bold statements with the benefit of hindsight. Today we have the luxury of high-resolution remotely sensed spatial data sets

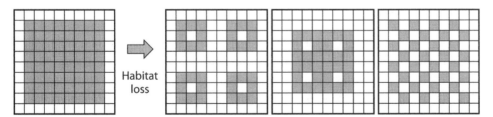

Figure 1.2. Conceptual mapping of habitat loss on a hypothetical landscape (*far left panel*) under three manifestations of a 50% reduction in habitat (*right panels*). Although each landscape has an equivalent composition of habitat, the configuration of habitat will influence population density differently depending on the wildlife species and its unique ecological responses to spatial heterogeneity on the landscape.

and technology to study ecology at virtually any desired scale, and it's easier than ever to fit models that were considered too computationally sophisticated a decade ago. Contemporary quail ecologists have theorized that composition of habitat on the landscape alone is only part of the dilemma for bobwhite (Guthery 1997, 1999). Several recent studies have identified critical relationships between landscape patterns and bobwhite populations (Veech 2006, Duren et al. 2011, Blank 2012, Parent et al. 2016, Miller et al. 2019). Thus it's clear that a landscape ecological perspective is needed to fully understand the ecology surrounding range-wide declines in bobwhite populations.

A landscape ecological perspective is justified when the factors that determine the distribution and abundance of wildlife populations are structured spatially and no longer operate at a single, local spatial scale, but rather multiple, broader scales (Fahrig 2005). From the standpoint of bobwhite, this means the ecological research results generated to inform management in the latter half of the 20th century may only be effective at reversing local, small-scale population declines as a result of proximate factors (Hernández et al. 2013). This issue is not unique to bobwhite; the same story could be told about prairie chicken (*Tympanuchus cupido*), sage-grouse (*Centrocercus urophasianus*), wild turkey (*Meleagris gallopavo*), moose (*Alces alces*), brown bear (*Ursus arctos*), and countless other species groups that command far less research attention. To affect wildlife populations at meaningful spatial scales, we need to understand the mechanisms that influence ecological population processes across large areas. Inherently, managers often assume that managing wildlife on a landscape entails manipulating every square meter of the landscape. This is not true, nor is it what landscape ecology advocates. Rather, the objective is to manage individual sites (i.e., a farmyard, pasture, section, region) as if they were a part of a landscape, which they are. Concepts from landscape ecology offer a framework to do this because the landscape ecological perspective necessarily entails

consideration of size and arrangement of habitats in a scale-dependent manner (Turner et al. 2001). An essential byproduct of the landscape ecological perspective is a blueprint for wildlife conservation and management in the 21st century. We suggest a framework for carrying out wildlife management on landscapes below.

A Comprehensive Framework for Managing Landscapes
Defining Concepts Central to Landscape Ecology

The first and second parts of this book (Chaps. 2–8) focus on linking habitat to landscape ecology and introducing landscape ecological ideas that are needed for managing landscapes. The material in these other parts of the book will be covered in much greater detail in their respective chapters. We do, however, need to introduce some of these concepts to facilitate discussion. Wildlife landscapes are defined in different ways. A general definition of a *landscape* might refer to an area in which spatial heterogeneity potentially influences ecological population processes (i.e., population size, vital rates, distribution) of interest (Turner et al. 2001). Conceptual models that define the manner in which spatial heterogeneity is translated to the landscape are necessary for landscape ecological analyses. These are formally referred to as *models of landscape structure* in the literature; for simplicity, we refer to them as *landscape models*. The patch-corridor-matrix model is used ubiquitously for this purpose (Fig. 1.3). The model treats the landscape as a mosaic of spatial patterns created from patches, corridors, and matrix. More specifically, habitat components exist in the form of patches that are connected to each other by corridors, all of which are embedded in a matrix of lands that do not contribute to habitat. A *habitat* is an area that contains resources (typically food, cover, and water), environmental conditions (abiotic and biotic), or other features that are conditional to occupancy for a specific species (Morrison et al. 2006).

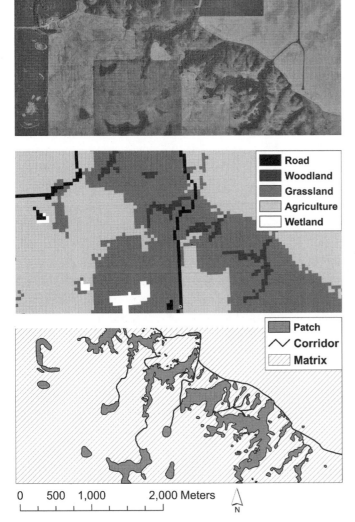

Road
Woodland
Grassland
Agriculture
Wetland

Patch
Corridor
Matrix

0   500   1,000   2,000 Meters

N

Figure 1.3. Examples of a patch-corridor-matrix model of landscape structure depicted as a landscape mosaic based on categorical data from the National Land Cover Database (*middle*) and as a set of patches and corridors embedded in a homogeneous matrix (*bottom*). An aerial image of the site from the National Agriculture Imagery Program is provided for reference (*top*).

Under this definition of habitat, vegetation is not the only habitat component (though it often is). Thus, under a model of landscape that is composed of habitat patches (i.e., the patch-corridor-matrix model), landscape also is a species-specific concept, similar to the habitat concept. Its area will depend on the ecological process being studied. This means that landscapes are not necessarily defined by size, but a large spatial extent is usually required to capture the full range of variability of an ecological process for multiple populations of an organism (Dunning et al. 1992).

Spatial heterogeneity occurs when environmental conditions on the landscape, such as vegetation composition or energy and nutrient flows, are discontinuous (Morrison et al. 2006). This discontinuity creates ecotones, which ecologists refer to as spatial heterogeneity on the landscape; the greater the variability and complexity of these ecotones, the greater the heterogeneity. Under the patch-corridor-matrix model, ecologists quantify the heterogeneity of habitat patches, corridors, and unusable background matrix across different scales. Heterogeneity is measured in terms of composition and configuration.

Composition refers to how much or the concentration of the habitat patches, and *configuration* refers to the structure or arrangement of patches of a particular type with respect to other patches on the landscape. In other words, the concept of heterogeneity is used to formally quantify the composition and configuration of patterns on the landscape, of which habitat is usually the primary focus. Quantifying heterogeneity is useful because a central tenet of landscape ecology is to understand interactions between ecological population processes and heterogeneity of habitat patches across scales. This concept is often referred to informally as the study of *pattern on process* (Turner et al. 1989). A convenient example that may demonstrate this a simple linear model. In landscape ecology parlance, we are basically saying that the spatial patterns of habitat are equivalent to our predictor variables in a linear regression, and that the ecological processes are equivalent to the response variables (Fig. 1.4).

Landscapes are spatially heterogeneous through time, across space, and hierarchically (O'Neill et al. 1989, Turner et al. 1989, Schneider 2001, Wu and Li 2006). Considerations of scale are important, so that

ecologists can more effectively and efficiently study the effect of pattern on process. *Scale* is described as the product of two components, extent and grain (Wiens 1989). *Extent* is defined as the physical size in space or duration of an ecological observation and often is what researchers imprecisely think of as scale (Wiens 1989). *Grain* defines the finest level of measurable detail for a variable that can be observed in a data set and is sometimes referred to as resolution. Because grain and extent define scale, altering either grain or extent results in a change of scale (Fig. 1.5) and therefore landscape heterogeneity (Turner et al. 2001). It follows, then, that changing the grain or extent will affect the outcomes of research and conclusions for management. Therefore any discussion of scale in wildlife ecology must define the spatial and temporal extent and grain of the pattern on process phenomenon being studied and, perhaps more importantly, the scale at which managers apply management.

Finally, while the patch-corridor-matrix model is used heavily throughout this book, there are a number of ways to conceptually model the structure of a landscape. Other models include point pattern

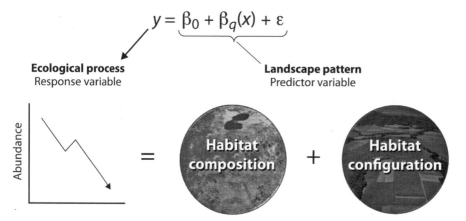

$$y = \beta_0 + \beta_q(x) + \varepsilon$$

Ecological process
Response variable

Landscape pattern
Predictor variable

Abundance

= **Habitat composition** + **Habitat configuration**

Figure 1.4. A primary goal of landscape ecology is to understand the effect of spatial patterns on population processes (i.e., pattern on process). Pattern on process evaluations often are translated to statistical models to facilitate evaluation. For example, wildlife ecologists might use a linear regression to evaluate the relationship between wildlife abundance (response, or dependent variable) and spatial patterns of habitat on the landscape (predictor, or independent variable).

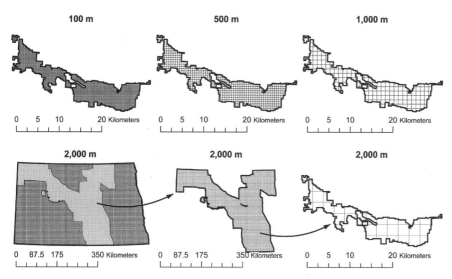

Figure 1.5. The components of scale include grain and extent. The Lonetree Wildlife Management Area in north central North Dakota, USA, is depicted at three spatial scales that change with grain sizes of 100 m, 500 m, and 1,000 m (*top*). A constant grain size of 2,000 m is applied to 3 different extents to depict 3 scales (*bottom*): a state scale (*left*), regional scale (*middle*), and wildlife management area scale (*right*). Grain becomes indistinguishable when extent is increased, as is the case with the two extents at left. Collecting or analyzing data at these scales is probably not feasible for most ecological processes, demonstrating why increasing extent is usually associated with an increase in grain.

models, where pattern and process variables are depicted through their *x,y* locations, and landscape gradient (sometimes called continuum) models created from spatial data depicting continuous, quantitative measurement of a variable across space. Ultimately, though, models of landscape structure are just like any other model; there are trade-offs associated with landscape models that attempt to reflect reality perfectly. Accordingly, ecologists and managers should think critically about the most appropriate landscape model to depict heterogeneity on their landscapes of interest. There are inherent benefits to the patch-corridor-matrix model, and we discuss these benefits below.

Linking Pattern and Process to Management

If we understand how the factors that determine habitat affect wildlife populations, then we should be able to predict how wildlife will respond to changes in habitat (Van Horne and Wiens 2015). Therefore linking pattern and process to management can be reduced to one basic goal: to develop a formal understanding about the relationship between habitat (quantified in terms of spatial heterogeneity) and wildlife populations to create beneficial landscapes for wildlife. There is no right way to link landscape ecological concepts to management. We would argue that wildlife ecologists have been performing some version of this linkage for years under the umbrella of wildlife–habitat relationships, the key distinction being that landscape ecologists are explicitly concerned with scales broad enough to capture the spatially heterogenous nature of the relationship. Concepts like scale and heterogeneity are increasingly becoming mainstays in the contemporary wildlife–habitat literature (Morrison et al. 2006). The boundary between what wildlife ecologists call wildlife–habitat relationships and what landscape ecologists call pattern on process is becoming blurry. We attempt to add clarity by suggesting a framework that merges the ecological underpinnings of pattern on process concepts from the landscape ecological literature and the management-oriented mindset of wildlife–habitat relationship concepts from the wildlife literature.

COLLECTING ECOLOGICAL DATA

Under a landscape perspective, it is typically desirable to link pattern and process to management at spatial scales that are meaningful to natural resource managers. This often means landscapes are represented by large areas such as ecoregions, hunting management units, or wildlife management areas, which necessitates the collection of ecological data for landscapes at corresponding spatial scales. Natural resource agencies collect ecological data in a variety of ways. Here, we review common sources and methods for collecting ecological data and discuss the relevance to linking pattern on process to management.

Ecological data are variables that describe the distribution (occupancy, presence-absence) and size (abundance, density) of a population along with relevant vital rates (survival, productivity). It is usually not feasible to observe ecological data for wildlife populations through censuses (i.e., complete counts), especially on a landscape, so wildlife ecologists sample key covariates from wildlife populations to permit the estimation of ecological data (Royle and Dorazio 2008). Common sampling protocols are based on capture-recapture sampling (e.g., band recovery, marking, removals), distance sampling, and occupancy sampling (Burnham and Anderson 1976, Otis et al. 1978, MacKenzie et al. 2006). A core feature of these sampling protocols are statistical methods that adjust population surveys for undetected individuals in the population (i.e., animals that were present but not counted during the survey). Accounting for detection requires the additional collection of covariates. For example, the perpendicular distance from a line transect is collected for observed individuals under the distance sampling protocol (Burnham and Anderson 1976). Such surveys often require a lot of sampling effort (e.g., large sample sizes, survey replicates) to obtain estimates of abundance or occupancy with reasonable levels of precision at meaningful spatial scales (Pollock et al. 2002, Pollock 2006).

Given the necessary collection of additional covariates for sampling protocols that adjust for detection, natural resource agencies often cannot justify the cost of obtaining absolute estimates of abundance, which often relegates natural resource managers to make decisions about wildlife management using indices (Pollock 2006). Much has been written about the use of indices in the ecological literature, primarily in terms of indices as a form of blasphemy, their defense, and validation of assumptions (Anderson 2001, Pollock et al. 2002, Johnson 2008). Absolute estimates of abundance are scientifically more defensible than indices (particularly for indices that go unvalidated). In reality, however, absolute estimates of abundance are hard to come by, and yet wildlife managers have been successfully managing wildlife populations in the absence of absolute estimates of abundance for many years. Suffice to say, if resources were available to design long-term monitoring programs focused on collecting the covariates needed to fit modern ecological models to estimate abundance or reconstruct vital rates at the landscape scale, wildlife managers would have already adopted such programs.

Selecting appropriate ecological population data is necessarily a collaborative effort. Managers need to make ecologists aware of the limitations in the types of ecological population data they're capable of collecting on landscapes. Ecologists need to help managers place results into context, depending on the types of ecological data that are used to link pattern and process to management. Specifically, ecological population data allow for direct linkages between the mechanisms that influence habitat on the landscape and population processes. In contrast, population indices are not population processes. Although pattern on process relationships using population indices can inform management, they cannot be used to evaluate the effect that a habitat manipulation may have on wildlife populations. At best, they can be viewed as factors related to relative changes in abundance over time.

HYPOTHESIS DEVELOPMENT

Much has been written about the philosophy of wildlife science and the role of hypotheses (Romesberg 1981, Shenk and Franklin 2001, Burnham and Anderson 2002, Guthery 2008, Sells et al. 2018). A critical first step prior to the analysis of any data is the development of research hypotheses about the mechanisms responsible for the ecological process under consideration. This critical thinking often is considered one of the hardest parts of wildlife research (Guthery 2008). If one struggles with this idea, we suggest that a good place to begin is by stepping outside. Once again, bobwhite offer a convenient case study.

Suppose you were transported in time to the quail heaven described by Leopold earlier in this chapter. If you walked through this landscape during the spring nesting season, you might observe that bobwhite flush from specific types of bunch grasses. The vertical structure of the vegetation offers concealment from predators but also permits young broods to move freely. From this you might conclude there are structural features of bunch grasses that convey some benefit to nesting and brood rearing for bobwhite (there are; see Lehman 1984, Roseberry and Klimstra 1984, Hernández and Peterson 2007). Further, suppose you also viewed the landscape from a helicopter (effectively, you are viewing these features at a different scale). You might observe nesting bobwhite flush to a nearby tree row or brush motte. As with bunch grasses, these patches of brush provide useful habitat features because they provide escape cover from predators (Hernández and Peterson 2007). If you traveled the rest of this landscape, either on foot or in a helicopter, no doubt a number of relationships between the various aspects of the bobwhite's life history and the habitat features at different scales would begin to emerge. What becomes clear is that if the availability, composition, or configuration of resources on which bobwhite rely should change (i.e., if the spatial heterogeneity of habitat patches changes), then naturally the survival, reproduction, and abundance of bobwhite populations on that landscape also will change.

The next step is where critical thinking becomes important, because you need to translate the patterns you observed about quail in the field into some meaningful, testable research hypothesis. Foundational concepts from ecology can help guide this process. We suggest a good place to begin is with the concept of limiting factors (but other relevant concepts might be equally useful, e.g., niche, assembly rules, ecological succession, competition, density dependence, metapopulation). Recall that limiting factors refer to resources that, below some minimum, will limit the growth, abundance, or distribution of an individual or population (Krebs 2001). With respect to bobwhite on the hypothetical landscape above, it would be straightforward to generate a set of research hypotheses relevant to the type of ecological data available from population monitoring surveys. For example, a reasonable hypothesis to evaluate is that landscapes with greater composition of landcovers containing bunch grasses will support larger populations of bobwhite. Because there is evidence that bobwhite select bunch grasses for nesting (Hernández and Peterson 2007), it follows that there would be a population response as a result of greater nest success and more sites for more bobwhite to nest.

HABITAT DATA COLLECTION

Measuring habitat in a landscape ecological context implicitly means quantifying heterogeneity, which requires methodology to objectively measure spatial heterogeneity of landscapes and habitats, for which there are many (Gustafson 1998). Metrics to quantify an index of spatial heterogeneity are applied to habitat patches (and potentially corridors and matrix, if desired) in the patch-corridor-matrix landscape model. Categorical map data sets such as the National Land Cover Database (Multi-Resolution Land Characteristics Consortium 2016) supply the raw spatial data to perform these calculations. Indices of spatial heterogeneity can be classified into six categories: area and edge, shape, core area, nearest neighbor, diversity, and contagion and interspersion. Documentation for a variety of metrics, such as the equation, units of measurement, theoretical minimum and maximum values, and assumptions are discussed extensively by McGarigal and Marks (1995). The number of metrics available to quantify heterogeneity has proliferated since landscape ecologists first started measuring heterogeneity in the 1980s. Today, programs dedicated to housing comprehensive suites of metrics have been created to streamline the quantification of spatial heterogeneity (e.g., FRAGSTATS v4, University of Massachusetts, Amherst, Massachusetts, USA). Such programs have been a boon (and some would say a bane) to landscape ecologists because they have made landscape ecological ideas more accessible (Li and Wu 2004).

Li and Wu (2004) outline appropriate uses, and misuses, of heterogeneity metrics. Their review emphasizes the notion that landscapes are complex, intractable spaces, and it is unlikely that a single index can meaningfully capture the dynamics of an ecological process (Li and Wu 2004). Rather, they suggest identifying multiple metrics that take into consideration the specific ecological process under investigation and the scale at which the process operates. Moreover, landscape ecologists must have a nuanced understanding of which metrics can adequately measure the form of heterogeneity described by research hypotheses. This is important because heterogeneity is a species-specific, scale-dependent concept that can be measured using a plethora of metrics, some of which have multiple meanings (Li and Wu 2004).

Because the ultimate goal of pattern on process investigations in our framework is to translate landscape ecological research results to management, additional consideration needs to be given to the transferability of a metric to on-the-ground management applications. A good rule of thumb is the bar napkin test. If the general form and appearance of a metric are too complex to be easily sketched on the back of a bar napkin, and too abstract to be easily interpreted, then we shouldn't expect such a pattern could be transferred onto the landscape. That is to say, so long as habitat on the landscape continues to be modified by plows, drip torches, livestock, and herbicide, complex metrics with abstract interpretations likely cannot be efficiently "sketched" onto the landscape to positively influence wildlife populations. Finally, other environmental factors might be considered beyond metrics of spatial heterogeneity. Climate, weather, soil type, and population density are just a few examples of environmental factors that exert profound influences on wildlife populations at broad scales, either directly, indirectly, or in an interaction with spatial heterogeneity (Parent et al. 2016).

USING MODELS TO EVALUATE RESEARCH HYPOTHESES
It is difficult to pick up a piece of ecological literature without encountering topics on modeling. Volumes have been written about modeling, and there are even entire disciplines of ecology that are based exclusively on the subject (e.g., population viability analysis). Accordingly, this book is not (and should not be) a how-to for developing, interpreting, validating, handling uncertainty, and applying models of wildlife–habitat or pattern on process relationships. Instead, it is more appropriate to view this book as a forum for ecologists and natural resource managers to get on the same page so that meaningful resource management can occur on landscapes. Here, we briefly highlight the role of modeling in wildlife–habitat or pattern on process relationships.

In the framework we present, ecologists use models to evaluate hypothesized relationships between spatially explicit patterns on the landscape and wildlife populations so that results can be used to inform management. Interpreting these relationships will vary depending on the objective of the model. Although ecologists have thought of many creative uses for models, when it comes to using models to studying pattern on process, there are three general objectives: understanding, prediction, and control (Table 1.1; Morrison et al. 2006, Collier and John-

Table 1.1. Plausible Research Hypotheses to Evaluate under Various Objectives for Modeling

Model Objective	Possible Hypothesis Evaluated
Understand	1. Has heterogeneity of habitat patches changed over time?
	2. Is heterogeneity of habitats similar between 2 or more landscapes, and if so, precisely how?
	3. Which hypothesized measure of heterogeneity has the greatest effect size on wildlife populations, and is the effect positive or negative?
	4. How do multiple measures of heterogeneity interact to drive an ecological population process?
Predict	1. Where on the landscape will wildlife be distributed?
	2. If habitat change occurs on the landscape, how will wildlife populations change? Will the effect be positive or negative?
	3. How sensitive are wildlife populations to changes in heterogeneity of habitat on the landscape?
Control	1. If we understand the effect of heterogeneity on wildlife populations, how will wildlife populations respond if we attempt to re-create this heterogeneity on the landscape through management?
	2. If we have limited resources for management, where is the most efficient place on the landscape to use management resources?

son 2015). Models used for understanding are done in service of inferring relationships between wildlife and habitat variables. The ability to detect a relationship is of little value if the direction and magnitude of the relationship are not also of interest (Johnson 2002); therefore ecologists seek to quantify effect sizes between wildlife population processes and patterns of heterogeneity. Models used for predictive purposes help us understand how wildlife populations will respond to a set of model inputs. For example, when exploring sensitivity of response (the population process) and predictor variables (the habitat patterns) in regression analyses, it is common to hold predictor variables at their mean to predict how the response variable changes. Models used to control reflect an implicit desire to manipulate inputs to the model (i.e., predictor variables) to achieve a desired result (Collier and Johnson 2015). Because the predictor variables used in pattern on process models are spatially explicit (i.e., they have a specific, known location on the landscape), they can be used to spatially prioritize management manipulations, which is particularly useful to natural resource managers. The intended purpose of a model obviously will depend on the research hypotheses developed earlier in the framework. Clearly, the purpose of a model can be applied to accomplish all three tasks; that is, a model used for prediction can be useful for understanding and also controlling wildlife population processes. Ultimately, the important point to keep in mind is that the fundamental objective of modeling pattern on process is to link landscape ecology to management in a useful, accessible way for natural resource managers.

Summary

Traditionally, wildlife ecologists and natural resource managers have applied research and management at human-centered scales: usually political boundaries that encompass a plantation, pasture, ranch, or management area. These scales are adequate if the full range of variability of a population process under consideration can be observed within the extent of the study area. But wildlife ecologists and managers could not have foreseen the impending problem that loss and fragmentation of habitat at a landscape scale would present. The traditional research and management strategies that target small-scale sources of population change are not effective at addressing landscape-scale changes in habitat, to which a vast array of wildlife is particularly sensitive. Thus a landscape ecological perspective is particularly relevant when it comes to developing management objectives aimed at impacting wildlife populations across their geographic distribution.

Landscape ecology became a popular area of wildlife research starting in the 1980s (Turner et al. 2001). The availability of technologies like geographic information systems to store and view the growing sources of spatial data provided a means to ask how broader spatial patterns on the landscape influenced wildlife ecology. Accordingly, identifying the relationship between the spatial pattern on the landscape and population processes in wildlife ecology is a major emphasis of landscape ecology (i.e., pattern on process). Landscape ecologists accomplish this goal through the use of tools that quantify spatial heterogeneity and link heterogeneity to wildlife population data through models. This approach to studying pattern on process that occurs at a landscape scale resembles closely wildlife–habitat relationships that generally occur at a local scale and therefore represents a solid foundation for integrating landscape ecology into natural resource management. Successful management of habitat on landscapes, however, cannot occur without a concerted, collaborative effort by ecologists and managers to develop hypothesis-driven models about pattern on process.

LITERATURE CITED

Anderson, D. R. 2001. The need to get the basics right in wildlife field studies. Wildlife Society Bulletin 29:1294–1297.

Blank, P. J. 2012. Northern bobwhite response to conservation reserve program habitat and landscape attributes. Journal of Wildlife Management 77:68–74.

Brennan, L. A. 1991. How can we reverse the northern bobwhite population decline? Wildlife Society Bulletin 19:544–555.

Brennan, L. A. 2012. The disconnect between quail research and management. Pages 119–128 in J. P. Sands, S. J. DeMaso, M. J. Schnupp, et al., editors. Wildlife science: connecting research and management. CRC Press, Taylor and Francis Group, Boca Raton, Florida, USA.

Brennan, L. A., and W. P. Kuvlesky. 2005. North American grassland birds: an unfolding conservation crisis? Journal of Wildlife Management 69:1–13.

Burnham, K. P., and D. R. Anderson. 1976. Mathematical models for nonparametric inferences form line transect data. Biometrics 32:325–326.

Burnham, K. P., and D. R. Anderson. 2002. Model selection and multi-model inference: a practical information–theoretical approach. Second edition. Springer-Verlag, New York, New York, USA.

Collier, B. A., and D. H. Johnson. 2015. Thoughts on models and prediction. Pages 117–127 in M. L. Morrison and H. A. Mathewson, editors. Wildlife habitat conservation: concepts, challenges, and solutions. Johns Hopkins University Press, Baltimore, Maryland, USA.

Dunning, J. B., B. J. Danielson, and H. R. Pulliam. 1992. Ecological processes that affect populations in complex landscapes. Oikos 65:169–175.

Duren, K. R., J. J. Buler, W. Jones, and C. K. Williams. 2011. An improved multi-scale approach to modeling habitat occupancy of northern bobwhite. Journal of Wildlife Management 75:1700–1709.

Errington, P. L., and F. N. Hammerstrom. 1936. The bobwhite's winter territory. Iowa Agricultural Experiment Station Research Bulletin 201:302–443.

Fahrig, L. 1997. Relative effects of habitat loss and fragmentation on population extinction. Journal of Wildlife Management 61:603–610.

Fahrig, L. 2005. When is a landscape perspective important? Pages 3–11 in J. Wiens and M. Moss, editors. Issues and perspectives in landscape ecology. Cambridge University Press, New York, New York, USA.

Gibbs, K. E., R. L. Mackey, D. J. Currie. 2009. Human land use, agriculture, pesticides and losses of imperiled species. Diversity and Distributions 15:242–253.

Gustafson, E. J. 1998. Quantifying landscape spatial pattern: what is the state of the art? Ecosystems 1:143–156.

Guthery, F. S. 1997. A philosophy of habitat management for northern bobwhites. Journal of Wildlife Management 61:291–301.

Guthery, F. S. 1999. Slack in the configuration of habitat patches for northern bobwhites. Journal of Wildlife Management 63:245–250.

Guthery, F. S. 2002. The technology of bobwhite management: the theory behind the practice. Iowa State University Press, Ames, Iowa, USA.

Guthery, F. S. 2008. A primer on natural resource science. Texas A&M University Press, College Station, Texas, USA.

Guthery, F. S., and R. L. Bingham. 1992. On Leopold's principle of edge. Wildlife Society Bulletin. 20:340–344.

Hernández, F., L. A. Brennan, S. J. DeMaso, J. P. Sands, and D. B. Wester. 2013. On reversing the northern bobwhite population decline: 20 years later. Wildlife Society Bulletin 37:177–188.

Hernández, F., and M. J. Peterson. 2007. Northern bobwhite ecology and life history. Pages 40–64 in L. A. Brennan, editor. Texas quails: ecology and management.

Texas A&M University Press, College Station, Texas, USA.

Johnson, D. H. 2002. The role of hypothesis testing in wildlife science. Journal of Wildlife Management 66:272–276.

Johnson, D. H. 2008. In defense of indices: the case of bird surveys. Journal of Wildlife Management 72:857–868.

Krausman, P. R. 2012. Strengthening the ties between wildlife research and management. Pages 13–26 in J. P. Sands, S. J. DeMaso, M. J. Schnupp, et al., editors. Wildlife science: connecting research and management. CRC Press, Taylor and Francis Group, Boca Raton, Florida, USA.

Krebs, C. J. 2001. Ecology: the experimental analysis of distribution and abundance. Benjamin Cummings, San Francisco, California, USA.

Lehmann, V. W. 1984. The bobwhite in the Rio Grande plain of Texas. Texas A&M University Press, College Station, Texas, USA.

Leopold, A. 1932. Report of the Iowa game survey, chapter one: the fall of the Iowa game range. Outdoor America 11:7–9.

Leopold, A. 1933. Game management. Charles Scribner's Sons, New York, New York, USA.

Levin, S. A. 1992. The problem of pattern and scale in ecology. Ecology 73:1943–1967.

Li, H., and J. Wu. 2004. Use and misuse of landscape indices. Landscape Ecology 19:389–399.

MacKenzie, D. I., J. D. Nichols, J. A. Royle, K. H. Pollock, L. L. Bailey, and J. E. Hines. 2006. Occupancy estimation and modeling: inferring patterns and dynamics of species occurrence. Elsevier/Academic Press, Burlington, Massachusetts, USA.

McGarigal, K., and B. J. Marks. 1995. FRAGSTATS: spatial pattern analysis program for quantifying landscape structure. US Department of Agriculture General Technical Report PNW-GTR-351, Portland, Oregon, USA.

Merkle, J. A., N. J. Anderson, D. L. Baxley, M. Chopp, L. C. Gigliotti, J. A. Gude, T. M. Harms, H. E. Johnson, E. H. Merrill, M. S. Mitchell, et al. 2019. A collaborative approach to bridging the gap between wildlife managers and researchers. Journal of Wildlife Management 83:1644–1651.

Miller, K. S., L. A. Brennan, H. L. Perotto-Baldivieso, F. Hernández, E. D. Grahmann, A. Z. Okay, X. B. Wu, M. J. Peterson, H. Hannush, J. Mata, et al. 2019. Correlates of habitat fragmentation and northern bobwhite abundance in the Gulf Prairie Landscape Conservation Cooperative. Journal of Fish and Wildlife Management 10:3–18.

Morrison, M. L., B. G. Marcot, and R. W. Mannan. 2006.

Wildlife–habitat relationships: concepts and applications. Third edition. Island Press, Washington, DC, USA.

Multi-Resolution Land Characteristics Consortium. 2016. National Land Cover Database (NLCD) 2016. https://www.mrlc.gov/national-land-cover-database-nlcd-2016. Accessed 1 September 2019.

O'Neill, R. V., A. R. Johnson, and A. W. King. 1989. A hierarchical framework for the analysis of scale. Landscape Ecology 3:193–205.

Otis, D. L., K. P. Burnham, G. C. White, and D. R. Anderson. 1978. Statistical inference from capture data on closed animal populations. Wildlife Monographs 62:3–135.

Parent, C. J., F. Hernández, L. A. Brennan, D. B. Wester, F. C. Bryant, and M. J. Schnupp. 2016. Northern bobwhite abundance in relation to precipitation and landscape structure. Journal of Wildlife Management 80:7–18.

Pollock, J. F. 2006. Detecting population declines over large areas with presence-absence, time-to-encounter, and count survey methods. Conservation Biology 20:882–892.

Pollock, K. H., J. D. Nichols, T. R. Simons, G. L. Farnsworth, L. L. Bailey, and J. R. Sauer. 2002. Large scale wildlife monitoring studies: statistical methods for design and analysis. Environmetrics 13:105–119.

Romesberg, H. C. 1981. Wildlife science: gaining reliable knowledge. Journal of Wildlife Management 45:293–313.

Roseberry, J. L., and W. D. Klimstra. 1984. Population ecology of the bobwhite. Southern Illinois University Press, Carbondale, Illinois, USA.

Royle, J. A., and R. M. Dorazio. 2008. Hierarchical modeling and inference in ecology: the analysis of data from populations, metapopulations, and communities. Elsevier/Academic Press, Burlington, Massachusetts, USA.

Sauer, J. R., D. K. Niven, J. E. Hines, K. L. Pardieck, J. E. Fallon, W. A. Link, and D. J. Ziolkowski Jr. 2017. The North American breeding bird survey, results and analysis 1966–2015. Version 12.23.2015. US Geological Survey Patuxent Wildlife Research Center. https://www.mbr-pwrc.usgs.gov/bbs/bbs2015.html. Accessed 8 Aug 2019.

Schipper, J., J. S. Chanson, F. Chiozza, N. A. Cox, M. Hoffmann, V. Katariya, J. Lamoreaux, A. S. L. Rodrigues, S. N Stuart, H. J. Temple, et al. 2008. The status of the world's land and marine mammals: diversity, threat, and knowledge. Science 322:225–230.

Schneider, D. C. 2001. The rise of the concept of scale in ecology. BioScience 51:545–553.

Sells, S. N., S. B. Bassing, K. J. Barker, S. C. Forshee, A. C.

Keever, J. W. Goerz, and M. S. Mitchell. 2018. Increased scientific rigor will improve reliability of research and effectiveness of management. Journal of Wildlife Management 82:485–494.

Shenk, T. M., and A. B. Franklin. 2001. Models in natural resource management: an introduction. Pages 1–10 *in* T. M. Shenk and A. B. Franklin, editors. Modeling in natural resource management: development, interpretation, and application. Island Press, Washington, DC, USA.

Stoddard, H. L. 1931. The bobwhite quail: its habits, preservation and increase. Scribner's Sons, New York, New York, USA.

Thomas, J. W. 1979. Introduction. Pages 10–21 *in* J. W. Thomas, editor. Wildlife habitats in managed forests: the blue mountains of Oregon and Washington. Agricultural Handbook No. 553, US Department of Agriculture Forest Service, Washington, DC, USA.

Turner, M. G., R. H. Gardner, and R. V. O'Neill. 2001. Landscape ecology in theory and practice: pattern and process. Springer-Verlag, New York, New York, USA.

Turner, M. G., R. O'Neill, R. Gardner, and B. Milne. 1989. Effects of changing spatial scale on the analysis of landscape pattern. Landscape Ecology 3:153–162.

Van Horne, B., and J. A. Wiens. 2015. Managing habitats in a changing world. Pages 34–46 *in* M. L. Morrison and H. A. Mathewson, editors. Wildlife habitat conservation: concepts, challenges, and solutions. Johns Hopkins University Press, Baltimore, Maryland, USA.

Veech, J. A. 2006. Increasing and declining populations of northern bobwhites inhabit different types of landscapes. Journal of Wildlife Management 70:922–930.

Wiens, J. A. 1989. Spatial scaling in ecology. Functional Ecology 3:385–397.

Wilcove, D. S., D. Rothstein, J. Dubow, A. Phillips, and E. Losos. 1998. Quantifying threats to imperiled species in the United States: assessing the relative importance of habitat destruction, alien species, pollution, overexploitation, and disease. BioScience 48:607–615.

Williams, C. K., F. S. Guthery, R. D. Applegate, and M. J. Peterson. 2004. The northern bobwhite decline: scaling our management for the twenty-first century. Wildlife Society Bulletin 32:861–869.

Wu, J., and H. Li. 2006. Concepts of scale and scaling. Pages 3–15 *in* J. Wu, K. B. Jones, H. Li, and O. L. Loucks, editors. Scaling and uncertainty analysis in ecology: methods and applications. Springer, Dordrecht, Netherlands.

2

James A. Martin and
John M. Yeiser

Wildlife Management and the Roots of Landscape Ecology

Introduction

Wildlife management is an applied interdisciplinary field that endeavors to manage wildlife populations in the context of biology, societal needs, and wants (Conroy and Peterson 2013). Wildlife biologists and managers are typically versed in foundational biology, ecology, statistics, physical sciences, zoology, botany, humanities, and human dimensions. It is a difficult job to manage something as complex as wildlife populations, which vary across space and time, as societal pressures ebb and flow. To accomplish such a task, one calls upon the aforementioned training in all the disciplines, but a strong thread connecting the population, habitat, and people triad (Krausman and Cain 2013) is interactions across landscapes (Liu and Taylor 2002). Simply put, management of the parcel (e.g., a wildlife management area) requires consideration of landscape context.

Wiens et al. (2002) described three basic challenges that confront ecologically sound management while integrating ecological principles across heterogeneous space. First, parcels of land are treated in isolation under the auspices of human constructs (i.e., ownership, landowner objectives), but not with regards to ecosystem processes that are inherently landscape oriented. Second, even though manage-

ment decisions are often restricted to shortened timescales or small spatial scales, ecological systems operate on much broader scales. Finally, it is important to integrate ecology with factors that are needed to make decisions (e.g., economics, land ownership, politics, sociology; Johnson et al. 1999, Daily and Walker 2000, Dale et al. 2000). Wildlife managers are familiar with these challenges, especially sociology. A state biologist with the task of setting harvest regulations for an exploited species (e.g., black bears [*Ursus americanus*] in Florida, USA) is challenged to make the correct decision.

Our objective in this chapter is to present a broad understanding to facilitate integration of landscape ecology into everyday wildlife management and essentially confront the challenges elucidated by Wiens et al. (2002) as well as a realization that there are many existing similarities between common wildlife management practices and landscape ecological concepts. An entire book could be written on just these topics, and there have been several on landscape ecology (Bissonette 2012, Turner and Gardner 2015) and wildlife management (Krausman and Cain 2013, Fryxell et al. 2014), but the integration of both is less common (Bissonette and Storch 2003). We first focus on the history of these disciplines and how the idea of spatial pattern and process is implicit in wildlife management and ecology. We then illustrate the similarities between wildlife management principles and landscape ecological concepts, and establish merit for the adoption of landscape perspectives among wildlife managers. We frame our discussion using four common wildlife management concepts: island biogeography, habitat, niche, and edge.

The Interplay of Landscape Ecology and the Lineage of Wildlife Management

Wildlife management as a discipline has evolved since its formal genesis in the early 20th century. The founding fathers of wildlife management (e.g., A. Leopold, H. L. Stoddard) developed and instigated the initial concepts and principles of the discipline. Shaped by the societal values of the time, they were solidly centered on the management of game species. Pointedly, the first sentence in Leopold's (1933:3) textbook states, "Game management is the art of making land produce sustained annual crops of wild game for recreational use." Leopold as a person was much more holistic, valuing game and nongame species (Leopold 1966), but his technical works were primarily about game: life history, population dynamics, range, monitoring, hunting, predator control, limiting factors, disease, economics, and policy (Leopold 1933). Leopold's famous tools for wildlife management (i.e., axe, cow, plow, fire, gun) still underpin the active management sector of contemporary wildlife management.

Concomitantly, Stoddard (1931) was establishing methods to maintain and enhance northern bobwhite (*Colinus virginianus*) populations. Stoddard was a habitat manager at heart and tended to think of things at the scale of a property, but he did think about spatial processes, albeit simplistically. Stoddard (1931) made numerous observations about the distances between captures of individual bobwhites. He also concluded that movement equalizes abundance. In this context, movement refers to dispersal

events that shape population structure. Stoddard specifically noted that some quail moved up to several kilometers, allowing them to occupy available habitat. This idea of spatial movements affecting demography is a central tenet in many foundational landscape ecological concepts (e.g., ideal free distribution and source-sink dynamics; see Fretwell and Calver 1969 and Pulliam 1988, respectively).

Leopold was more explicit with his landscape thinking. Silbernagel (2003) makes a strong case that Leopold applied spatial theory (i.e., landscape ecology) through many avenues in his conservation design work (Table 2.1). Leopold determined that game range must have the essential environmental types represented. This idea that landscape composition was important to wildlife was novel to North America at the time. Grinnell (1917) introduced the niche concept decades prior, but did not formally include the idea that abiotic resources needed by a species could come from the landscape instead of a single patch. Leopold implicitly generalized the concept of niche without even knowing it!

Leopold also developed the law of interspersion to describe the positive association between animal density and heterogeneity of resources in the landscape (Leopold 1933). Thus not only was composition important, but also juxtaposition affected movements and density: "The game must usually be able to reach each of the essential types each day. The maximum population of any given piece of land depends, therefore, not only on its environmental types or composition, but also on the *interspersion* of these types in relation to the cruising radius of the species. Composition and interspersion are thus the two principal determinants of potential abundance on game range" (Leopold 1933:128–129). Leopold's classic patchwork landscape is crystalized in the memory of most wildlife biologists and managers (Fig. 2.1).

Silbernagel (2003) makes a case that the creation of interspersion as a concept led to the concept of edge. Others have questioned Leopold's law in the

Table 2.1. Relationship of Leopold's Spatial Terms to Contemporary Landscape Ecology Concepts

Term	Leopold's (1933) Definition	Related Contemporary Concept/Term	Turner et al.'s (2001) Definition
Composition	Not defined explicitly but suggested as the mix of environmental types on a particular landscape/range.	Heterogeneity	Quality or state consisting of dissimilar elements, as with mixed habitats or cover types occurring on a landscape.
Cover	Vegetative or other shelter for game.	Cover type	Category within a classification scheme defined by the user that distinguishes among the different habitats, ecosystems, or vegetation types on a landscape.
Covert	A geographic unit of game cover.	Patch	Surface area that differs from its surroundings in nature or appearance.
Cruising radius	The distance between locations at which an individual animal is found at various hours of the day, or at various seasons, or during various years.	—	Related concepts are used but not defined here; e.g., dispersal distance, foraging range, and the like.
Edge	Not defined explicitly but occurs where the types of food and cover needed by game species come together.	Edge	Length of adjacency between cover types.
Interspersion	The degree to which environmental types are intermingled or interspersed on a game range.	Configuration	Specific arrangement of spatial elements; also spatial, or patch, structure.
Mobility	The tendency of the individual animal to change location during the day or between seasons or years.	Connectivity	Relates to spatial continuity of a habitat or cover type across a landscape.
Game range	Not defined explicitly but suggests a piece of land suitable for a given species of game; includes food and cover; optimum range composition.	Landscape	Area that is spatially heterogeneous in at least one factor of interest.
—	Optimum range composition.	Matrix	Background cover type in the landscape.
Remise	European term for an artificially established game bird covert. Sometimes includes food as well.	—	Relates indirectly to patch, corridor, matrix, habitat.

Source: Adapted from Silbernagel (2003:table 1).

context of the definition of edge or type peripheries (Guthery and Bingham 1992), but Silbernagel (2003) argues that Leopold had a hypothesis that tied spatial configuration of the landscape to wildlife populations, which was likely the first of its kind in the literature. Not to be stopped there, Leopold (1933) suggested that some species (i.e., those with high type requirements and low radius) were positively affected by edge (e.g., northern bobwhite), whereas forest and range game (e.g., white-tailed deer [*Odocoileus virginianus*]) are best suited for a matrix dominated by forest with minimal interspersion. Over time, in parallel with advances in landscape ecolog-

ical concepts such as functional edges (McCollin 1998) and spatial scale (Kie et al. 2002), our understanding of the importance of interspersion to wildlife has progressed. Nonetheless, Leopold's species-specific view of landscape heterogeneity was novel at the time, and prominent landscape ecologists still argue the continued need for such landscape thinking (Fahrig et al. 2011).

Troll formulated landscape ecology as a discipline in 1939 (Troll 1939, Turner and Gardner 2015). Without question, the technology of the time kept A. Leopold and C. Troll from knowing about each other's work. It is clear, however, that concepts cen-

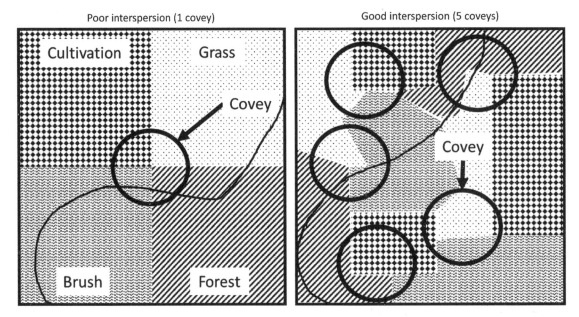

Figure 2.1. Leopold illustrated the concept of interspersion through a pair of diagrams that represent poor interspersion (*left*) versus good interspersion (*right*). Note that both diagrams have the same types and total area.   Modified from Leopold (1933).

tral to landscape ecology shaped the trajectory of wildlife management as a discipline.

Commonalities of Wildlife Ecology and Management with Landscape Ecology

Wildlife management can be simplified into four options for wildlife populations: make it increase, make it decrease, manage the population for sustainable yield, or do nothing (Leopold 1933, Krausman and Cain 2013, Fryxell et al. 2014). Biologists accomplish these goals using population management, habitat management, people management, or a combination of all three. In all cases, biologists take specific management actions (e.g., prescribed fire) that are based on a set of underlying principles and assumptions. The remainder of this chapter describes how landscape ecology contributes to the underlying principles associated with common management paradigms and how management at the parcel scale is implicitly landscape management. There are reoc-

curring themes in our discussion linking landscape ecology to wildlife management, particularly patch dynamics and heterogeneity.

Population Structure

To discuss the ties between landscape ecology and wildlife management, we first set the stage for how management of land parcels scales up to landscape or regional scales. How are groups of subpopulations within or among land parcels connected to each other, and how does wildlife management influence these broad patterns? We use island biogeography theory (and, by extension, the patch mosaic model) to explain how individuals and populations are connected across space. Although there are ways to classify landscapes without using patches (e.g., the continua-umwelt model; Manning et al. 2004) and there are alternative (yet related) ways to describe population structure (Box 2.1), much of landscape ecology can be explained and related to wildlife management using island biogeography as a foundational concept.

ISLAND BIOGEOGRAPHY THEORY AND THE PATCH MOSAIC MODEL

MacArthur and Wilson's (1967) classic monograph is woven throughout modern landscape ecology and wildlife ecology and management. They described how the area of an island and its distance to the mainland affect biodiversity. At the core of island biogeography theory is the dynamic equilibrium model that predicts the number of species found on an island will be a function of opposing forces leading to colonization of new species (gains) and losses through local extinctions, creating a constant turnover of species on the island through time (MacArthur and Wilson 1967, Whittaker and Fernández-Palacios 2007).

Through natural extension, this theory has been applied to terrestrial systems and led to the formulation of the patch mosaic model (Forman 1995, Wiens 1995, Wu and Levin 1997, Pickett and White 2012). Landscape ecology defines patches technically as homogeneous areas different from their surroundings (Forman 1995) in nature or appearance (Forman 1995, Turner and Gardner 2015). These patches are typically embedded in a matrix or the background ecological system (Forman 1995). Together the patches and matrix create a pattern often described as the mosaic (Forman 1995). Patches in this model are not static; they fluctuate in existence and the resources they provide animals.

In the context of wildlife ecology and management, we simply ascribe function to patches based on landcover or land use. They are often defined by their resources (e.g., food) or as habitat patches (which is not entirely correct; see "Patch Mosaic Model" below and Morrison and Block, Chap. 3, this volume). Farmlands are examples of patchy landscapes that can be described similarly using a wildlife management approach or a landscape ecology approach. There are distinct, homogeneous patches of land (e.g., forest, row crops) that are embedded in a matrix of the most common land type (e.g., cattle operations), and these patterns across broad scales produce a mosaic of patch types. The management or exploitation of these patches over time defines the dynamic nature of the landscape mosaic.

PATCH MOSAIC MODEL: FOOD AS AN EXAMPLE RESOURCE

Most wildlife biologists and managers are comfortable with the idea of a patch, especially a food patch. Food availability is an ultimate factor that contributes to the reproduction and survival (i.e., fitness) of an individual (Hildén 1965). Wildlife managers that attempt to increase populations manage vegetation to promote food resources or vice versa to reduce the population. The patch mosaic model's linkages to food resources reside in optimal foraging theory (Charnov 1976, Pyke et al. 1977), where food is assumed to be patchily distributed across an area, which by definition creates a landscape. Thus if food occurs as patches surrounded by a matrix, then as wildlife biologists we must begin to think explicitly about larger scales to be effective managers.

Habitat

Habitat is an area with the amalgamation of resources (e.g., food, cover, and water) and environmental conditions (e.g., temperature, precipitation) that promote occupancy (see Fig. 2.2A, where the Grinellian Niche is analogous to habitat; Whittaker et al. 1973, Pulliam 2000) by individual animals that assure survival and reproduction (Hall et al. 1997, Morrison et al. 2012). In landscape ecology circles, habitat is often synonymous with habitat type or habitat patch, but in reality these are landcover types or patch types. Just as habitat is a species-centric concept, recent landscape ecology literature strives to measure landscapes from a species-centered approach (Betts et al. 2014).

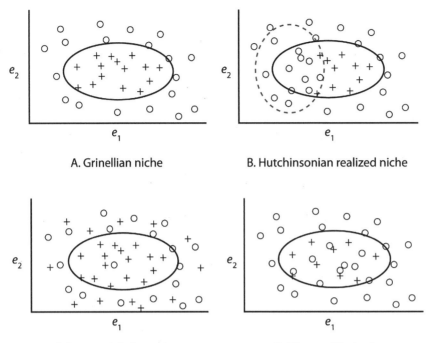

Figure 2.2. Four views of the relationship between niche and species distribution. In each diagram the solid oval refers to the fundamental niche or the combination of environmental factors (e_1 and e_2) for which the species has a finite rate of increase (λ) greater than or equal to 1.0. The pluses indicate the presence of the species in a patch of habitat characterized by particular values of e_1 and e_2, and the zeroes indicate the absence of the species in a patch of habitat. According to the Grinnellian niche concept (*A*), a species occurs everywhere that conditions are suitable and nowhere else. Hutchinson's realized niche concept (*B*) postulates that a species will be absent for those portions of the niche space that are utilized by a dominant competitor. According to source-sink theory (*C*), a species may commonly occur in sink habitat where $\lambda < 1.0$. Metapopulation dynamics and dispersal limitation (*D*) posit that species are frequently absent from suitable habitat because of frequent local extinctions and the time required to recolonize suitable patches. Reproduced from Pulliam (2000:fig. 1).

DISTURBANCE ECOLOGY AND HABITAT MANAGEMENT
Disturbance events alter ecosystem structure, community, or populations and can change resource availability because of an altered physical environment (White and Pickett 1985; cited in Turner and Gardner 2015). The pattern of disturbance over time within an ecosystem, its disturbance regime, is characterized by its frequency, intensity, and spatial scale, which in turn dictate severity and residual effects from a species perspective (Turner et al. 1998).

Habitat management usually consists of modifying existing patches or creating new patch types to manipulate species-specific resources (i.e., influencing composition and dynamicity of the patch mosaic landscape). Managers employ concepts of disturbance ecology (and thus landscape ecology) when performing habitat management; in fact, successful management of natural resources emulates natural disturbance regimes (Long 2009, Drapeau et al. 2016). Prescribed fire is an example of habitat management equating to a vector of disturbance that manipulates the density and spatial arrangement of resource patches (i.e., landscape heterogeneity). When managers compose a prescribed fire plan for a longleaf pine (*Pinus palustris*) landscape, they must make decisions about burn unit size (spatial scale of disturbance), rotation (frequency of disturbance), and acceptable burn conditions (intensity of disturbance). The manager dictates the severity of disturbance. Management decisions repeated across many landscapes over many years translate to a manipulation of the interactions between organisms and patchy resources over broad scales. Luckily, there is a wealth of literature on how landscape patterns influence population processes (Turner 1989, Dunning et al. 1992, Pickett and Cadenasso 1995, Tscharntke et al. 2012, Turner and Gardner 2015).

HABITAT FRAGMENTATION

Habitat fragmentation and habitat loss have been common themes of landscape ecology and wildlife ecology and management. Unfortunately, these terms have been used synonymously, creating a great deal of confusion among landscape ecologists and wildlife biologists. Fahrig (2003) synthesized these two concepts in a seminal review. The process of fragmentation results in four effects on habitat pattern: reduction in habitat amount (i.e., habitat loss), increase in number of patches, decrease in sizes of habitat patches, and increase in isolation of patches (Fahrig 2003). The reversal of habitat amount reductions or habitat loss is the underpinning of habitat management. But what are the commonalities with these other effects of fragmentation? Regrettably, given the ambiguity of the term *habitat fragmentation*, many biologists have dogmatic beliefs regarding the negative effects of fragmentation on wildlife populations. Habitat fragmentation concepts emerged from the theory of island biogeography (Haila 2002), which likely contributed to the idea that fragmentation resulted in an automatic negative effect because the distance from a patch to its neighbor was analogous to isolation. Not to say these negative effects do not exist, because they are real (discussed in "Edge," below). The implications of these effects are important to wildlife conservationists trying to conserve biodiversity in managed landscapes.

Niche Theory

Morrison and Block (Chap. 3, this volume) state that the distribution of animals is intimately tied to the concept of niche. The concept of niche (Grinnell 1917, Gause 1934, Elton 1946, Hutchinson 1957, Whittaker et al. 1973) has been around for almost a century and is fundamental to our understanding of how abiotic and biotic factors determine where a species can survive and reproduce. Hutchinson's n-dimensional hypervolume is perhaps the niche concept that has the closest commonality with landscape ecology (Hutchinson 1957). Hutchinson (1957:416; cited by Pulliam 2000) stated, "consider two independent environmental variables e_1 and e_2 which can be measured along ordinary rectangular coordinates . . . an area is defined, each point of which corresponds to a possible environmental state permitting the species to exist indefinitely." In this scenario, a species occurs everywhere that conditions are suitable and never occurs where conditions are unsuitable (Fig. 2.2A). Because of competition, however, the entire niche would not be filled or realized (i.e., competitive exclusion; Gause 1934, Hardin 1960; Fig. 2.2B). Hutchinsonian and contemporary niche concepts were essentially spatially implicit where resources were assumed to be homogeneous in a given space. Resources and populations are not, however, homogeneous.

HETEROGENEITY AFFECTING RESOURCE DISTRIBUTION

Environmental heterogeneity has been viewed as a determinant of species diversity (Hutchinson and MacArthur 1959, MacArthur and MacArthur 1961) and wildlife population size (Leopold 1933). Later chapters (see Part II of this volume) discuss how heterogeneity should be incorporated into wildlife management, but we introduce its roots in landscape ecology and wildlife management. Heterogeneity, and its usefulness in wildlife management, is bound by how you define landscape features (i.e., patches).

Box 2.1 Metapopulation Theory

Metapopulation theory is another way of describing the spatial structuring of populations and can be used to predict how species will respond to future changes in land use. Levin and Paine (1974), Levin (1976), and Hanski (1998) introduced the concept of a metapopulation as a system of local populations (found in patches) connected by dispersing individuals among patches, which together form a functioning population (Fig. 2.3). Abundance of individuals oscillates through time and space within subpopulations. Patches go through extinction events and are colonized by individuals from other subpopulations. For this to occur, dispersal between subpopulations is necessary.

Several models have been proposed to explain how individuals of a subpopulation disperse to another subpopulation. Slatkin (1985; see Fig. 2.4) postulated that dispersal may simply be a random walk in space with few, if any, factors influencing species dispersal. It is not likely, however, that most species perceive the environment in this manner, and this model is not well supported in the ecological literature. The ideal free distribution model (Fretwell and Lucas 1970; see Fig. 2.5) or balanced dispersal model (Doncaster et al. 1997) states that dispersal patterns are contingent upon the fitness of the individual in a given habitat and dispersal is not constrained by population density of the other patches.

Conversely, for general source-sink models (Holt 1985, Pulliam 1988; see Fig. 2.6), dispersal is constrained between patches, density-independent and -dependent dispersal are both possible, and habitat quality may vary greatly among patches. The source-sink dynamics also dictate the occupancy of niche space or non-niche space (Figs. 2.2C, D). Under strict source-sink dynamics, a species may occupy space beyond its fundamental niche (e.g., the population will decline). Also, as the name implies, the presence of a sink is assumed under source-sink models (Fig. 2.6).

Finally, Senar et al. (2002; see Fig. 2.7) proposed a model based on work with citril finches (*Serinus citronella*), whereby animals may disperse from low- to high-quality sites because the high-quality sites act as pools of genetic variability and are sources of higher food quality. Given the assumptions of each model,

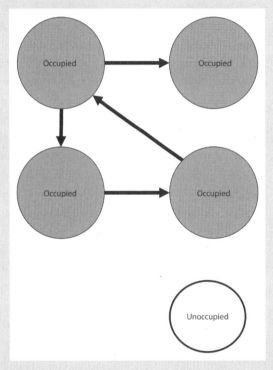

Figure 2.3. Metapopulation structure where patches are occupied (gray) with subpopulations and individuals disperse and colonize other patches. Unoccupied patches (white) are possible because of dispersal limitations.

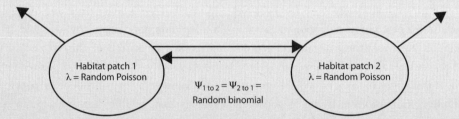

Figure 2.4. Slatkin (1985) postulated that dispersal may simply be a random walk-in space with few, if any, factors driving species dispersal. Ψ represents the probability that a species will transition from one patch to another, and λ represents population growth rate.

predictions may vary regarding dispersal. Further, the response of individual populations to human and environmental perturbations and land management actions will be dependent upon which model of dispersal is applicable in a given metapopulation system. For a given wildlife community, these management activities are likely to create multiple and potentially conflicting models of metapopulations, which then need to be incorporated into overall ecosystem management objectives.

The application of metapopulation theory to wildlife conservation is clear only in a few cases (e.g., pond breeding amphibians; Marsh and Trenham 2001). Although most wildlife population boundaries are not easily delineated, predator-prey dynamics may share common concepts with metapopulation ecology. Gause (1934) concluded that predator-prey systems were prone to extinctions, and Huffaker's (1958) work using mites in a microcosm provides evidence that the spatial arrangement of refugia facilitated dispersal and persistence of prey. These concepts have not matriculated to wildlife management. Managers have tried to directly create patches of refugia through predator exclusion (Conner et al. 2016) or predator removal (Beasom 1974), but it is uncommon for managers to dictate spatial heterogeneity of refugia to increase prey populations.

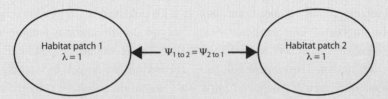

Figure 2.5. The ideal free distribution model (Fretwell and Lucas 1970), or balanced dispersal model (Doncaster et al. 1997), states that dispersal patterns are contingent upon the fitness of the individual in a given habitat type and dispersal is not constrained by population density of the other patches. Ψ is the probability that an individual transitions from one patch to another, and λ represents the population growth rate.

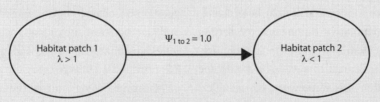

Figure 2.6. General source-sink models (Holt 1985, Pulliam 1988) whereby dispersal is constrained between patches and density-independent and -dependent dispersal are both possible. Additionally, habitat quality between patches may hinder or facilitate dispersal. A patch with a growth rate (λ) greater than 1 and at carrying capacity is considered a source.

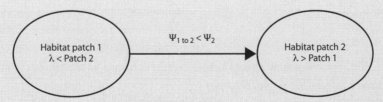

Figure 2.7. Senar et al. (2002) model based on work with Citril finches (*Serinus citronella*), whereby animals disperse from lower- to higher-quality patches because the high-quality sites act as pools of genetic variability and are sources of higher food quality, enhancing reproduction and survival rates. Ψ is the probability that an individual will transition from one patch to another, and λ represents the population growth rate.

By nature, the more specifically one defines patch types, the higher heterogeneity you will have within that landscape (Turner and Gardner 2015). This is an important consideration when managing landscape heterogeneity. Specifically, goals for increasing or decreasing the functionality of a landscape for a certain species, guild, or taxa of interest should be established before management action is taken. Patches should be defined by their function to the animal group of interest (Fahrig et al. 2011). For northern bobwhite, for example, woody cover may be classified as thermal and escape cover, and native grassland cover may be classified as nesting and brooding cover.

For practical purposes, heterogeneity can be dissolved into two main categories: compositional and configurational heterogeneity. Compositional heterogeneity is the number of landcover types within a landscape, and configurational heterogeneity is the complexity of the spatial pattern of patches within a landscape (Fahrig and Nuttle 2005). Pioneers of wildlife management understood the importance of variation in patch types to promote wildlife populations.

Indeed, most biologists and managers are familiar with the multitude of challenges when managing habitat. The concept of compositional heterogeneity provides a quantitative framework for setting landscape management goals to meet niche requirements of wildlife. Central to successful landscape management within this framework is an understanding of how individuals interact with certain landcover types and how these interactions change over time through predator-prey dynamics, resource selection, and as a result of intra- and interspecific competition (see Morrison and Block, Chap. 3, and Martin et al., Chap. 7, this volume).

EDGE

Tscharntke et al. (2012) reviewed how spatial configuration of landcover influences population processes, but we focus on a single important aspect of configurational heterogeneity that applies directly to wildlife management: edge. Increasing edge densities of certain landcover types can elicit either a positive or negative population response (Fahrig 2003). Returning to the example of northern bobwhite, woody cover (thermal and escape cover) and native grasslands (nesting and brooding habitat) can be classified more broadly as beneficial landcover. Within the class of beneficial landcover, increasing configurational heterogeneity can promote movement between landcover types that meet differing life history requirements (i.e., landscape complementation; Dunning et al. 1992). Increasing configuration heterogeneity can be a double-edged sword, however, as implementing beneficial landcover within a matrix of detrimental landcover (e.g., closed canopy forest) may benefit northern bobwhite only temporarily and may promote lower fitness over time. This is common sense to many, but without the theoretical framework of landscape heterogeneity, these considerations are potentially ignored.

Summary

Broadscale management of wildlife populations is a complex task because of spatial variability in climate, geomorphic characteristics, and vegetative structure and composition; changing sociopolitical undertones; and limited monetary and logistical resources. We provided a cursory overview of how principles of landscape ecology mirror those of wildlife management and how prevailing landscape ecological theories can help practitioners anticipate broadscale responses of populations to management activities.

Foundational concepts of wildlife management have built-in spatial considerations, and approaching management problems from a landscape ecology perspective can help practitioners make decisions that will help expand the consequences of habitat management from a local scale to the landscape scale.

LITERATURE CITED

Beasom, S. L. 1974. Relationships between predator removal and white-tailed deer net productivity. Journal of Wildlife Management 38:854–859.

Betts, M. G., L. Fahrig, A. S. Hadley, K. E. Halstead, J. Bowman, W. D. Robinson, J. A. Wiens, and D. B. Lindenmayer. 2014. A species-centered approach for uncovering generalities in organism responses to habitat loss and fragmentation. Ecography 37:517–527.

Bissonette, J. A. 2012. Wildlife and landscape ecology: effects of pattern and scale. Springer Science and Business Media, Berlin, Germany.

Bissonette, J. A., and I. Storch. 2003. Landscape ecology and resource management: linking theory with practice. Island Press, Washington, DC, USA.

Charnov, E. L. 1976. Optimal foraging, the marginal value theorem. Theoretical Population Biology 9:129–136.

Conner, L. M., M. J. Cherry, B. T. Rutledge, C. H. Killmaster, G. Morris, and L. L. Smith. 2016. Predator exclusion as a management option for increasing white-tailed deer recruitment. Journal of Wildlife Management 80:162–170.

Conroy, M. J., and J. T. Peterson. 2013. Decision making in natural resource management: a structured, adaptive approach. John Wiley and Sons, Hoboken, New Jersey, USA.

Daily, G. C., and B. H. Walker. 2000. Seeking the great transition. Nature 403:243–245.

Dale, V. H., S. Brown, R. Haeuber, N. Hobbs, N. Huntly, R. Naiman, W. Riebsame, M. Turner, and T. Valone. 2000. Ecological principles and guidelines for managing the use of land. Ecological Applications 10:639–670.

Doncaster, C. P., J. Clobert, B. Doligez, E. Danchin, and L. Gustafsson. 1997. Balanced dispersal between spatially varying local populations: an alternative to the source-sink model. The American Naturalist 150:425–445.

Drapeau, P., M. A. Villard, A. Leduc, and S. J. Hannon. 2016. Natural disturbance regimes as templates for the response of bird species assemblages to contemporary forest management. Diversity and Distributions 22:385–399.

Dunning, J. B., H. R. Pulliam, and B. J. Danielson. 1992. Ecological processes that affect populations in complex landscapes. Oikos 65:169–175.

Elton, C. 1946. Competition and the structure of ecological communities. Journal of Animal Ecology 15:54–68.

Fahrig, L. 2003. Effects of habitat fragmentation on biodiversity. Annual Review of Ecology, Evolution, and Systematics 34:487–515.

Fahrig, L., J. Baudry, L. Brotons, F. G. Burel, T. O. Crist, R. J. Fuller, C. Sirami, G. M. Siriwardena, and J. Martin. 2011. Functional landscape heterogeneity and animal biodiversity in agricultural landscapes. Ecology Letters 14:101–112.

Fahrig, L., and W. K. Nuttle. 2005. Population ecology in spatially heterogeneous environments. Pages 95–118 in G. M. Lovett, M. G. Turner, C. G. Jones, et al., editors. Ecosystem function in heterogeneous landscapes. Springer, New York, New York, USA.

Forman, R. T. 1995. Land mosaics: the ecology of landscapes and regions. Springer, New York, New York, USA.

Fretwell, S. D., and J. S. Calver. 1969. On territorial behavior and other factors influencing habitat distribution in birds. Acta Biotheoretica 19:37–44.

Fretwell, S. D., and H. L. Lucas, Jr. 1970. On territorial behavior and other factors influencing habitat distribution in birds. I. Theoretical development. Acta Biotheoretica 19:16–36.

Fryxell, J. M., A. R. Sinclair, and G. Caughley. 2014. Wildlife ecology, conservation, and management. John Wiley and Sons, Hoboken, New Jersey, USA.

Gause, G. F. 1934. The struggle for existence. Science 79:16–17.

Grinnell, J. 1917. The niche-relationships of the California thrasher. Auk 34:427–433.

Guthery, F. S., and R. L. Bingham. 1992. On Leopold's principle of edge. Wildlife Society Bulletin 20:340–344.

Haila, Y. 2002. A conceptual genealogy of fragmentation research: from island biogeography to landscape ecology. Ecological Applications 12:321–334.

Hall, L. S., P. R. Krausman, and M. L. Morrison. 1997. The habitat concept and a plea for standard terminology. Wildlife Society Bulletin 25:173–182.

Hanski, I. 1998. Metapopulation dynamics. Nature 396:41–49.

Hardin, G. 1960. The competitive exclusion principle. Science 131:1292–1297.

Hildén, O. 1965. Habitat selection in birds: a review. Annales Zoologici Fennici 2:53–75.

Holt, R. D. 1985. Population dynamics in two-patch environments: some anomalous consequences of an optimal habitat distribution. Theoretical Population Biology 28:181–208.

Huffaker, C. B. 1958. Experimental studies on predation: dispersion factors and predator-prey oscillations. Hilgardia 27:343–383.

Hutchinson, G. E. 1957. Concluding remarks. Cold Spring Harbor Symposia on Quantitative Biology 22:415–427.

Hutchinson, G. E., and R. H. MacArthur. 1959. A theoretical ecological model of size distributions among species of animals. American Naturalist 93:117–125.

Johnson, K. N., J. Agee, R. Beschta, V. Dale, L. Hardesty, J. Long, L. Nielsen, B. Noon, R. Sedjo, and M. Shannon.

1999. Sustaining the people's lands: recommendations for stewardship of the national forests and grasslands into the next century. Journal of Forestry 97:6–12.

Kie, J. G., R. T. Bowyer, M. C. Nicholson, B. B. Boroski, and E. R. Loft. 2002. Landscape heterogeneity at differing scales: effects on spatial distribution of mule deer. Ecology 83:530–544.

Krausman, P. R., and J. W. Cain. 2013. Wildlife management and conservation: contemporary principles and practices. Johns Hopkins University Press, Baltimore, Maryland, USA.

Leopold, A. 1933. Game management. Charles Scribner's Sons, New York, New York, USA.

Leopold, A. 1966. A Sand County almanac: and sketches here and there. Oxford University Press, Oxford, UK.

Levin, S. A. 1976. Population dynamic models in heterogeneous environments. Annual Review of Ecology and Systematics 1:287–310.

Levin, S. A., and R. T. Paine. 1974. Disturbance, patch formation, and community structure. Proceedings of the National Academy of Sciences 71:2744–2747.

Liu, J., and W. W. Taylor. 2002. Integrating landscape ecology into natural resource management. Cambridge University Press, Cambridge, UK.

Long, J. N. 2009. Emulating natural disturbance regimes as a basis for forest management: a North American view. Forest Ecology and Management 257:1868–1873.

MacArthur, R. H., and J. W. MacArthur. 1961. On bird species diversity. Ecology 42:594–598.

MacArthur, R. H., and E. Wilson. 1967. The theory of island biogeography. Princeton University Press, Princeton, New Jersey, USA.

Manning, A. D., D. B. Lindenmayer, and H. A. Nix. 2004. Continua and umwelt: novel perspectives on viewing landscapes. Oikos 104:621–628.

Marsh, D. M., and P. C. Trenham. 2001. Metapopulation dynamics and amphibian conservation. Conservation Biology 15:40–49.

McCollin, D. 1998. Forest edges and habitat selection in birds: a functional approach. Ecography 21:247–260.

Morrison, M. L., B. Marcot, and W. Mannan. 2012. Wildlife–habitat relationships: concepts and applications. Island Press, Washington, DC, USA.

Pickett, S. T. A., and M. L. Cadenasso. 1995. Landscape ecology: spatial heterogeneity in ecological systems. Science 269:331–334.

Pickett, S. T. A., and P. S. White. 2012. The ecology of natural disturbance and patch dynamics. Academic Press, Cambridge, Massachusetts, USA.

Pulliam, H. R. 1988. Sources, sinks, and population regulation. American Naturalist 132:652–661.

Pulliam, H. R. 2000. On the relationship between niche and distribution. Ecology Letters 3:349–361.

Pyke, G. H., H. R. Pulliam, and E. L. Charnov. 1977. Optimal foraging: a selective review of theory and tests. Quarterly Review of Biology 52:137–154.

Senar, J. C., M. J. Conroy, and A. Borras. 2002. Asymmetric exchange between populations differing in habitat quality: a metapopulation study on the citril finch. Journal of Applied Statistics 29:425–441.

Silbernagel, J. 2003. Spatial theory in early conservation design: examples from Aldo Leopold's work. Landscape Ecology 18:635–646.

Slatkin, M. 1985. Gene flow in natural populations. Annual Review of Ecology and Systematics 16:393–430.

Stoddard, H. 1931. The bobwhite quail: its habits, preservation and increase. Charles Scribner's Sons, New York, New York, USA.

Troll, C. 1939. Luftbildplan und okologische Bodenforschung. Zeitschrift der Gesellschaft fur Erdkunde, Berlin, Germany.

Tscharntke, T., J. M. Tylianakis, T. A. Rand, R. K. Didham, L. Fahrig, P. Batáry, J. Bengtsson, Y. Clough, T. O. Crist, C. F. Dormann, et al. 2012. Landscape moderation of biodiversity patterns and processes—eight hypotheses. Biological Reviews 87:661–685.

Turner, M. G. 1989. Landscape ecology: the effect of pattern on process. Annual Review of Ecology and Systematics 20:171–197.

Turner, M. G., W. L. Baker, C. J. Peterson, and R. K. Peet. 1998. Factors influencing succession: lessons from large, infrequent natural disturbances. Ecosystems 1:511–523.

Turner, M. G., and R. H. Gardner. 2015. Landscape ecology in theory and practice: pattern and process. Springer, New York, New York, USA.

Turner, M. G., R. H. Gardner, and R. V. O'Neill. 2001. Landscape ecology in theory and practice: pattern and process. Springer, New York, New York, USA.

White, P. S., and S. T. A. Pickett. 1985. The ecology of natural disturbance and patch dynamics. Academic Press, Orlando, Florida, USA.

Whittaker, R. H., S. A. Levin, and R. B. Root. 1973. Niche, habitat, and ecotope. American Naturalist 107:321–338.

Whittaker, R. J., and J. M. Fernández-Palacios. 2007. Island biogeography: ecology, evolution, and conservation. Oxford University Press, Oxford, UK.

Wiens, J. A. 1995. Landscape mosaics and ecological theory. Pages 1–26 in L. Harrison, L. Fahrig, and G. Merriam, editors. Mosaic landscapes and ecological processes. Springer, New York, New York, USA.

Wiens, J. A., B. Van Horne, and B. R. Noon. 2002. Integrating landscape structure and scale into natural resource management. Pages 23–67 *in* J. Liu and W. M. Taylor, editors. Integrating landscape ecology into natural resource management. Cambridge University Press, Cambridge, UK.

Wu, J., and S. A. Levin. 1997. A patch-based spatial modeling approach: conceptual framework and simulation scheme. Ecological Modelling 101:325–346.

3 — Wildlife–Landscape Relationships

Michael L. Morrison and
William M. Block

A Foundation for Managing Habitats on Landscapes

Introduction

Numerous journal articles, conference proceedings, books, and symposia have been devoted to the topic of habitat; this is a fact that requires no long list of citations for support (see review by Morrison et al. 2006). Yet there remains immense confusion throughout all corners of the ecological profession on the definition of habitat. This confusion arises largely because habitat is a concept and not a testable theory per se. The enormous proliferation of terms attempting to describe habitat is clear testimony to the vagueness of the concept. When such a central concept causes such consternation among scientists, there is no possibility that more data will solve the problem or result in substantial improvement in the way we manage and ultimately conserve animal populations.

One of our objectives is for readers to gain added insight and knowledge in three areas. First, to understand how the habitat concept can be used as a means of drawing on the wide array of concepts and tools in landscape ecology to address wildlife management. Second, to apply habitat as the common concept linking wildlife management to landscape ecology to improve communication between wildlife managers and landscape ecologists. Finally, to apply

concepts and tools of landscape ecology to reframe habitat management as a part of the landscape.

To meet these objectives requires more than a description of habitat and a plea for clarity in terminology. Thus a primary goal of this chapter is to integrate multiple ecological concepts into a coherent approach that ultimately leads to improving how we maximize the likelihood of long-term persistence of animals. We review and synthesize previously covered material, but we do so as a foundation for a comprehensive strategy for managing animal populations. We first review traditional approaches for studying wildlife–habitat relationships. To ensure a common understanding, we provide definitions for key terms related to habitat and population and build upon these definitions by incorporating the niche concept. Implicit to incorporating niche is consideration of biotic and abiotic factors that influence the distribution and habitat of a species across the landscape, which leads to discussion of assembly rules and the development of species distribution models.

The State of Wildlife–Habitat Relationships

In his review of habitat studies, Morrison (2012) concluded that no major change has occurred in the way we approach studies of wildlife–habitat relationships. Most advances since the 1970s have come in the form of improved technology and more sophisticated statistical analyses. We are not criticizing such advances, however, as reviews of most journal manuscripts focus on matters of technology and statistics, and largely downplay concerns about sampling design and the relevance thereof to the biological population (Block 2012). Because most habitat studies collect additional examples of phenomena that were already well studied, Morrison (2012) concluded that what we call wildlife–habitat relationships are failing to advance understanding and risk becoming outmoded.

Typical habitat studies examine a convenient study area, collect samples of the usual vegetation variables and other environmental parameters (often well more than needed), conduct a series of statistical analyses, and compare results to studies done at different times and usually different locations, with publication then justified by extrapolating findings to some unspecified larger area. In most cases the animals studied are only a part of the biological population, and any inferences that are drawn apply only to the place and time of study (Hurlbert 1984). The investigator may measure dozens of variables on thousands of plots, but the reality is that they have a sample size of one arbitrary study area, and extrapolation beyond that location is not justifiable. Even studies that incorporate measures of reproductive success or body condition usually fail to sample from a known segment of the population. Are we sampling from the middle or an edge of the range of the species? Are we sampling from only a small portion of the conditions that the species can occupy within a portion of their range? Is what we are doing of relevance to the biology of the species? The problem is that our local studies might yield results that are quickly swamped by interactions happening outside the boundaries of our artificial study area and are being modified by the presence, abundance, and behavior of multiple other species.

We seem to have things reversed. It seems logical that we would first determine the biological population of interest and then design a way to adequately sample characteristics of interest from that population, such as features of the environment being used (i.e., habitat). As noted above, however, with most of our wildlife studies we first designate a sampling area, find some individuals of interest, and measure things about them. It would be like going to the local shopping mall (because it is close and convenient) and asking the assembled mass of people (good sample size) if they thought the health benefits of outdoor recreation was overstated.

Thus we believe that a broader approach to habitat studies is needed: an approach that steps through various descriptions of the environment but does so under a comprehensive framework that also recognizes the closely linked factors of the animal population (in the biological sense); interactions with other animal species; and a number of key factors that ultimately determine the distribution, abundance, and persistence of animals. In the following sections we begin by defining relevant and useful habitat terminology; this is necessary so readers clearly understand our approach. We discuss habitat in its classical sense and in a hierarchical sense, integrate related and core concepts (e.g., niche), and then describe the link between habitat and landscape (which itself requires careful description). These next few sections set the stage for our description of what we think is a comprehensive way to study wildlife.

Definitions

There are only a few habitat-related terms that need definition, because only a few key terms and concepts are actually needed to understand, quantify, and manage a species' habitat (our terminology follows Morrison et al. 2006 and Morrison 2009). Unfortunately, there has been a proliferation of habitat terms, likely because concepts by nature are vague in ecology. Numerous authors have applied modifiers to the basic term *habitat*—including high-quality, microhabitat, macrohabitat, marginal, transitional, optimal, used, unused, and suitable—and the list seems to keep growing (Hall et al. 1997). Many of these modified terms arise from a failure to assign habitat to a single species (Rountree and Able 2007), or they are inappropriately used when population demographics are not considered (Hall et al. 1997). There cannot be unsuitable habitat because a lack of suitability simply means that it is not habitat in the first place. Attempts to divide the environment into habitat subcomponents is likewise inappropriate. For example, by definition, a habitat patch is nonsensical because it means there is an identifiable habitat that is separate from the animal. Yet we see this term used frequently. Because of this confusion and misuse of terms even within the same paper, Guthery and Strickland (2015) suggested that, to forestall overuse and misuse of the word, authors do a word search for "habitat" after finishing a paper to ensure that any particular usage matches the author's stated definition. A key concept to remember is that habitat must be linked to features associated with animal-use patterns and not refer to specific, fixed locations.

There are few appropriate terms to use in conjunction with habitat. Here, we briefly discuss those terms to set the stage for what follows in this chapter and in other chapters in this book. What we provide here has been published on previous occasions (Block and Brennan 1993, Hall et al. 1997, Morrison et al. 2006, Morrison 2009, Mathewson and Morrison 2015), and the fact we feel the need to repeat this discussion here indicates that the message is only slowly breaking through into the scientific literature.

Habitat use is the way an animal uses physical and biological components (i.e., resources) in an area. Habitat use focuses on how an organism uses their environment; habitat use is either a description of how resources are used or often it is quantified as the proportion of time that an animal spends within various components of the environment (Beyer et al. 2010).

Habitat selection is the hierarchical process of innate and learned decisions by an animal about where it should be across space and time to persist (Johnson

1980, Hutto 1985, Piper 2011). Hildén (1965) parsed habitat selection into proximate and ultimate factors. Proximate factors are cues that elicit a settling response by an individual to use an area. Ultimate factors are those that contribute to reproduction and survival, such as nest sites, food availability, and the like. Habitat selection is thus an evolutionary process based on fitness consequences that vary with differential resource use (Morris 2011).

Habitat preference is restricted to the consequence of the habitat–selection process, resulting in the disproportional use of some resources over others. Habitat use and habitat preference differ from habitat selection in that the former are based on quantifiable patterns using measures of availability (whatever and however availability is defined). Authors oftentimes confuse these concepts and use the term habitat selection to represent the outcome of statistical interpretation of habitat use compared to habitat availability (i.e., habitat–selection models; Mannan and Steidl 2013). Rather, habitat selection should be restricted to the behavioral or evolutionary process (Hutto 1985, Morris 2003, Beyer et al. 2010).

Habitat quality is the ability of the environment to provide conditions appropriate for individual and population persistence (Morrison 2009). Habitat quality is thus an assessment based on demographics (reproduction, survival, immigration, emigration) of individuals or populations, which fluctuate over space and time. Thus habitat per se has no inherent quality. Habitat quality is an outcome that can change through time over the same area. Throughout the literature are papers that assign quality to an area when they really mean the probability of an area to result in quality (i.e., the production of young or survival of individuals).

Guthery and Strickland (2015) reviewed uses of the term habitat and summarized its primary uses to mean where an organism lives, which includes biotic and abiotic requisites and time; an arbitrary area of interest (habitat) that contains demographically or behaviorally unique subareas (habitats) that in turn contain descriptively unique subareas (habitats);

and gibberish, that is, the meaning is apparently in-decipherable.

They then broke these meanings into three classes: classical, hierarchical, and irrational. The classical outlook includes definitions with meanings like a place to live or natural abode. Classical habitat is an existential matter: it either exists or it does not. They noted that the classical version of habitat is log-ically a dead-end concept except for its application to descriptive science. They defined the hierarchical outlook as an arbitrary area (level 1) with (demo-graphically, behaviorally) meaningful subdivisions (level 2) that are in turn subdivided (level 3), which may in turn be further subdivided (level 4). The hierarchical view of habitat represents sequential parsing, which is the study of an entity composed of parts that are in turn composed of parts and so on. Lastly, our favorite, the irrational outlook, includes any use of habitat in an indecipherable manner. Guthery and Strickland (2015) provided examples of the often tautological and multiple meanings ap-plied to habitat throughout the scientific literature. Guthery and Strickland (2015:17) asked if the clas-sical definition held sway over the hierarchical defi-nition, or vice versa, as the appropriate model for research and management. They commented that both seemed to have a place and concluded, in appar-ent exasperation, that "both will be with us forever." These comments highlight the plight that research-ers delving into habitat relationships find themselves stuck within.

The hierarchical concept of habitat selection has been promoted as a way to help ecologists under-stand how an individual decides (using innate and learned behaviors) where to settle; we then mea-sure this settlement to describe habitat. Areas (often circular in shape) of various sizes are constructed around the settlement location (e.g., foraging loca-tion, nest, den site), with increasingly gross mea-sures of the environment measured with distance from the settlement location. Some studies also at-tempt to provide a measurement of the presence or abundance of a few key competitors or predators of the focal species of interest. There is nothing incor-rect with proceeding in this manner, as the results give us a picture of where the animal was located at the time of measurement and some notion of condi-tions surrounding it. What these studies lack, how-ever, is much, if any, knowledge of how the focal species will respond to its surroundings as biotic and abiotic conditions change through time; this statement holds even when a study is repeated for multiple seasons or years. This is because habitat de-scriptions per se do not capture anything about the various dynamic processes that interact directly and indirectly with our focal species or group of species. Hierarchies should be defined by structures where the lower levels are bounded and enclosed by higher levels; for example, forests contain trees that contain leaves that contain cells that contain chloroplasts that contain DNA that contain base pairs and so on. Nested circles do not describe a hierarchy; rather, they represent discrete points along a scaling gradi-ent. This is shown by the fact that the same features (e.g., areal proportion of a specific landcover type) are generally measured within all of the nested cir-cles and compared across circle radii. In actual hier-archies, you cannot measure the same structures at lower levels because attributes of higher hierarchal levels are hidden; that is, you can look down but you cannot look up (e.g., you can measure photosynthe-sis at the tree, leaf, or cellular level, but you cannot measure tree geometry by studying cells). Thus, in true hierarchal analyses, the selection analysis at each smaller scale is bounded by the results of the selection at the next larger scale.

A key point here is that habitats are not static in time. Vegetation changes by succession or distur-bance, as does the rest of the biotic community. Thus habitats are dynamic in that they may become more beneficial to a species or may change to the point that they are no longer habitat for a focal species. Descriptions of habitat use are widely available for most species. It is a worthwhile endeavor to fill gaps in knowledge when certain areas of a species range have been ignored. Nevertheless, such studies only give us a description of habitat use at a point in time (or several points at best) and do little, if anything, to inform us of how the species of interest is occupying and persisting at a meaningful biological level. We must be clear: using a hierarchical approach to habitat description is not equivalent to a landscape approach.

We understand that defining the *biological population* is complicated because of discontinuities in distribution that are difficult to recognize, and we often know little about individual animal movements across any spatial scale. Regardless, we must first understand how the animals are assorted into meaningful, interacting groups of individuals before we can gather a proper sample. Population demography and viability have not been emphasized in most wildlife habitat studies; abundance, density, or occupancy has been the focus. Measures of abundance or density are emphasized because they do not require that we know anything about the underlying population structure. The area needed to support viable demographic units, however, changes as we change the size of our study areas.

The number of species across space is constantly changing and is simultaneously interacting with various numbers of individuals within each of those species. As we change spatial extents, these relationships likewise change. There are, of course, additional and more subtle interactions occurring, such as changes in sex and age ratios. Smallwood (2001) reported how increasing the spatial extent of a sampling area increasingly captured functionally significant demographic units of different taxa, including predators and prey (Fig. 3.1). Although this is a simple case, it clearly shows how the demographic relationships being expressed (e.g., density) change as the sampling area changes in size. Because population viability considers how long a population will persist into a specified future time, we must know how a species is structured into populations if we are to appropriately design our studies. The primary reason we study animals and where they live (habitat) is to ensure persistence into the future.

Morrison (2012:fig. 1) expanded the example given by Van Horne (2002:fig. 24.2) and Morrison (2009:fig. 3.5) that depicted how the abundance of a species could decline toward edges of its range (Fig. 3.2). Where we sample along environmental gradients relative to geographic location of the species will give us different relationships between a parameter and the response of the species (i.e., positive, negative, neutral). We also see that the response of a

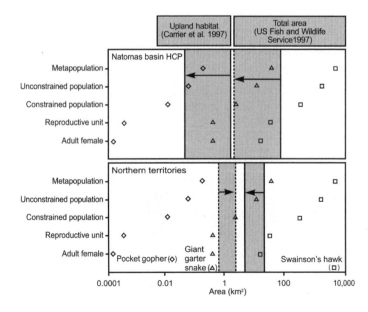

Figure 3.1. The areas needed to support functionally significant demographic units of animals vary with project spatial extent (size). The vertical lines correspond to habitat area that are upland (dashed lines) and both upland and wetland (solid lines) at the start and end of projects, as indicated by the respective origins and directions of the arrows. Arrows pointing left indicate loss of habitat space, and arrows pointing right indicate gains. Reproduced from Smallwood (2001:fig. 4).

species can change along a gradient even when distributions overlap (Fig. 3.2). Three major messages emerge. First, averaging across the gradient without regard to the underlying population structure (e.g., ecotypes) would result in wide variance in parameter estimates and give a false impression of how the species is responding to environmental conditions across space. Second, considering variation in how the species responds to environmental conditions is an important consideration, especially if source-sink dynamics (Pulliam 1988) come into play. Here, a species may be able to reproduce in one area (source) and their offspring disperse to another area where they are unable to breed (sink). Averaging abundance or reproductive output across a broader landscape would provide misleading information on how a species is performing within a smaller area. A study occurring in just a source or sink area would not represent the overall status of the population. And third, we would not know how the individuals inhabiting our snapshot in space (study area) were being influenced by adjacent individuals, or how relationships would change through time as individuals move across our study area. This neighborhood effect as described by Dunning et al. (1992) depends on the juxtaposition of patches on the landscape, relationships of species with other species in adjoining patches, and how species move among those patches. Simply declaring that the study population was closed for the duration of the study does nothing to relieve these real-world dynamics.

By not accounting in any manner for these spatial and temporal dynamics, our studies become simple case studies that document, perhaps precisely, habitat relationships at a specific time and place. Because we do not capture the broader dynamics, we cannot accurately predict how habitat relationships will change given future environmental conditions. Metareplication has been recommended as a means of piecing together knowledge across many different studies, thus potentially circumventing some of the problems associated with such local studies. Metareplication uses the results of studies of different

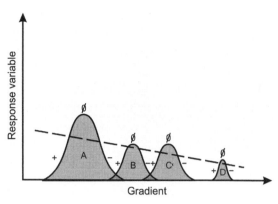

Figure 3.2. The response (e.g., density, productivity) of a species can vary across environmental gradients through space (*A–D*). Additionally, the portion of the gradient from where samples are taken can result in differing relationships (positive, +; negative, –; or neutral, Ø) with the response variable. The dashed line represents the hypothetical relationship when the underlying responses of different ecotypes are ignored. Reproduced from Morrison (2012:fig. 1).

sampling intensities and spatial extents to increase confidence that a relationship can be generalized across time and space (Johnson 2002). If conducted in a planned manner, repeating studies across space and time certainly has merit. Regardless of planning, however, such efforts do not account for any of the issues we have raised concerning broader-scale dynamics. Below we develop a more holistic approach to studying wildlife with the goal of maintaining species viability.

Moving beyond Habitat: The Niche Concept

Habitat is widely used because it is easy to see and to quantify its primary components; namely, vegetation and other obvious structures (e.g., rocks, water, topography). The distribution of animals, however, is intimately tied to the concept of niche. As reviewed in Morrison et al. (2006:chap. 1), Grinnell (1917) introduced the term niche when explaining the distribution of a single species of bird and included spatial considerations (e.g., reasons for a close association with a vegetation type), dietary dimensions, and constraints placed by the need to avoid predators. Thus in this view the niche included both positional (i.e., habitat) and functional roles in the community. Elton (1927) later described the niche as the status of an animal in the community and focused on trophic position and diet. A key advance in the niche concept was distinguishing the fundamental niche from the realized niche. The fundamental niche is basically the suite of resources (biotic and abiotic) a species could use in absence of interference by other species. The realized niche, however, is a reduced space used by the species considering the influence of other species and the environment. Other, more complicated views of the niche also arose following the pioneering work by Grinnell (1917), Elton (1927), and Hutchinson (1957).

We need to recognize that habitat can only provide limited insight into factors responsible for animal survival and fitness, and thus population responses to changing environments. Mathewson and Morrison (2015) concluded that the proliferation of habitat terms occurs largely because of a failure to think about the niche concept when studying how and why animals are distributed as they are. The habitat concept alone usually cannot describe the underlying mechanisms that determine survival and fecundity. Researchers have reported that a number of environmental factors can restrict survival and productivity of a species (across its range or a portion thereof), and the influence of any single factor is not necessarily additive to the influence of any other factor. That is, usually only one factor is limiting in any particular location and time, and it is unlikely that the same factor will always be limiting because natural variation causes shifting of the quality and quantity of resources.

As developed by Morrison (2009), focusing on habitat alone is problematic because the environmental features we measure can stay the same, while use of important resources by an animal within that habitat can change. For example, consider changes in the species or size of prey taken by a bird forag-

ing on shrubs. The shrubs (habitat) do not need to change in physical dimensions or appearance. If we describe habitat only as structural or floristic aspects of vegetation, we will often fail to predict organism health because we did not recognize constraints on use of other resources that are limiting factors (Dennis et al. 2003).

Wildlife Habitat and Landscape Ecology

Landscape ecology focuses on how elements or patches in the environment are configured relative to one another in an overall mosaic, and in turn how such landscape structure influences ecological patterns and processes (Wiens and Milne 1989). Although the public attaches a human dimension to landscape, in landscape ecology there is no specific spatial extent. Rather, landscape must be viewed from the perspective of the animal under study; studies of multiple species are clearly complicated and involve multiple spatial scales. As noted by Wiens and Milne (1989), a landscape that is heterogeneous from the perspective of a ground-dwelling insect may only contain one or two seemingly homogeneous patches from the perspective of a foraging ungulate. When considering a landscape, we should think not only of the patch being used, but also of adjoining patches and connectivity among patches as influencing use. The function and existence of a location (patch) as habitat are entirely conditional on the attributes of the landscape in which they are embedded; describing them as habitat or not in isolation is clearly problematic because it is the nexus of patches on the landscape that enhance fitness, not any one patch. Mitchell and Powell (2003) reviewed the concept of linking landscapes to fitness of the organism. Here, habitat is mapped and modeled as a response surface depicting how conditions contribute to fitness of the organism. This approach does not rely on the tenuous assumptions that patches are discrete and homogeneous but incorporates concepts of heterogeneity, matrix areas, and corridors. Such a map represents a testable hypothesis about the relationship between habitat features on a landscape and fitness that can be tested with empirical data.

Turner (2005) reviewed the development of landscape ecology in North America and found that concepts emerging from landscape ecology now permeate ecological research. She reviewed definitions of landscape ecology, noting that they all shared the explicit focus on the importance of spatial heterogeneity for ecological processes. Of particular relevance to considerations of wildlife, she noted that the scale at which characteristics of the surrounding landscape influence a local response has been demonstrated for a variety of taxa. Turner mentioned habitat but used the term in two distinct contexts. First in the sense of a landcover, such as habitat fragmentation, habitat abundance, and habitat configuration, and second in the species-specific sense when discussing habitat suitability and habitat use. Thus we think that our research focus should shift from rather arbitrary hierarchically arranged measurements to an approach that focuses on scale-dependent examinations of the ecological processes responsible for the patterns we observe.

Approaching habitat analysis from a hierarchical perspective is not equivalent to conducting a landscape analysis. As reviewed above, the typical hierarchical approach simply increases the area under study to determine what conditions (i.e., habitat) prevail in areas of increasing size around an animal (or group of animals) under study. This hierarchy does place the specific location of an organism into a broader context, such as noting that the gopher (*Thomomys* sp.) is in a meadow surrounded by a deciduous woodland that is part of a conifer forest; all of these spatial areas can be described using any number of variables. Such information can be used, for example, to talk about how many meadows like this one occur throughout the forest. Likewise, we can (and have thousands of times in the literature) describe the nest site, the breeding territory, the home range, and so forth across increasingly larger spatial extents for individual animals. Such descriptions can provide useful information, for example,

in predicting the potential distribution of the species across space. But these analyses are static and not dynamic, and they are not reflective of population dynamics in space or over temporal scales that are meaningful to the persistence of a population. The nature of population dynamics is particularly important when considering the annual cycle of a species during which their needs and distribution can change over time. For example, many migratory birds select an area to breed, establish a territory, and nest. Once the nestlings fledge, the area used may expand beyond the territory as the adults mentor the fledglings on where and how to forage. As fall approaches, birds migrate along pathways that may include different conditions than those found within their breeding grounds and ultimately settle in a wintering area that may be different from where they bred or were reared.

A Broader Approach: Assembly Rules and Habitat Analysis

As mentioned above, many wildlife habitat studies focus on one species found in one location and measure a set of presumed habitat attributes they deem important to the species under study. Variables often relate to vegetation structure and composition and perhaps topography to describe the species habitat. Data can be analyzed to simply describe the habitat or compare that used with that available to infer preference. If done well, these studies might describe what the species is using, but not why.

Thus it should be clear that our studies of habitat are misleading at best and simply incorrect under multiple and often undescribed circumstances. It is the evaluation of status and trend of biological populations through time and space as conditions change that determines persistence. Studies of habitat, be they classical or hierarchical, are necessary but insufficient in understanding populations and managing them in the long term. Habitat studies are a component of a multipart approach to wildlife management and conservation. What we really need to know is the type and intensity of key factors that ultimately determine settlement. Traditional autecological studies fail to get at the root of the matter, as they often ignore the suite of biotic interactions that influence the space used by a species and how they use it. Diamond (1975) developed assembly rules to describe broad patterns of species co-occurrence. Assembly rules have received substantial attention in the ecological literature and have been adapted to the field of restoration ecology (Weiher and Keddy 1999, Temperton et al. 2004). The role of competition in development of assembly rules has been controversial (Connor and Simberloff 1979, Gotelli 1999). We are not evoking any particular process in outlining how species might assemble in space and time. Rather, we are describing a method that can be applied to better understand how and why species occur where they do. Assembly of species, as a concept, provides a framework for a more holistic way to look at habitat because it forces an integration of various biotic and abiotic filters and constraints of studies across space and time, and thus delves into niche factors of which classical habitat is a part. Thus, while the hierarchical view and analysis of habitat selection describe a point in time, the framework of assembly rules starts by trying to understand how animals were able to settle in the first place, given certain key filters and constraints. Characteristics of the selected environment, habitat, can then be quantified across the various spatial and temporal scales relevant to the focal species and those species it interacts with in ways that modify its behavior.

The practice of sampling habitat across spatial scales is best viewed, then, as a parallel complement to the more holistic and primary approach related to the assembly of species. We describe patterns of habitat use and seek the processes the assemble species in time and space. The measurements of habitat will change as the focal species (or group of species) change in location owing to changes in their biotic and abiotic environment (niche factors). Setting habitat analysis in this framework forces the observer to consider a wider range of factors influencing behavior, including intra- and interspecific interactions, in a context broader than a user-defined and convenient study area.

As reviewed by Temperton and Hobbs (2004), the study of how to assemble species encompasses various approaches to finding rules that govern how ecological communities develop. Various schools of thought have emphasized biotic interactions (e.g., competition), whereas others have emphasized a more holistic approach in which the interaction of the environment with the organisms of a community, and the interactions among organisms, restricts how a community is structured and develops.

Keddy and Weiher (1999) described the procedure for finding assembly rules as a four-step process: define and measure a property of assemblages, describe patterns in this property, explicitly state the rules that govern the expression of the property, and determine the mechanism that caused the patterns. Temperton and Hobbs (2004) concluded that no real consensus existed about what assembly rules constitute. Keddy and Weiher (1999) concluded that the search for assembly rules will benefit from a multitude of perspectives and approaches. Morrison (2009) concluded that ecologists would benefit by focusing on quantifying the relative contributions of biotic and abiotic factors and other constraints to avoid debates about the role of competition in structuring assemblages and if an ecological community actually exists.

Terminology

As reviewed by Morrison and Hall (2002), ecological terminology is not well standardized and often poorly used in the literature. One term that is directly involved with assemblage rules is *community*. Morrison and Hall (2002) surveyed definitions of community and reported that the co-occurrence of individuals of several species in time and space is common to most; the definitions usually stressed the role of interdependences among the species (populations) under study. Wiens (1989) concluded that multispecies assemblages do occur in nature, and we should focus on identifying interactions among these species. Wiens (1989) also noted that ecologists cannot hope to understand such groupings of species and interactions among them if we choose study areas arbitrarily. Thus we cannot use arbitrary and artificial boundaries because those very boundaries influence what we call a population or a community. Thus the interactions we see among species will reflect our decisions and usually will not be real.

Here we follow Morrison and Hall (2002) and use the term *species assemblage* to denote the group of species that are present and potentially interacting. Although such an assemblage could be part of a larger community, there is no need to even invoke the community concept. Here, we focus on identifying the filters and constraints that will modify the species present in an area throughout a successional pathway. To avoid confusion, however, we do retain the original terminology used in literature that we cite herein.

Thus our approach includes the classical and hierarchical notions of habitat within the overall concept of the niche (factors) but does so as part of a much broader framework for managing wildlife. Under this framework we can incorporate classical, descriptive habitat studies; sample across a hierarchy of spatial scales; and study elements of the niche. We want, however, to do these studies after we have developed a more encompassing framework that centers on species assembly, as we develop next.

Species Pool

Van Andel and Grootjans (2006) summarized the *species pool* concept, where a regional species pool occurs within a biogeographic region, thus encompassing multiple local species assemblages. The species pool becomes increasingly smaller as we move from the watershed extent down to the stream reach extent (Fig. 3.3). By passing through a series of filters, the local assemblage emerges from the larger pool. Van Andel and Grootjans (2006) defined *regional species pool* as the set of species occurring in a certain biogeographic or climatic region that are potential members of the target assemblage, *local species pool* as the set of species occurring in a subunit of the biogeographic region (e.g., a valley segment), and *community species pool* as the set of species present in a site within the target community (community and assemblage would be synonymous).

The concept of *ecological filters* forms one of the main approaches in assembly rules theory (Hobbs and Norton 2004). Of the total pool of potential colonists, only those that are adapted to the abiotic and biotic conditions present at a local site will be able to establish themselves. Thus a process of deletion or filtering of species occurs. Hobbs and Norton (2004) applied assembly rules to restoration ecology because of its application at the beginning of restoration projects in determining what factors were likely limiting membership in the community. Likewise, we can apply the concept of species assembly to initially hypothesize the key parameters (filters and constraints), spatial extent, and interactions of potential influence in the system we are studying. This becomes our broad framework that incorporates habitat. This construct will apply regardless of the taxa of primary interest. Assembly rules range from the obvious (e.g., predators without prey will starve) to more complex (e.g., the abundance of prey necessary for a new predator to enter a community; Temperton and Hobbs 2004). It is also critical to recognize that the action of filters will change over space and time.

Morrison (2009) identified abiotic major filters (modified from Hobbs and Norton 2004) to include climate including rainfall and temperature gradients; substrate including fertility, soil water availability, and toxicity; and landscape structure including landscape position, previous land use, patch size, and isolation. Major biotic filters will include competition with preexisting and potentially invading species and between planted or introduced species; predation-trophic interactions from preexisting and potentially invading species, and predation between

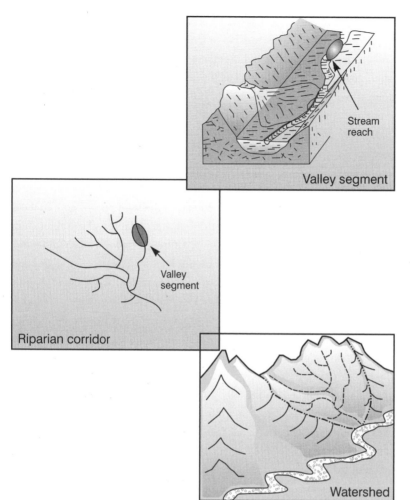

Figure 3.3. Schematic of the hierarchical scales used in stream reach, valley segment, riparian corridor, and watershed. Reproduced from Chambers and Miller (2004:fig. 1).

reintroduced animal species; propagule availability (dispersal) such as bird perches, proximity to seed sources, and presence of seed banks; mutualisms including mycorrhizae, rhizobia, pollination and dispersal, defense, and so forth; disturbance including the presence of previous or new disturbance regimes; the order of species arrival and successional model (i.e., facilitation, inhibition, dispersal capabilities, and tolerance); and current and past composition and structure (biological legacy).

At least seven figures in Temperton et al. (2004) depicted the generalized pathway from the potential pool of species, through various abiotic and biotic filters and other constraints, to the realized species pool in a location. Morrison (2009:fig. 5.2) synthe-sized those figures into a single diagram depicting pathways and filters, and how species fit into available niche space throughout the course of succession (Fig. 3.4).

As shown for fish assemblages by Mouillot et al. (2007), the latitudinal gradient is the most influential factor explaining richness distribution. At the regional scale, assemblage diversity is predictable based on a limited number of factors. Finally, at the local scale, fish richness is determined by abiotic filters and biotic interactions acting simultaneously. As further reviewed by Mouillot et al. (2007), the niche filtering hypothesis states that coexisting species are more similar to one another than would be expected by chance because of the filtering effect of

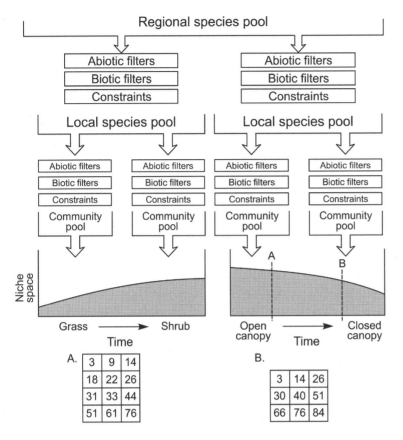

Figure 3.4. The species present at any area (community pool) are those remaining after the filtering process occurring at the local and regional levels. The figure represents two local species pools drawn from the same regional pool and co-occurring in time in two vegetation types. The community pool associated with a seral stage is drawn from the local pool that is specific to that more general vegetation type. The type and number of species present across the seral stage will also reflect the size of the target area and niche space available. The cross-sectional cut depicted as *A* and *B* indicates how species (numbered squares) change in type and total number. Reproduced from Morrison (2009:fig. 5.2).

environmental conditions that only allow species with similar traits to occur. Additionally, biotic interactions further limit species presence at the local (pool) scale because of factors such as competitive exclusion and limiting similarity principles.

Time and Succession

The succession concept emerged from plant ecology, whereas the assembly rules concept emerged from studies of animal communities in the context of island biogeography. Integration of paradigms such as these is critical in moving toward a better understanding of habitat across time and space (Pickett et al. 1994). These approaches are intertwined because filters, assembly rules, disturbance, and succession are part of a larger concept; namely, community assembly. Studies of disturbance and succession

contribute to our understanding of the processes that shape communities (White and Jentsch 2004). Nuttle et al. (2004) identified another relationship between assembly rules and succession; whereas succession describes the dynamics of changes in species composition, assembly rules deal with the interactions between organisms that determine the trajectory of those changes. Thus we witness the distinction between pattern (succession) and process (assembly rules).

The regional species pool of Van Andel and Grootjans (2006) is filtered through various abiotic and biotic factors to result in the local species pool. This process of filtering out species that are not adapted to a specific set of conditions must include this temporal component because the actual habitat or niche space available to species will vary through time as succession proceeds. Thus we quickly see how this assembly approach incorporates time, which is missing except in a static sense (e.g., breeding season in 2013 and 2014) from most habitat studies.

A Step Forward

To this point we have provided a framework for studying and managing species across a landscape. The foundation is a common understanding of the habitat concept, incorporating biotic and abiotic factors that influence a species distribution. Below, we outline steps to follow in application of this framework, with the goal of substantially advancing our knowledge of how animals are distributed in space and time.

First, define the population or a relevant biological segment thereof (e.g., subspecies, ecotype, isolated segment). This will likely become a study in and of itself, but doing so would provide data that are important to all subsequent work. Develop a study design that samples from across the biological distribution of the focal species, rather than simply selecting an area of convenience. If funding or other constraints dictate the study area, at a minimum this procedure means that you are aware of where your study animals reside relative to the overall range; that is, on an edge or near the center (Van Horne 2002).

Second, determine animal species that could potentially influence the behavior of the focal species (or group of species); this is the regional species pool. The regional species pool will often include species that occupy much larger areas than those used by the focal species, but this determination is a needed step in developing a valid landscape context. It is an initial list that will be reduced in a subsequent step.

Third, determine additional potential filters and constraints that could modify the behavior, abundance, movements, and other aspects of the biology of the regional species pool. Much of this work will involve literature reviews, along with expert opinion as needed.

Fourth, given the current ecological state of the sampling region, apply the filters and constraints and determine the resulting local species pool. This process could involve several likely scenarios based on your knowledge of the environment under study. You have now developed a list of species likely to occur across the biologically relevant area under study and an understanding of those factors in need of quantification. This is a point-in-time estimate and could be the ending point needed to finish the design of your study.

Finally, if you are interested in projecting how the distribution and abundance of your focal species could change through time, then you will need to determine the likely scenarios for natural disturbance, climate change, management actions, and plant succession over a reasonable and relevant timeframe, such as 50 to 100 years. Such scenarios could include various outcomes based on planned management activities (e.g., burning, clearing). Modeling changes in vegetative conditions through time would be a useful procedure, with a subsequent reevaluation of the regional species pool, filters, and constraints that might be related to the future conditions (e.g., based on 10-year intervals). This step is important to forecast changes in landscape configuration and concomitant changes in animal distribution and abundance because it includes related changes in other species that are influential on the focal species (or, if the animal assemblage is the focus, how it changes through time).

At a minimum, the above general procedure results in predictions of how the focal species (or assemblage of species), and the larger pools of species, will change in distribution and abundance (given the methods used) as the environment changes through time. You will also generate what we currently would consider a typical description of habitat use, but within a much more biologically relevant spatial extent and with knowledge of why changes in use are occurring (e.g., succession, plant species diversity, predator-competitor occurrence). Because this procedure is based on a spatial extent that is relevant to the biological population and other species influencing the focal species, you have also generated a more realistic representation of the landscape of the focal species. And you will have followed a process that is clearly defined and justifiable at each step. An additional advantage is that the researcher or researcher could modify your input (e.g., species pools, filters, constraints) as new information is obtained.

What we propose is not entirely new. There are a variety of analyses available that generally fall under species distribution models (SDMs). As thoroughly reviewed by Guisan and Thuiller (2005) and Elith and Leathwick (2009), these models relate field observations (e.g., presence or absence, abundance) to environmental predictor variables based on statistically or theoretically derived response. Environmental predictors are chosen from the three main types of influences on the species (i.e., limiting factors, disturbances affecting the environmental systems [natural or human-induced], resources). A core characteristic of SDMs is their relationship to the niche concept. The SDMs are thus directly applicable as tools for studying and developing management applications in applied ecology and wildlife biology.

The SDM models can be constructed such that relationships between species and their overall environment are evident at different spatial scales. The distribution observed over a large spatial extent and at coarse resolution is likely controlled by climatic regulators, whereas a patchy spatial distribution observed over a smaller area and at fine resolution is likely related to a patchy distribution of resources caused by natural topographic variation or human-induced fragmentation (Fig. 3.5). Among the many applications provided by Guisan and Thuiller (2005:table 1) were studies that (1) quantified the environmental niche of species; (2) tested biogeographical, ecological, and evolutionary hypotheses; (3) assessed species invasion and proliferation; (4) assessed the effect of climate, land use, and other environmental changes on species distributions; (5) supported appropriate management plans for species recovery, mapping suitable sites for species reintroduction; (6) supported conservation planning and reserve selection; and (7) modeled species assemblages (biodiversity, composition) from individual species predictions.

An output of SDMs are habitat maps, which have clear and direct utility in describing animal habitat

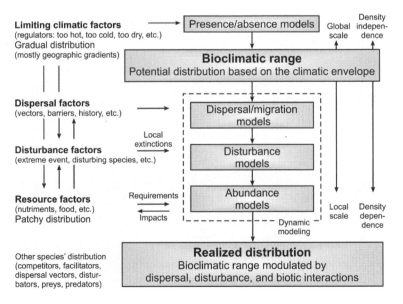

Figure 3.5. General hierarchical modeling framework that depicts a means of integrating disturbance, dispersal, and population dynamics within species distribution models. Reproduced from Guisan and Thuiller (2005:fig. 1).

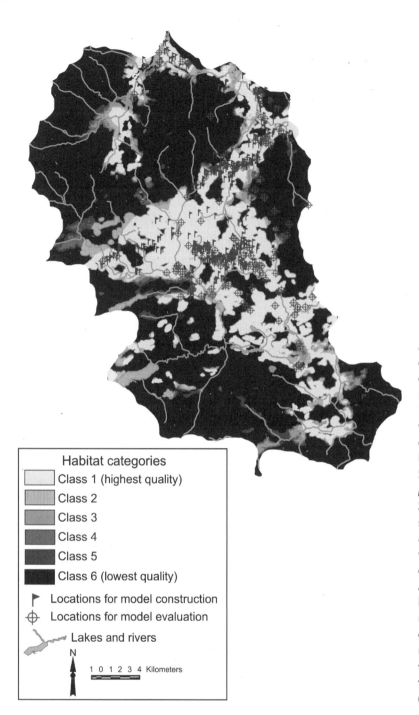

Figure 3.6. Example of a species distribution map for woodland caribou based on ecological niche models using variables representing the distance of caribou locations from the nearest patch, the density of landcover patches, or both patch density and distance. Species distribution maps were generated at a cell resolution of 25 × 25 m. A six-class ranking scheme was used, where it was assumed that class 1 habitats were of the highest quality and class 6 were of the lowest quality. A validation set of caribou locations was overaid, and the number of locations within each of the six habitat classes normalized by the area of that class was determined. Reproduced from Johnson and Gillingham (2005:fig. 2).

use along with changes in use through space and time, and lay the foundation for analyzing the effect of potential natural and human-caused environmental changes in the future. Species distribution for woodland caribou (*Rangifer tarandus*) based on ecological niche models using variables that represent the distance of caribou locations from the nearest patch, the density of landcover patches, or both patch density and distance provides an example (Fig. 3.6). There are numerous statistical packages directly related to SDM analysis that use a variety of statistical methods, including artificial neural networks, Bayesian approaches, classification and regression trees, ecological niche factor analysis, generalized additive and linear models, and other techniques (Guisan and Thuiller 2005:table 3).

As Guisan and Thuiller (2005) noted, a lack of knowledge of the focal species and assemblage of species substantially complicate development of a study of species distributions. Of particular relevance here is their call for information on home range sizes and how the species uses resources in the landscape. They noted that the choice of the geographical extent likely depends on existing knowledge of environmental gradients in the study area and how different ages and sexes of animals use resources across seasons. The concept and application of species assembly we developed above provide the framework necessary to fulfill these requirements outlined by Guisan and Thuiller (2005). They called specifically for the collection of preliminary field observations, sensitivity analyses, and conducting experiments to quantify the fundamental range of tolerance of organisms to predictor variables. Stepping through development of species assembly provides an overall framework for any study, identifies the data needed and the species for which the data are needed (e.g., as gathered through existing information, expert opinion, preliminary studies), and identifies the appropriate spatial scales.

Application of SDMs has emphasized plant and fish, with fewer applications to other wildlife species. Many SDMs use combinations of focal species distributions, measures of vegetation and topography, and climate in development of models (Lyet et al. 2013, Pikesley et al. 2013, Shirley et al. 2013, Bucklin et al. 2015, Young and Carr 2015). Pikesley et al. (2013) used telemetry data from olive ridley turtles (*Lepidochelys olivacea*) and information on the physical and biological environment to develop an SDM for this species. Lyet et al. (2013) combined species location data with climate and landcover data to develop an SDM for the endangered Orsini's viper (*Vipera ursinii*), and Shirley et al. (2013) evaluated various remote sensing tools to model distributions of 40 land birds in western Oregon. Missing from many of the models is inclusion of other species and how they might influence the distribution of the focal species. We envision this as the next logical step at improving the predictive capabilities of SDMs. These models offer a path forward in advancing our understanding of how animals are distributed in time and space.

Fundamentally, we are suggesting that we do not need to reinvent the wheel; for decades, ecologists from related disciplines have been delving into the factors responsible for species occurrences. Wildlife scientists simply need to more fully embrace the extensive work on these topics that have been conducted by theoretical ecologists, aquatic and fisheries scientists, and others.

Summary

Habitat and landscape remain two of the most misused terms throughout all corners of the ecological profession by both researcher and manager. Unfortunately, this vagueness inhibits advancement of knowledge about animal ecology and the subsequent translation of research findings into successful management applications. In this chapter we have pleaded for clarity in terminology. There are only a few habitat-related terms that need definition, because only a few key terms and concepts are actually needed to understand, quantify, and manage a species' habitat. A major first step in improving how we approach studies of animal ecology is to understand the geographic extent of the actual biological population(s) of interest. Additionally, landscape must be viewed from the perspective of the animal under study, and not that of the human observer.

LITERATURE CITED

Beyer, H. L., D. T. Haydon, J. M. Morales, J. L. Frair, M. Hebblewhite, M. Mitchell, and J. Matthiopoulosl. 2010. The interpretation of habitat preference metrics under use-availability designs. Philosophical Transactions of the Royal Society B: Biological Sciences 365:2245–2254.

Block, W. M. 2012. Analysis paralysis. Journal of Wildlife Management 76:875–876.

Block, W. M., and L. A. Brennan. 1993. The habitat concept in ornithology: theory and applications. Current Ornithology 11:35–91.

Bucklin, D. N., M. Basille, A. M. Benscoter, L. A. Brandt, F. J. Mazzotti, S. S. Romanach, C. Speroterra, and J. I. Watling. 2015. Comparing species distribution models constructed with different subsets of environmental predictors. Diversity and Distributions 21:23–35.

Carrier, W. D., K. S. Smallwood, and M. L. Morrison. 1997. Natomas Basin Habitat Conservation Plan: narrow channel marsh alternative wetland mitigation. Report to Northern Territories, Inc., Sacramento, California, USA.

Chambers, J. C., and J. R. Miller. 2004. Restoring and maintaining sustainable riparian ecosystems: the Great Basin Ecosystem Management Project. Pages 1–23 in J. C. Chambers and J. R. Miller, editors. Great basin riparian systems: ecology, management, and restoration. Island Press, Washington, DC, USA.

Connor, E. F., and D. Simberloff. 1979. The assembly of species communities: chance or competition? Ecology 60:1132–1140.

Dennis, R. L. H., T. G. Shreeve, and H. V. Dyck. 2003. Towards a functional resource-based concept for habitat: a butterfly biology viewpoint. Oikos 102:416–426.

Diamond, J. M. 1975. Assembly of species communities. Pages 342–444 in M. L. Cody and J. M. Diamond, editors. Ecology and evolution of communities. Harvard University Press, Cambridge, Massachusetts, USA.

Dunning, J. B., B. J. Danielson, and H. R. Pulliam. 1992. Ecological processes that affect populations in complex landscapes. Oikos 65:169–175.

Elith, J., and J. R. Leathwick. 2009. Species distribution models: ecological explanation and prediction across space and time. Annual Review of Ecology, Evolution, and Systematics 40:677–697.

Elton, C. S. 1927. Animal ecology. Sidgwick and Jackson, London, UK.

Gotelli, N. J. 1999. How do communities come together? Science 286:1684–1685.

Grinnell, J. 1917. The niche relationships of the California thrasher. Auk 34:427–433.

Guisan, A., and W. Thuiller. 2005. Predicting species distribution: offering more than simple habitat models. Ecology Letters 8:993–1009.

Guthery, F. S., and B. K. Strickland. 2015. Exploration and critique of habitat and habitat quality. Pages 9–18 in M. L. Morrison and H. A. Mathewson, editors.

Wildlife habitat conservation: concepts, challenges, and solutions. Johns Hopkins University Press, Baltimore, Maryland, USA.

Hall, L. S., P. R. Krausman, and M. L. Morrison. 1997. The habitat concept and a plea for standard terminology. Wildlife Society Bulletin 25:173–182.

Hildén, O. 1965. Habitat selection in birds: a review. Annals Zoologica Fennici 2:53–75.

Hobbs, R. J., and D. A. Norton. 2004. Ecological filters, thresholds, and gradients in resistance to ecosystem assembly. Pages 72–95 in V. M. Temperton, R. J. Hobbs, T. Nuttle, and S. Halle, editors. Assembly rules in restoration ecology: bridging the gap between theory and practice. Island Press, Washington, DC, USA.

Hurlbert, S. H. 1984. Pseudoreplication and the design of ecological field experiments. Ecological Monographs 54:187–211.

Hutchinson, G. E. 1957. Concluding remarks. Cold Springs Harbor Symposium on Quantitative Biology 22:415–422.

Hutto, R. L. 1985. Habitat selection by nonbreeding land birds. Pages 455–476 in M. L. Cody, editor. Habitat selection in birds. Academic Press, Orlando, Florida, USA.

Johnson, C. J., and M. P. Gillingham. 2005. An evaluation of mapped species distribution models used for conservation planning. Environmental Conservation 32:1–12.

Johnson, D. H. 1980. The comparison and usage and availability measurements for evaluating resource preference. Ecology 61:65–71.

Johnson, D. H. 2002. The importance of replication in wildlife research. Journal of Wildlife Management 66:919–932.

Keddy, P., and E. Weiher. 1999. The scope and goals of research on assembly rules. Pages 1–23 in E. Weiher and P. Keddy, editors. Assembly rules: perspectives, advances, and retreats. Cambridge University Press, Cambridge, UK.

Lyet, A., W. Thuiller, M. Cheylan, and A. Besnard. 2013. Fine-scale regional distribution modelling of rare and threatened species: bridging GIS Tools and conservation in practice. Diversity and Distributions 19:651–663.

Mannan, R. W., and R. J. Steidl. 2013. Habitat. Pages 229–245 in P. R. Krausman and J. W. Cain III, editors. Wildlife management and conservation: contemporary principles and practices. Johns Hopkins University Press, Baltimore, Maryland, USA.

Mathewson, H. A., and M. L. Morrison. 2015. The misunderstanding of habitat. Pages 3–8 in M. L. Morrison and H. A. Mathewson, editors. Wildlife habitat conservation: concepts, challenges, and solutions. Johns Hopkins University Press, Baltimore, Maryland, USA.

Mitchell, M. S., and R. A. Powell. 2003. Linking fitness landscape with the behavior and distribution of animals. Pages 93–124 in J. A. Bissonette and I. Storch, editors. Landscape ecology and resource management: linking theory with practice. Island Press, Washington, DC, USA.

Morris, D. 2003. Toward an ecological synthesis: a case for habitat selection. Oecologia 136:1–13.

Morris, D. W. 2011. Adaptation and habitat selection in the eco-evolutionary process. Proceedings of the National Academy of Sciences 278:2401–2411.

Morrison, M. L. 2009. Restoring wildlife: ecological concepts and practical applications. Island Press, Washington, DC, USA.

Morrison, M. L. 2012. The habitat sampling and analysis paradigm has limited value in animal conservation: a prequel. Journal of Wildlife Management 76:438–450.

Morrison, M. L., and L. S. Hall. 2002. Standard terminology: toward a common language to advance ecological understanding and application. Pages 43–52 in J. M. Scott, P. J. Heglund, M. L. Morrison et al., editors. Predicting species occurrences: issues of scale and accuracy. Island Press, Washington, DC, USA.

Morrison, M. L., B. G. Marcot, and R. W. Mannan. 2006. Wildlife–habitat relationships: concepts and applications. Third edition. Island Press, Washington, DC, USA.

Mouillot, D., O. Dumay, and J. A. Tomasini. 2007. Limiting similarity, niche filtering and functional diversity in coastal lagoon fish communities. Estuarine, Coastal and Shelf Science 71:443–456.

Nuttle, T., R. J. Hobbs, V. M. Temperton, and S. Halle. 2004. Assembly rules and ecosystem restoration: where to from here? Pages 410–422 in V. M. Temperton, R. J. Hobbs, T. Nuttle, et al., editors. Assembly rules and restoration ecology: bridging the gap between theory and practice. Island Press, Washington, DC, USA.

Pickett, S. T. A., J. Kolasa, and C. G. Jones. 1994. Ecological understanding. Academic Press, San Diego, California, USA.

Pikesley, S. K., S. M. Maxwell, K. Pendoley, D. P. Costa, M. S. Coyne, A. Formia, B. J. Godley, W. Klein, J. Makanga-Bahouna, S. Maruca, et al. 2013. On the front line: integrated habitat mapping for olive ridley sea turtles in the southeast Atlantic. Diversity and Distributions 19:1518–1530.

Piper, W. H. 2011. Making habitat selection more "familiar": a review. Behavioral Ecology and Sociobiology 65:1329–1351.

Pulliam, H. R. 1988. Sources, sinks, and population regulation. American Naturalist 132:652–661.

Rountree, R. A., and K. W. Able. 2007. Spatial and temporal habitat use patterns for salt marsh nekton: implications for ecological functions. Aquatic Ecology 41:25–45.

Shirley, S. M., Z. Yang, R. A. Hutchinson, J. D. Alexander, K. McGarigal, and M. G. Betts. 2013. Species distribution modelling for the people: unclassified Landsat TM imagery predicts bird occurrence at fine resolutions. Diversity and Distributions 19:855–866.

Smallwood, K. S. 2001. Linking habitat restoration to meaningful units of animal demography. Restoration Ecology 9:253–261.

Temperton, V. M., and R. J. Hobbs. 2004. The search for assembly rules and its relevance to restoration ecology. Pages 34–54 in V. M. Temperton, R. J. Hobbs, T. Nuttle, et al., editors. Assembly rules in restoration ecology: bridging the gap between theory and practice. Island Press, Washington, DC, USA.

Temperton, V. M., R. J. Hobbs, T. Nuttle, M. Fattorini, and S. Hallel. 2004. Introduction: why assembly rules are important to the field of restoration ecology. Pages 1–8 in V. M. Temperton, R. J. Hobbs, T. Nuttle, et al., editors. Assembly rules in restoration ecology: bridging the gap between theory and practice. Island Press, Washington, DC, USA.

Turner, M. G. 2005. Landscape ecology: what is the state of the science? Annual Review of Ecology, Evolution, and Systematics 36:319–344.

US Fish and Wildlife Service. 1997. Natomas Basin Habitat Conservation Plan. US Fish and Wildlife Service Sacramento Field Office, Sacramento, California, USA.

Van Andel, J., and A. P. Grootjans. 2006. Concepts in restoration ecology. Pages 16–28 in J. van Andel and J. Aronson, editors. Restoration ecology. Blackwell, Oxford, UK.

Van Horne, B. 2002. Approaches to habitat modelling: the tensions between pattern and process and between specificity and generality. Pages 63–72 in J. M. Scott, P. J. Heglund, M. L. Morrison, et al., editors. Predicting species occurrences: issues of accuracy and scale. Island Press, Washington, DC, USA.

Weiher, E., and P. Keddy. 1999. Assembly rules as general constraints on community composition. Pages 251–271 in E. Weiher and P. Keddy, editors. Ecological assembly rules: perspectives, advances, and retreats. Cambridge University Press, Cambridge, UK.

White, P. S., and A. Jentsch. 2004. Disturbance, succession and community assembly in terrestrial plant communities. Pages 342–366 in V. M. Temperton, R. J. Hobbs, T. Nuttle, et al., editors. Assembly rules in restoration ecology: bridging the gap between theory and practice. Island Press, Washington, DC, USA.

Wiens, J. A. 1989. Spatial scaling in ecology. Functional
    Ecology 3:385–397.
Wiens, J. A., and B. T. Milne. 1989. Scaling of "landscapes"
    in landscape ecology, or, landscape ecology from a
    beetle's perspective. Landscape Ecology 3:87–96.
Young, M., and M. H. Carr. 2015. Application of species
    distribution models to explain and predict the distribu-
    tion, abundance and assemblage structure of nearshore
    temperate reef fishes. Diversity and Distributions
    21:1428–1440.

PART II • Establishing a Landscape Foundation for Wildlife Managers

4

Essential Concepts in Landscape Ecology for Wildlife and Natural Resource Managers

HUMBERTO L. PEROTTO-BALDIVIESO

Introduction

As a plane takes off and gains altitude, its passengers enjoy watching the surrounding landscapes from their airplane windows. If it is departing from London Heathrow (UK), people will observe an urban landscape composed of houses, streets, parks, and trees. If it is heading northwest from Corpus Christi International Airport (Texas, USA), the landscape is dominated by cropland, then by rangeland. Finally, if it is taking off from El Alto International Airport (Bolivia), below are landscapes dominated by small houses around the airport, followed by large open pastures and then the Tropical Andes mountain range (Fig. 4.1).

These landscapes all have something in common: they are composed of patches (e.g., crop fields, urban blocks), corridors (e.g., streets, rivers, hedgerows), and a matrix. The spatial distribution of these patches creates patterns that drive ecological processes (e.g., livestock grazing, animal movements, water and nutrient flows). In turn, these ecological processes can affect existing patterns and modify or create patterns through flooding, seed dispersal, or corridor development. As a plane gains altitude, one can see that the patches seem to coalesce, and less detailed patterns (e.g., streets, parks, croplands, pas-

tures) emerge in the much larger matrix; the scale is changing. At these larger landscape scales it may no longer be relevant to discuss species' individual movements, but rather their populations and habitats; instead of individual trees and grasses, one may be looking at the continuity of rangelands and other features of habitat important to wildlife. Both patterns and processes are present, but these are different from patches at small scales to landscapes at large scales.

So how do pattern, process, and scale fit within wildlife management? To answer this question, we need to provide an understanding of how spatial patterns affect life history processes of wildlife (ecological processes), a fundamental idea of effective management and conservation. Therefore the goal of this chapter is to provide an introduction to spatial heterogeneity, the relationship between pattern and process at different scales, and their potential applications to manage wildlife habitat. The information presented in this chapter should provide a good basis for designing a conservation strategy to maintain and enhance ecological diversity and ecosystem function that will benefit wildlife species. To achieve this goal, this chapter is divided into three sections: Spatial Heterogeneity, Scale, and Models for Assessing Landscape Structure. Finally, there is

Figure 4.1. Landsat 8 (15-m-resolution panchromatic; geographical scale 1:200,000; 2017) images showing the landscape matrix around London Heathrow Airport, England, UK (*top*); Corpus Christi International Airport, Texas, USA (*middle*); and El Alto International Airport, Bolivia (*bottom*).

a brief concluding section about landscape ecology and wildlife management.

Spatial Heterogeneity

Spatial heterogeneity can be defined as the distribution of biologically important resources on the landscape. This distribution depends on factors such as topography, soils, and climate, as well as processes driven by wildlife species (e.g., species dispersal) and humans (e.g., pasture management, agricultural practices). Spatial heterogeneity has significant effects on the recruitment, demographic variability, and dispersal of plant and animal species (Miller et al. 1995). Dispersal in heterogeneous landscapes is key to maintaining or enhancing population persistence, as spatial heterogeneity can provide stability among population interactions. Thus spatial patterns (Fig. 4.2) can have important effects on the stability and size of populations (Turner and Gardner 2015). Spatial heterogeneity can also provide benefits for higher-order ecological processes such as species richness (McGranahan et al. 2012), community structure (Wijesinghe et al. 2005), and ecosystem processes (Dubois et al. 2015). Conversely, loss of spatial heterogeneity may reduce long-term persistence of ecosystems (Turner and Gardner 2015). The relationship between ecological processes and spatial heterogeneity is not unidirectional. Ecological processes such as population interactions can also modify spatial heterogeneity in areas where landscape patterns, particularly of resources, are homogeneous. The biotic feedbacks between vegetation and herbivory are thought to be major drivers of spatial heterogeneity (Okayasu et al. 2012).

Rangelands represent one landscape where those drivers shape spatial heterogeneity. In rangelands, water, soil, nutrients, herbivory, and grazing are the main factors affecting ecosystem structure and function at multiple scales (Otieno et al. 2011). While water and nutrients influence productivity across the landscape, herbivory and fire determine water and nutrient availability. Changes in vegetation communities within rangelands are regulated by herbivory with an effect on soil structure and function that feedback to plant productivity. Within savanna grasslands the presence of *termaria* (termite mounds) and trees provides good insight into understanding how spatial heterogeneity is affected by ecological processes. Termaria modify surrounding soil composition and structure by increasing the percentage of soil nitrogen, organic and colloidal matter, and soil moisture. These changes in soil are beneficial to plant growth and plant diversity. Similar effects can be observed under the canopy of trees and their surrounding areas. The islands of fertility created by the

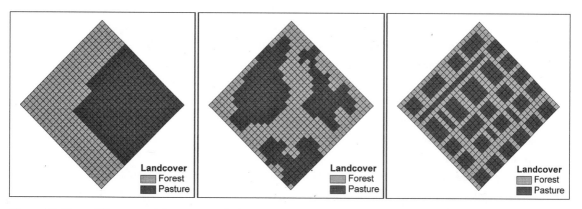

Figure 4.2. Three different spatial patterns with the same proportion of pasture of forest and pasture. These three spatial patterns will provide habitat for different wildlife species owing to their spatial configuration in the landscape.

processes described above can then affect large herbivory (e.g., deer, cattle) dispersal, and pasture use. Areas with palatable forage tend to be used repeatedly owing to possible dietary preferences and postgrazing regrowth of new and more palatable forage (Okayasu et al. 2012, Fulbright and Ortega-Santos 2013). The preference for specific grazing areas can generate changes in vegetation cover and composition toward unpalatable forage, brush encroachment, and pasture degradation, which in turn can change landscape spatial structure. These changes have effects on predator foraging efficiency where areas with high spatial heterogeneity may diminish predators' ability to identify prey habitat (Gulsby et al. 2017).

The impacts of spatial heterogeneity need to be assessed on the basis of the conservation or management goals to be achieved. While there are benefits for some species, others may be negatively affected by spatial heterogeneity. This negative effect can be true for species that prefer undisturbed habitat for movement and dispersal (Baggio et al. 2011, Zemanova et al. 2017). Fragmentation by deforestation can increase landscape spatial heterogeneity and generate significant changes on habitat availability and quality with direct effects on species survival, abundance, and dispersal (Dixo et al. 2009, Gao et al. 2013, Zemanova et al. 2017). These changes tend to be detrimental to large carnivores that prefer large tracts of undisturbed habitat. Some of these species (e.g., jaguars [*Panthera onca*]) tend to be sensitive to human activities such as hunting (e.g., through prey depletion) and perceived threats to humans and livestock (Cavalcanti and Gese 2009, Horev et al. 2012, Zeilhofer et al. 2014). Alternatively, Turner et al. (2013) concluded that spatial heterogeneity in Yellowstone National Park contributes to wildlife habitat by providing a variety of habitat through forest regeneration and timber production. This provision will increase patchiness and selection of microhabitats for certain species, while negatively affecting habitat for species requiring large undisturbed areas such as northern spotted owl (*Strix occidentalis cau-*

rina) and pine marten (*Martes martes*). Whether it is plant distribution around termaria, the spatial distribution of those termaria, or large-scale spatial distribution, the impacts of spatial heterogeneity are different as scale changes. Therefore understanding the role of scale in the distribution of ecological and biological resources and their related ecological processes is fundamental in landscape ecology.

Scale
The Importance of Scale

What is *scale*, and why it is so important? Scale is a critical concept in the natural sciences (Turner et al. 2001) and has been defined in various ways in different disciplines (Schneider 2001, Turner and Gardner 2015). For landscape ecologists, the term scale refers to the spatial and temporal extent of ecological processes and their spatial interpretation (de Knegt et al. 2011, Schneider 2001, Turner and Gardner 2015). Bissonette (2017:193) considers scale a "metric that refers to the spatial or temporal dimension of an object, pattern or process." Although the term has been used for several decades, it was in the 1970s and 1980s that the concept of scale became widely recognized.

Scale is characterized by two components: *grain* and *extent*. "Grain refers to the finest spatial resolution within a given data set," and "extent refers to the size of the overall study area" (Turner and Gardner 2015:17). In other words, grain can be considered as the minimum resolution at which data are being observed and analyzed, while extent is the domain or size of the landscape. When using digital imagery, the pixel size will represent the grain of the imagery. For example, the National Agriculture Aerial Photography Program provides imagery at 1-m resolution, and the Landsat program offers imagery ranging from 15- to 90-m resolution (Loveland and Dwyer 2012, Roy et al. 2014). When using geographic information vector databases (i.e., shapefiles), the size of the polygon or minimum mapping unit will be considered the grain. The extent, being the domain of the data, represents the population to be sampled from a statistical standpoint. Therefore grain represents the lower limit of resolution and extent the upper limit of resolution. Patterns cannot be detected outside the boundaries of grain (finer resolution) or extent (coarser resolution). Selecting the adequate grain and extent in wildlife studies is fundamental to making acceptable inferences about pattern-process relationships (Fig. 4.3). If the grain is too fine, the extent of the landscape may introduce unnecessary noise into the analysis, or it may be a limiting factor for the extent of the spatial analysis owing to file size constraints. If the grain is too coarse, however, some important information may be missed or misinterpreted. For example, Perotto-Baldivieso et al. (2009) used IKONOS imagery (1-m resolution) to map depression forest patches, which are important habitat features for the Mona rock iguana (*Cyclura cornuta stejnegeri*). Mona rock iguanas are widely distributed across Mona Island (Puerto Rico, USA) but have small home ranges. The use of high-resolution imagery allowed for the production of more detailed maps than previously reported for depression forest maps. The resulting map using the original IKONOS pixel resolution, with a minimum mapping unit (MMU) equal to 5 m², contained 9,220 depression forest patches of habitat versus the 218 patches reported previously (Cintrón and Rogers 1991). The higher number of depression forest patches in this study arose from the detection of small patches, which Mona rock iguanas used for dispersal, while large patches tend to be used for nesting. Further analysis comparing the original resolution map (MMU = 5 m²) to lower resolutions (MMU = 100 m², 500 m², and 1,000 m²) showed that the results of metrics describing spatial structure and connectivity varied significantly. In particular, the integral index of connectivity (Pascual-Hortal and Saura 2006, Saura and Pascual-Hortal 2007) decreased by 60% when assessed with the lowest resolution (MMU = 1,000 m²).

Therefore, when studying landscape processes or wildlife species, it is crucial that the grain and extent selected embody these processes or species under study. When it is not possible to identify the desired grain and extent, Thompson and McGarigal (2002) suggest the use of the finest grain that may be representative of the species or process being studied, which provides the opportunity to reanalyze spatial patterns at coarser resolution until the most representative scale is found. This approach was reported useful to represent and analyze bald eagle habitat in New York, USA (Thompson and McGarigal 2002). The interpretation of landscape metrics can have a significant impact when utilizing them for planning and decision making, as the interpretation of ecological data is fundamentally a function of scale (Wheatley 2010). In this context it is important to differentiate scale from levels of organization, which identify a place within a biotic hierarchy. In wildlife studies, for example, levels of organization could be grouped into individuals, herds or flocks, populations, or metapopulations where applicable. Each of these levels of organizations is characterized by different ecological properties. As these levels of organization are aggregated into higher levels (e.g., individuals to herds), new ecological properties emerge, which could provide a hierarchical structure for scale and the patterns and processes that need to be studied at each scale. Ecological systems are complex, however, as their dynamics can follow large spatial and temporal extents (Schneider 2001), and common levels of organization are not necessarily scale dependent (Allen and Hoekstra 1991). Hierarchy theory provides a useful framework to set the spatial and temporal dimension to study ecological systems.

Hierarchy Theory

Hierarchy theory deals with the ecological consequences of levels of organization within ecological systems. One of the foundations for hierarchy theory was presented by Delcourt et al. (1983) by showing the positive correlation between spatial and temporal scales for environmental disturbance regimes, biotic responses, and vegetation patterns (Turner and

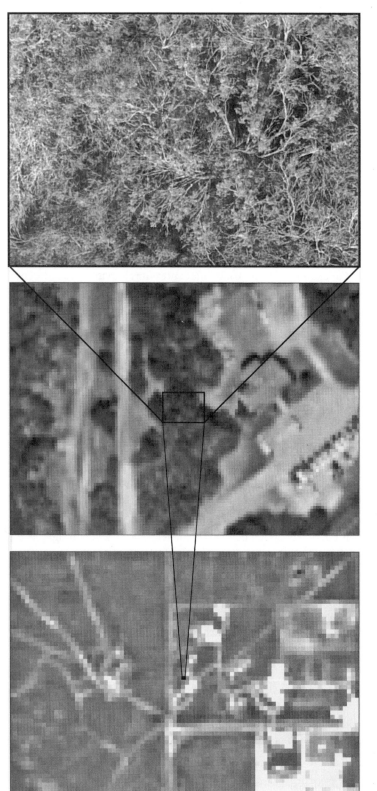

Figure 4.3. Grain and extent differences from remote sensing imagery. The top panel is a mesquite patch acquired with an unmanned aerial vehicle (0.02-m grain resolution and 1:10 image geographical scale; 2017). The middle panel shows the same mesquite patch acquired from National Agricultural Imagery Program aerial photography (1-m grain resolution and 1:100 image geographical scale; 2016). The bottom panel shows the extent of the mesquite patch derived from Landsat 8 panchromatic imagery (15-m grain resolution and 1:1000 image geographical scale; 2017). Note the difference in scale and resolution for the same mesquite patch.

Gardner 2015). Delcourt et al.'s (1983) framework has been widely used to explain the linkages between scale and hierarchy, which are intricately interrelated. A hierarchy is defined as "a system of interconnections wherein the higher levels constrain the lower levels to various degrees, depending on the time constraints of the behavior" (Turner and Gardner 2015:21). Ecological organizations in nature tend to show a hierarchical structure; thus the levels of a landscape hierarchical system are separated by the spatial and temporal dimensions of the process within each level. Within the hierarchical framework, levels are not isolated but exist between other levels or scales. It is important to understand this concept because processes may or may not change, but the direction or the relative importance of the variable will change at different scales. The application of the hierarchy theory framework requires the identification of three hierarchical levels. The level at which the question is addressed is the focal level or level of interest. The below level provides details that are important to understand the processes and patterns at the focal level. The above level sets the controls and constraints for the focal level (Turner and Gardner 2015). Hierarchy theory therefore provides a similar framework for pattern analysis at different scales. The focal scale is the scale at which the phenomenon of interest is being studied. The broader scale will set the constraints, and the finer scale will provide details to explain the observed pattern. When processes are operating at different scales, they are referred to as cross-scale interactions. Studies across scales and cross-scale processes can provide insights into analyses of emergent patterns with linear or nonlinear dynamics that could identify thresholds in spatial patterns and processes within the focal scales (Turner and Gardner 2015, Zemanova et al. 2017). Although this framework is exciting conceptually, it is still challenging to identify the appropriate scale at which a problem needs to be addressed.

What is the best scale? Or what is the right scale? These are two common questions in wildlife man-

agement studies. The answers to these questions rely on the ecological process being studied. It is now well understood that there is no single scale that fits all ecological problems (Turner et al. 2001). Therefore selecting an adequate scale is fundamental to obtaining reliable data (Bissonette 2017). This selection requires a good understanding of the scale at which the ecological processes occur, the scale of the sampling protocol, and the scale of the statistical methods used in the study (Dungan et al. 2002, Bissonette 2017). Additionally, the scale of management has important implications for the outputs of the ecological questions being asked in wildlife management studies. Management practices are often conducted at different extents than the ecological processes being manipulated. Therefore the larger scale should be used to provide the constraints for the focal scale being studied. In some cases it is not possible to define the scale a priori, and several approaches have been proposed to identify relevant scales. For example, Thompson and McGarigal (2002) suggested the use of the finest grain that may be representative of the species or process being studied with further spatial pattern reanalysis at coarser resolution until the most representative scale is found. Alternatively, Wheatley (2010) proposed the use of domains of scale for the interpretation of landscape metrics when utilizing them for planning and decision making, as the interpretation of ecological data is fundamentally a function of scale. In recent years the number of studies using the term multiple scales has significantly increased as an alternative to single-scale studies.

Multiple Scales and Domains of Scale

Habitat selection is a key component of wildlife management. It has been well established that habitat selection is species specific, scale dependent (space or time), and context dependent (e.g., life history stage, location, season), and it can be studied at single or multiple scales (Mayor et al. 2009, McGarigal et al. 2016). One of the main focuses of multiscale habitat selection is identifying the adequate scale(s) based on biologically meaningful information (Miller et al. 2019) or empirical methods (Thompson and McGarigal 2002). "Multi-scale habitat selection modeling refers to any approach that seeks to identify the scale or scales (in space and time), at which the organism interacts with the environment to determine it being found in, or doing better in, one place (or time) over another" (McGarigal et al. 2016:1162). Although there is a significant amount of research on multiscale habitat selection, McGarigal et al. (2016) report that most of these studies do not have an adequate multiscale framework. Their review classifies multiscale studies into 11 classes (Table 4.1) based on level of organization, spatial scale (grain and extent), covariate scale, temporal scale, and scale optimization. As a result of this study, the following key points are highlighted: (1) there is an amalgamation in the use of the terms level and scale emphasizing spatial scale rather than level of organization; (2) en-

vironmental variables are not usually assessed across different grains, which can significantly affect the interpretation of results in habitat–selection studies; (3) scale selection or optimization through statistical methods combined with empirical approaches can significantly improve the outputs of habitat–selection studies; and (4) the use of multiple levels and scales may be more appropriate given the complexity and nature of habitat selection (McGarigal et al. 2016). In this context the concept of domains of scales has emerged as a potential framework to understand multiple-scale studies (Wheatley 2010).

The term *domain of scale* was coined by Wiens (1989:392) as a "portion of the scale spectrum within which process–pattern relationships are consistent regardless of scale." This concept provides a good theoretical basis to identify what scales need to be analyzed. Wheatley (2010) used a scale continuum to identify domains and transitions across scales. The author used this framework in west-central Al-

Table 4.1. Classification of Multilevel, Multiscale Studies According to Level of Organization and Spatial, Temporal, and Covariate Scale

Classification	Level of Organization	Spatial Scale: Grain	Spatial Scale: Extent	Covariate Scale	Temporal Scale	Scale Optimization
Multilevel in space	Multiple	Single	Single	Single	None	None
Multilevel in time	Multiple	Single	Single	Single	Multiple	None
Multilevel in space and time	Multiple	Single	Multiple	Single	Multiple	None
Multiscale in space	Single	Single	Multiple	Single	Single	None
Multiscale in time	Single	Single	Single	Single	Multiple	None
Multiscale in space and time	Single	Multiple	Multiple	Multiple	Multiple	None
A priori single scale	Single	Single	Single	Single	Single	Scale preselected
A priori separate scales	Single	Single or multiple	Single or multiple	Single or multiple	Single or multiple	Multiple scales selected by user but analyzed separately
A priori multiple scales	Single	Single or multiple	Single or multiple	Single or multiple	Single or multiple	Single multivariable multiscale model
Pseudo-optimized single scale	Single	Single	Single	Multiple	Single or multiple	Analysis of range of scales for a single-scale selection
Pseudo-optimized multiple scales	Single or multiple	Single or multiple	Single or multiple	Single or multiple	Single or multiple	Prespecified scales. Covariates evaluated separately. Single-scale optimization and covariate combination for a multi-variable, multiscale model.

Source: Adapted from McGarigal et al. (2016).

berta to calculate landscape metrics and empirically identify these domains of scales using statistical differences. This approach has also been used by Fisher et al. (2011) to predict scale based on body size of 12 sympatric mammal species in west-central Alberta and by Little et al. (2016) to assess nest site selection by eastern wild turkeys (*Meleagris gallopavo*) in Georgia. While the use of metrics to identify domains of scale has provided useful insight on the behavior of landscape metrics and the interpretation of spatial patterns at different sampling scales, it is fundamental to consider that as patterns change, the processes influencing species will change at different spatial and temporal scales. A good example is the study of northern bobwhite (*Colinus virginianus*) across their distribution range. While studies at local spatial scales focus on individual home ranges and their relationship to vegetation spatial structure, they are inherently constrained to small temporal scales (monthly or seasonal studies; Miller et al. 2019). Alternatively, range-wide distribution studies focus on processes occurring at much larger spatial and temporal scales. Changes in abundance may be related to human population changes, agricultural practices and policies, and human infrastructure development (Okay 2004). These processes and their corresponding spatial patterns can be identified over large time periods spanning decades.

Models for Assessing Landscape Structure

Landscape structure is defined by a specific spatial pattern in the landscape. Two main components are used to assess landscape structure: composition (nonspatial component; number and abundance of elements considered) and configuration (spatial component; character, arrangement, and context of the elements in the landscape; McGarigal et al. 2012). There are numerous models that provide a good understanding of landscape structure depending on the research objectives, the patterns to be analyzed, the process to be studied, and the type of available data (McGarigal and Cushman 2005). Point pattern, patch mosaic,

landscape gradient, and graph theory are the most frequently used models.

Point Pattern Models

Point pattern models are based on points for individuals or group of individuals with specific (x, y) locations. Examples in wildlife studies include locations of wildlife obtained through telemetry or global positioning system collars and count surveys. These locations are usually stored in vector format, with each individual location being assigned a unique identification as well as x, y coordinates. If these are the only attributes in the table, then locations are considered to be unweighted. When locations contain additional attributes that can be used for further analysis, however, they are considered weighted. All events recorded during a study are considered the mapped point pattern, but if samples are collected from different areas, we refer to them as sampled point patterns.

Point data can be used to determine whether there is a tendency of events to occur in a pattern across a landscape rather than a random distribution of points. Complete spatial randomness of points assumes that the density of points is constant across the landscape and that events follow a homogeneous Poisson distribution (Wiegand et al. 2013). Point distributions follow one of these three patterns: random, uniform (each point is as far for its neighbors as possible), or clustered (points are concentrated close together, and large areas may or may not contain any points).

Point data can be analyzed using first-order properties such as intensity and density of data or second-order properties to measure spatial dependence based on separation distance among locations (Bailey and Gatrell 1995). First-order methods to analyze point patterns include quadrat analysis and kernel density estimators. The quadrat method divides the landscape into regions of equal size and counts the number of events in each subregion (Shiode 2008). Based on the total number of events, the intensity of events in each subregion is calculated. One weakness of this method is that if the quadrat size is too small, intensity values will be low; conversely, if quadrat sizes are too large, intensity values will be too high. The quadrat method is a measure of dispersion, and not really pattern, because it only explains the density of points and not the spatial relationship among them. The kernel density estimator is a nonparametric statistical analysis designed to estimate a probability density function from a point variable (e.g., individual locations; Rosenblatt 1956). The kernel density estimator is estimated by calculating the density of locations within a specified search radius using a three-dimensional function (the kernel). The addition of individual kernels for the landscape produces a smoothed surface (raster). The most common and widely used application of this type of analysis in wildlife is home range estimation (Worton 1989, Moorcroft et al. 1999, Mitchell and Powell 2004, Van Moorter et al. 2016).

Second-order properties (Bailey and Gatrell 1995) include the nearest neighbor analysis (G-function) and K-function (Ripley 1976). The G-function is the simplest measure, and it is similar to a mean estimation. The G-function quantifies the distance to the nearest neighbor and helps examine the cumulative frequency distribution of the nearest neighbor event (Baddeley and Gill 1997). Values are then compared to the intensity or expected number of points per unit area (λ). Deviations from the empirical data and theoretical G-curves may suggest spatial clustering or spatial regularity. If the values of the empirical curve increase rapidly at short distances, patterns may be clustered. Conversely, if the G-values increase slowly up to a distance where all events are evenly spaced, and then G-values increase rapidly, events may be evenly distributed. One of the limitations of the G-function is that it uses the nearest distance and considers only the shortest scale of variation. Ripley (1976) proposed the K-function, which provides an estimate of spatial dependence over a wider range of scales than the G-function. The K-function is based on all distances among events

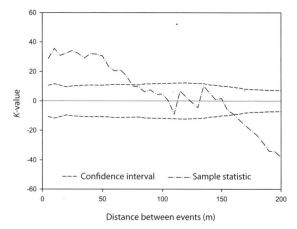

Figure 4.4. Ripley-*K* function diagram. The *x*-axis represents the distance between events, and the *y*-axis represents the *K*-values at specific distances between events. The dashed lines indicated the upper and lower 95% intervals. The dot-dashed line indicates *K*-values derived from event data. When *K*-values are above the confidence interval, data are significantly clustered. When *K*-values are below the confidence interval, data are significantly overdispersed. When *K*-values are within the confidence interval, data are randomly distributed.

and assumes isotropy across the landscape (Haase 1995, Feagin and Wu 2007). The K-function is estimated by quantifying the distance between events within circles of different radii and comparing them to a random distribution of points at the same scale. The result is a graph showing the K-value compared to a 95% confidence interval (Fig. 4.4). If K-values are above the confidence interval, events at those scales are assumed to be clustered. When K-values are within the confidence interval, patterns are considered random. If K-values are below the confidence interval, spatial patterns are considered overdispersed (Dickins et al. 2013).

The Patch Mosaic Model

The patch mosaic model is one of the most common models used to assess landscapes. Data are represented by a mosaic of discrete patches that can be identified by their edges that create discontinuities from their surroundings. Examples of patch mosaic models include land use and landcover maps, vegetation maps, parcel maps, the soil survey database, or any map that consists of classified polygons (vector format) or raster databases (pixels). The patch mosaic model has three main elements: patch, corridor, and matrix. A patch is a homogeneous area that differs from its surroundings in nature or appearance (Collinge 2009). Key properties of a patch are area, perimeter, and shape. Patches occur at different spatial and temporal scales, and the internal structure of a path will define patchiness at finer scales (Gergel and Turner 2002). *Patches* are defined in relation to the scale of the phenomenon being studied. *Corridors* are defined as "narrow strips of land which differ from the matrix on either side" (Forman and Godron 1986:123). This definition is focused on the structural character of corridors, and Forman and Godron (1986) classify them into three categories: line, strip, and stream corridors. The *matrix* is the background cover class in the landscape, and it is defined on the basis of the most abundant patch class and its connectivity. The matrix is scale dependent and affects both patches and corridors (Prevedello and Vieira 2010). In recent years the matrix has become more important in research, as it is becoming a critical component of landscape context, particularly in human-dominated landscapes (Rodríguez-San Pedro and Simonetti 2015).

Landscape structure in the patch mosaic model is usually assessed using metrics that quantify area, shape, contrast, and aggregation at the patch (i.e., individual patches), class (all patches of the same class), and landscape (all patches in the landscape) level. FRAGSTATS is one of the most widely used software packages to quantify metrics in the patch mosaic model. FRAGSTATS can quantify 15 metrics at the patch level, 37 metrics at the class level, and 46 metrics at the landscape level. Additionally, FRAGSTATS provides 7 sampling strategies for quantifying landscape metrics (McGarigal et al. 2012).

The use of landscape metrics to quantify landscape structure and its potential effects on wildlife is well documented. For example, Perotto-Baldivieso et al. (2011) used aerial photography from the National Agriculture Imagery Program to develop categorical maps of woody and herbaceous vegetation to quantify changes in landscape structure in wild turkey habitat. Zemanova et al. (2017) used classified imagery derived from Landsat between 1986 and 2005 (Killeen et al. 2007, 2008) to quantify landscape changes that affect jaguar habitat. Mata et al. (2018) used the National Agriculture Imagery Program to look at the progression of tanglehead (*Heteropogon contortus*) in south Texas between 2008 and 2014. The results of this study were used to develop a landscape gradient model to assess winter habitat characteristics for northern bobwhite in south Texas.

The Landscape Gradient Model

The gradient concept of landscape structure is a conceptual paradigm in landscape ecology that provides an additional framework for pattern-process relationships (Abdel Moniem and Holland 2013). This model, introduced by McGarigal and Cushman (2005), takes into consideration the inherent environmental variability that exists in the landscape. This variability has an important effect on ecological processes and their responses to changes along this gradient. The inherent continuous nature of ecological attributes has not been well represented with previous models, such as the patch mosaic model, which has been used by landscape ecologists to describe landscape spatial heterogeneity (Wu et al. 2000, Zemanova et al. 2017, Mata et al. 2018). Conceptually, the patch mosaic model represents spatial heterogeneity with categorical data (Turner and Gardner 2015). But this categorization results in a loss of critical ecological information, as it disregards the continuous nature of ecological factors (McGarigal et al. 2009). Although the patch mosaic model paradigm has significantly contributed to the understanding of patterns and processes occurring in a landscape and has been applied globally with success, discontinuities that account for environmental variation cannot be fully addressed using the patch mosaic model.

The gradient concept is landscape structure based and quantifies landscape patterns in a way that provides a better understanding of how an organism perceives its surrounding environment and its potential response to alterations (McGarigal and Cushman 2005). This concept has been used to study the influence of habitat heterogeneity on connectivity for pollinator beetles across Indiana, USA, using a combination of surface metrics and the gradient concept (Abdel Moniem and Holland 2013). Although there is little information on the application of this model in wildlife studies, the results of Abdel Moniem and Holland (2013) show potential for the assessment of wildlife habitat. With the landscape gradient model the loss of information is minimal, it provides contextual information about the surroundings, geostatistical and multivariate methods can be applied, and landscape metrics from the patch mosaic model can be easily converted into landscape gradient models (Lausch et al. 2015). But the use of the landscape gradient model requires good knowledge of geographic information systems and remote sensing, spatial data mining skills, and an understanding that the process is computationally intensive.

The Graph Theory Model

Graph theory refers to the study of mathematical structures used to model relations between objects (Biggs et al. 1986). In landscape ecology these graphs can represent landscape features of interest across the landscape. The graph theory model is composed of (1) nodes, also known as vertices or points that represent patches, and (2) links, edges, lines, or arcs. Nodes and links create mathematical structures (graphs) that help model pairwise relations between elements (Pascual-Hortal and Saura 2006, Saura and Pascual-Hortal 2007). Key node attributes are the spatial location (x, y), size, shape, and characteristics (functional attributes; Balkenhol et al. 2014). The links can be separated into aspatial theoretical connections (e.g., Euclidean distance between patches) and physical structures on the landscape (e.g., roads and rivers). Graphs can be classified as undirected or directed. Undirected graphs have flows that move in both directions through the links, whereas directed graphs have directional constraints based on the flow of elements across the landscape. The graph matrix model is being used in wildlife studies to assess habitat connectivity, evolution, and genetics (Perotto-Baldivieso et al. 2009, Pirnat and Hladnik 2016, Drake et al. 2017).

Summary

Understanding how spatial patterns and their related processes affect wildlife species is fundamental for habitat management. The rapid development of theoretical work in landscape ecology in the past 40 years, combined with significant technological advances such as computing power and satellite imagery platforms, is providing unique opportunities to assess and quantify the effects of land use change and management into species habitat. Decisions made at small scales will coalesce at large scales and potentially define the changes in abundance of wildlife species across their distribution range. Combined with the effect of human pressures through urban development, road infrastructure, and agricultural activities, the impacts will accumulate across the distribution ranges of wildlife, and though the impacts may not be observable immediately, they will likely set the trends for declines of wildlife species across regional scales (Miller et al. 2019). It is important to understand the scales at which species operate and to build a multiple-scale framework that provides an understanding of the drivers that affect processes and patterns at each scale. Providing and maintaining spatial heterogeneity in the landscape will be key to the sustainability of ecosystem function and consequently wildlife habitat. A significant amount of work has already been done in the theoretical and applied aspects of scale, spatial heterogeneity, and wildlife habitat use and interactions (Anderson et al.

2005, Stuber et al. 2017, Hegland and Hamre 2018). But there is still a lot of work to be done in terms of capturing the scales at which ecological processes operate and how these scales are reconciled with wildlife management. Bridging the biology and ecology of wildlife species with landscape ecology principles will be key to gaining insight on the landscape processes that affect movement, habitat selection, and demographic structures across wildlife populations.

LITERATURE CITED

Abdel Moniem, H. E. M., and J. D. Holland. 2013. Habitat connectivity for pollinator beetles using surface metrics. Landscape Ecology 28:1251–1267.

Allen, T. F. H., and T. W. Hoekstra. 1991. Role of heterogeneity in scaling ecological systems under analysis. Pages 47–68 in J. Kolasa and S. T. A. Pickett, editors. Ecological heterogeneity. Springer-Verlag, New York, New York, USA.

Anderson, P., M. G. Turner, J. D. Forester, J. Zhu, M. S. Boyce, H. Beyer, and L. Stowell. 2005. Scale-dependent summer resource selection by reintroduced elk in Wisconsin, USA. Journal of Wildlife Management 69:298–310.

Baddeley, A., and R. D. Gill. 1997. Kaplan-Meier estimators of distance distributions for spatial point processes. Annals of Statistics 25:263–292.

Baggio, J. A., K. Salau, M. A. Janssen, M. L. Schoon, and Ö. Bodin. 2011. Landscape connectivity and predator–prey population dynamics. Landscape Ecology 26:33–45.

Bailey, T. C., and A. C. Gatrell. 1995. Interactive spatial analysis. Routledge, New York, New York, USA.

Balkenhol, N., J. D. Holbrook, D. Onorato, P. Zager, C. White, and L. P. Waits. 2014. A multi-method approach for analyzing hierarchical genetic structures: a case study with cougars Puma concolor. Ecography 37:552–563.

Biggs, N., E. Lloyd, and R. Wilson. 1986. Graph theory 1739–1936. Oxford University Press, Oxford, UK.

Bissonette, J. A. 2017. Avoiding the scale sampling problem: a consilient solution. Journal of Wildlife Management 81:192–205.

Cavalcanti, S. M. C., and E. M. Gese. 2009. Spatial ecology and social interactions of jaguars (Panthera onca) in the southern Pantanal, Brazil. Journal of Mammalogy 90:935–945.

Cintrón, B., and L. Rogers. 1991. Plant communities of Mona Island. Acta Científica 5:10–64.

Collinge, S. K. 2009. Ecology of fragmented landscapes. Johns Hopkins University Press, Baltimore, Maryland, USA.

de Knegt, H. J., F. van Langevelde, A. K. Skidmore, A. Delsink, R. Slotow, S. Henley, G. Bucini, W. F. de Boer, M. B. Coughenour, C. C. Grant, et al. 2011. The spatial scaling of habitat selection by African elephants. Journal of Animal Ecology 80:270–281.

Delcourt, H. R., P. A. Delcourt, and T. Webb. 1983. Dynamic plant ecology: the spectrum of vegetational change in space and time. Quaternary Science Reviews 1:153–175.

Dickins, E. L., A. R. Yallop, and H. L. Perotto-Baldivieso. 2013. A multiple-scale analysis of host plant selection in Lepidoptera. Journal of Insect Conservation 17:933–939.

Dixo, M., J. P. Metzger, J. S. Morgante, and K. R. Zamudio. 2009. Habitat fragmentation reduces genetic diversity and connectivity among toad populations in the Brazilian Atlantic Coastal Forest. Biological Conservation 142:1560–1569.

Drake, J. C., K. L. Griffis-Kyle, and N. E. McIntyre. 2017. Graph theory as an invasive species management tool: case study in the Sonoran Desert. Landscape Ecology 32:1739–1752.

Dubois, L., J. Mathieu, and N. Loeuille. 2015. The manager dilemma: optimal management of an ecosystem service in heterogeneous exploited landscapes. Ecological Modelling 301:78–89.

Dungan, J. L., J. N. Perry, M. R. T. Dale, P. Legendre, S. Citron-Pousty, M. J. Fortin, A. Jakomulska, M. Miriti, and M. S. Rosenberg. 2002. A balanced view of scale in spatial statistical analysis. Ecography 25:626–640.

Feagin, R. A., and X. B. Wu. 2007. The spatial patterns of functional groups and successional direction in a coastal dune community. Rangeland Ecology and Management 60:417–425.

Fisher, J. T., B. Anholt, and J. P. Volpe. 2011. Body mass explains characteristic scales of habitat selection in terrestrial mammals. Ecology and Evolution 1:517–528.

Forman, R. T. T., and M. Godron. 1986. Landscape ecology. Wiley, New York, New York, USA.

Fulbright, T. E., and J. A. Ortega-Santos. 2013. White-tailed deer habitat: ecology and management on rangelands. Second edition. Texas A&M University Press, College Station, Texas, USA.

Gao, P., J. Kupfer, D. Guo, and T. Lei. 2013. Identifying functionally connected habitat compartments with a novel regionalization technique. Landscape Ecology 28:1949–1959.

Gergel, S. E., and M. G. Turner, editors. 2002. Learning landscape ecology: a practical guide to concepts and techniques. Springer-Verlag, New York, New York, USA.

Gulsby, W. D., J. C. Kilgo, M. Vukovich, and J. A. Martin. 2017. Landscape heterogeneity reduces coyote predation on white-tailed deer fawns. Journal of Wildlife Management 81:601–609.

Haase, P. 1995. Spatial pattern analysis in ecology based on Ripley's *K*-function: introduction and methods of edge correction. Journal of Vegetation Science 6:575–582.

Hegland, S. J., and L. N. Hamre. 2018. Scale-dependent effects of landscape composition and configuration on deer-vehicle collisions and their relevance to mitigation and planning options. Landscape and Urban Planning 169:178–184.

Horev, A., R. Yosef, P. Tryjanowski, and O. Ovadia. 2012. Consequences of variation in male harem size to population persistence: modeling poaching and extinction risk of Bengal tigers (*Panthera tigris*). Biological Conservation 147:22–31.

Killeen, T. J., V. Calderón, L. Soria, B. Quezada, M. K. Steininger, G. Harper, L. A. Solórzano, and C. J. Tucker. 2007. Thirty years of land-cover change in Bolivia. Ambio 36:600–606.

Killeen, T. J., A. Guerra, M. Calzada, L. Correa, V. Calderón, L. Soria, B. Quezada, and M. K. Steininger. 2008. Total historical land-use change in eastern Bolivia: who, where, when, and how much? Ecology and Society 13:36.

Lausch, A., T. Blaschke, D. Haase, F. Herzog, R.-U. Syrbe, L. Tischendorf, and U. Walz. 2015. Understanding and quantifying landscape structure—a review on relevant process characteristics, data models and landscape metrics. Ecological Modelling 295:31–41.

Little, A. R., N. P. Nibbelink, M. J. Chamberlain, L. M. Conner, and R. J. Warren. 2016. Eastern wild turkey nest site selection in two frequently burned pine savannas. Ecological Processes 5:1–10.

Loveland, T. R., and J. L. Dwyer. 2012. Landsat: building a strong future. Remote Sensing of Environment 122:22–29.

Mata, J. M., H. L. Perotto-Baldivieso, F. Hernández, E. D. Grahmann, S. Rideout-Hanzak, J. T. Edwards, M. T. Page, and T. M. Shedd. 2018. Quantifying the spatial and temporal distribution of tanglehead (*Heteropogon contortus*) on south Texas rangelands. Ecological Processes 7:2.

Mayor, S. J., J. A. Schaefer, D. C. Schneider, and S. P. Mahoney. 2009. The spatial structure of habitat selection: a caribou's-eye-view. Acta Oecologica 35:253–260.

McGarigal, K., and S. A. Cushman. 2005. The gradient concept of landscape structure. Pages 112–119 *in* J. A. Wiens and M. R. Moss, editors. Issues and perspectives in landscape ecology. Cambridge University Press, Cambridge, UK.

McGarigal, K., S. A. Cushman, and E. Ene. 2012. FRAGSTATS v4: spatial pattern analysis program for categorical and continuous maps. http://www.umass.edu/landeco/research/fragstats/fragstats.html. Accessed 10 Aug 2018.

McGarigal, K., S. Tagil, and S. A. Cushman. 2009. Surface metrics: an alternative to patch metrics for the quantification of landscape structure. Landscape Ecology 24:433–450.

McGarigal, K., H. Y. Wan, K. A. Zeller, B. C. Timm, and S. A. Cushman. 2016. Multi-scale habitat selection modeling: a review and outlook. Landscape Ecology 31:1161–1175.

McGranahan, D. A., D. M. Engle, S. D. Fuhlendorf, S. J. Winter, J. R. Miller, and D. M. Debinski. 2012. Spatial heterogeneity across five rangelands managed with pyric-herbivory. Journal of Applied Ecology 49:903–910.

Miller, K. S., L. A. Brennan, H. L. Perotto-Baldivieso, F. Hernández, E. D. Grahmann, A. Z. Okay, X. B. Wu, M. J. Peterson, H. Hannusch, J. Mata, et al. 2019. Correlates of habitat fragmentation and northern bobwhite abundance in the Gulf Prairie Landscape Conservation Cooperative. Journal of Fish and Wildlife Management 10:3–18.

Miller, R. E., J. M. ver Hoef, and N. L. Fowler. 1995. Spatial heterogeneity in eight central Texas grasslands. Journal of Ecology 83:919–928.

Mitchell, M. S., and R. A. Powell. 2004. A mechanistic home range model for optimal use of spatially distributed resources. Ecological Modelling 177:209–232.

Moorcroft, P. R., M. A. Lewis, and R. L. Crabtree. 1999. Home range analysis using a mechanistic home range model. Ecology 80:1656–1665.

Okay, A. Z. 2004. Spatial pattern and temporal dynamics of northern bobwhite abundance and agricultural land use, and potential causal factors. Dissertation, Texas A&M University, College Station, Texas, USA.

Okayasu, T., T. Okuro, U. Jamsran, and K. Takeuchi. 2012. Degraded rangeland dominated by unpalatable forbs exhibits large-scale spatial heterogeneity. Plant Ecology 213:625–635.

Otieno, D., G. K'Otuto, B. Jákli, P. Schröttle, J. Maina, E. Jung, and J. Onyango. 2011. Spatial heterogeneity in ecosystem structure and productivity in a moist Kenyan savanna. Plant Ecology 212:769–783.

Pascual-Hortal, L., and S. Saura. 2006. Comparison and development of new graph-based landscape connectivity indices: towards the priorization of habitat patches and corridors for conservation. Landscape Ecology 21:959–967.

Perotto-Baldivieso, H. L., X. Ben Wu, M. J. Peterson, F. E. Smeins, N. J. Silvy, and T. Wayne Schwertner. 2011.

Flooding-induced landscape changes along dendritic stream networks and implications for wildlife habitat. Landscape and Urban Planning 99:115–122.

Perotto-Baldivieso, H. L., E. Meléndez-Ackerman, M. A. García, P. Leimgruber, S. M. Cooper, A. Martínez, P. Calle, O. M. Ramos Gónzales, M. Quiñones, C. A. Christen, et al. 2009. Spatial distribution, connectivity, and the influence of scale: habitat availability for the endangered Mona Island rock iguana. Biodiversity and Conservation 18:905–917.

Pirnat, J., and D. Hladnik. 2016. Connectivity as a tool in the prioritization and protection of sub-urban forest patches in landscape conservation planning. Landscape and Urban Planning 153:129–139.

Prevedello, J. A., and M. V. Vieira. 2010. Does the type of matrix matter? a quantitative review of the evidence. Biodiversity and Conservation 19:1205–1223.

Ripley, B. D. 1976. The second-order analysis of stationary point processes. Journal of Applied Probability 13:255–266.

Rodríguez-San Pedro, A., and J. Simonetti. 2015. The relative influence of forest loss and fragmentation on insectivorous bats: does the type of matrix matter? Landscape Ecology 30:1561–1572.

Rosenblatt, M. 1956. Remarks on some nonparametric estimates of a density function. Annals of Mathematical Statistics 27:832–837.

Roy, D. P., M. A. Wulder, T. R. Loveland, C. E. Woodcock, R. G. Allen, M. C. Anderson, D. Helder, J. R. Irons, D. M. Johnson, R. Kennedy, et al. Landsat-8: science and product vision for terrestrial global change research. Remote Sensing of Environment 145:154–172.

Saura, S., and L. Pascual-Hortal. 2007. A new habitat availability index to integrate connectivity in landscape conservation planning: comparison with existing indices and application to a case study. Landscape and Urban Planning 83:91–103.

Schneider, D. C. 2001. The rise of the concept of scale in ecology. Bioscience 51:545–553.

Shiode, S. 2008. Analysis of a distribution of point events using the network-based quadrat method. Geographical Analysis 40:380–400.

Stuber, E. F., L. F. Gruber, and J. J. Fontaine. 2017. A Bayesian method for assessing multi-scale species-habitat relationships. Landscape Ecology 32:2365–2381.

Thompson, C. M., and K. McGarigal. 2002. The influence of research scale on bald eagle habitat selection along the lower Hudson River, New York (USA). Landscape Ecology 17:569–586.

Turner, M. G., D. Donato, and W. Romme. 2013. Consequences of spatial heterogeneity for ecosystem services in changing forest landscapes: priorities for future research. Landscape Ecology 28:1081–1097.

Turner, M. G., and R. H. Gardner. 2015. Landscape ecology in theory and practice. Second edition. Springer-Verlag, New York, New York, USA.

Turner, M. G., R. H. Gardner, and R. V. O'Neill. 2001. Landscape ecology in theory and practice. Springer-Verlag, New York, New York, USA.

Van Moorter, B., C. M. Rolandsen, M. Basille, and J. M. Gaillard. 2016. Movement is the glue connecting home ranges and habitat selection. Journal of Animal Ecology 85:21–31.

Wheatley, M. 2010. Domains of scale in forest-landscape metrics: implications for species-habitat modeling. Acta Oecologica 36:259–267.

Wiens, J. A., 1989. Spatial scaling in ecology. Functional Ecology 3:385–397.

Wiegand, T., F. He, and S. P. Hubbell. 2013. A systematic comparison of summary characteristics for quantifying point patterns in ecology. Ecography 36:92–103.

Wijesinghe, D. K., E. A. John, and M. J. Hutchings. 2005. Does pattern of soil resource heterogeneity determine plant community structure? An experimental investigation. Journal of Ecology 93:99–112.

Worton, B. J. 1989. Kernel methods for estimating the utilization distribution in home-range studies. Ecology 70:164–168.

Wu, X. B., T. L. Thurow, and S. G. Whisenant. 2000. Fragmentation and changes in hydrologic function of tiger bush landscapes, south west Niger. Journal of Ecology 88:790–800.

Zeilhofer, P., A. Cezar, N. M. Tôrres, A. T. de Almeida Jácomo, and L. Silveira. 2014. Jaguar *Panthera onca* habitat modeling in landscapes facing high land-use transformation pressure—findings from Mato Grosso, Brazil. Biotropica 46:98–105.

Zemanova, M. A., H. L. Perotto-Baldivieso, E. L. Dickins, A. B. Gill, J. P. Leonard, and D. B. Wester. 2017. Impact of deforestation on habitat connectivity thresholds for large carnivores in tropical forests. Ecological Processes 6:21.

5 — Using Landscape Ecology to Inform Effective Management

Joseph A. Veech

Introduction

Throughout this book the landscape perspective has been presented as an important framework for managing populations of wildlife species and their habitats. Wildlife species respond to factors on the landscape that exist at different spatial scales and in various patterns. Ecologists and managers are interested in understanding these relationships to more effectively conserve wildlife and biodiversity in general. Individuals and populations reside within and among landscapes and use the resources within those landscapes. Understanding the ecological and environmental conditions that give rise to patterns of resources on the landscape is an important prerequisite for the development of any wildlife–habitat or wildlife–landscape relationship. But there lies a significant challenge. Designing research projects on landscapes is inherently intractable; there are seldom (if ever) suitable control landscapes, and baseline measurements are not common (Morrison et al. 2006). Hence creating, modifying, and maintaining habitat at landscape scales are logistically difficult and sometimes not well informed by theory or knowledge. Yet management actions at broad, landscape extents are likely to have the greatest measurable effect on wildlife populations precisely because

such actions consider ecological relationships across large areal extents and thus affect a large portion of the population. The landscape perspective also needs to extend temporally to cover the daily, seasonal, and annual resource needs of individuals and populations. Satisfying the needs of individuals and populations over large spatial scales and extended temporal periods naturally leads to greater reproduction and population persistence. A first step in achieving such broadscale management is to recognize that landscapes are not simple or static.

One of the most important emerging principles of landscape ecology is that landscapes are naturally heterogeneous and dynamic (Turner 1989, Forman 1995, Pickett and Cadenasso 1995, Shugart 2005, Turner 2010). That is, at some spatial scale and typically multiple spatial scales, landscapes are mosaics of different landcover types, vegetation associations, stages of succession, surface topographies, waters (e.g., ponds, lakes, streams), and other discrete structural elements. Landscapes consist of patches of various sizes, shapes, and spatial arrangements. Patches are internally consistent in that any discrete point within the patch has a common attribute (e.g., shortgrass prairie in which grass height is 5–15 cm) shared with all other points in the patch. Areas of the mosaic where common attributes are no longer consistent are considered separate patches (e.g., a patch of shortgrass prairie with a high density of brush mottes). Patches can vary in size from a few hundred square meters to tens of hectares or more, although the scale that is relevant will depend on the wildlife species. Landscapes are dynamic in the sense that the mosaic is always in flux, changing at various rates through either the relatively slow process of succession or the comparatively rapid seasonal changes encompassed by the phenology of vegetation and anthropogenic changes (Pickett and Cadenasso 1995). Even patches of resources within the mosaic that would appear at first glance to be homogeneous and static may have substantial spatial heterogeneity (i.e., constantly changing through time and space). For example, an area of tallgrass prairie may appear to be structurally homogeneous (grass of a constant height and thickness), but the composition of plant species might actually vary from place to place depending on surface topography, soil type, and previous disturbance events. In this scenario, there is substantial spatial heterogeneity but without any discrete patches. Thus spatial heterogeneity is a concept that can be measured by patch dynamics, fragmentation, or edge and not just by patch type. Throughout this chapter, I use an inclusive definition that considers the composition and configuration of patches and gradients in environmental characteristics (e.g., soil moisture) within a landscape.

A second important emerging principle of landscape ecology is that the spatial heterogeneity matters to organisms be they animal, plant, or microbe, and it matters in a species-specific way (Johnson et al. 1992, Lindenmayer and Hobbs 2004, Murwira and Skidmore 2005, Whitehead et al. 2005, Leyequién et al. 2010). In essence, spatial heterogeneity of habitat features (i.e., components of habitat) and other landcover types necessarily entails that a landscape is composed of resource patches that are usable to specific species (to another species, the spatial heterogeneity of these resource patches may be completely unusable; Fig. 5.1). For example, a small-bodied organism such as a ground-dwelling

beetle might be influenced by heterogeneity (in this case, measured in terms of patchiness) on a scale of 1 m² or less when moving across the substrate (Johnson et al. 1992). To fully understand its habitat use, the heterogeneity must be measured and studied at the relevant spatial scale. Highly mobile wildlife species that access and use habitats spaced far apart (e.g., >10 km) are not likely to experience heterogeneity in the same way that a beetle would. There is no need for the wildlife manager to be concerned with such fine-grained heterogeneity of patches if the goal for management is to manage habitats on landscapes.

Various natural and anthropogenic processes can act to create and maintain the spatial heterogeneity of patches on a landscape. These influences determine the configuration and composition of patches, and in turn they may be influenced by the patterning (Fig. 5.2). The challenge for the wildlife manager is to understand how these processes affect the availability and quality of habitat, at spatial scales relevant to the managed species, and within the context of a much larger landscape. This chapter focuses on the influences that produce and maintain spatial heterogeneity of habitats within a landscape. Within any landscape altered by humans, there can also be substantial spatial patchiness represented by how humans use the landscape (e.g., urbanization, transportation, agricultural production), and this patterning is important because it leads to habitat fragmentation and loss (see Martin et al., Chap. 7, this volume). But these anthropogenic processes are not directly involved in determining the type of vegetation remaining in the unaltered natural areas of a landscape that are primarily used by wildlife.

Factors Affecting Spatial Heterogeneity

In the context of wildlife and habitat management the distinction between natural and anthropogenic factors is important because some processes affecting spatial heterogeneity of habitat can be initiated and controlled by humans, whereas others cannot.

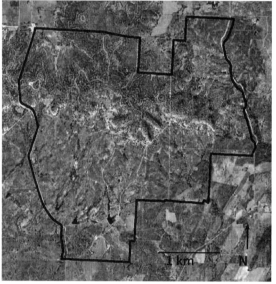

Figure 5.1. Mason Mountain Wildlife Management Area (WMA; 30.842°N, 99.215°W) in Mason County, Texas, USA. The maps depict spatial heterogeneity in landcover types within the property and surrounding area. The top panel shows distribution of forest (dark gray), shrubland (light gray), and grassland (white). The heavy black line is the property boundary. Black polygons outside of the property boundary (southeast and northwest corners) are cropland. Two small black polygons near the southern boundary of property are ponds. Black lines in the south-east corner are roads and highways. The map indicates a patch mosaic structure in which forest and grassland form patches in a landscape that is otherwise continuous shrubland. Map resolution is 30 × 30 m. The bottom panel shows an aerial image from January 2016. In both maps the spatial heterogeneity in landcover types arises from underlying heterogeneity in soil type, topography, and management practices. Mason Mountain WMA is notable for having 14 species of exotic ungulates (mostly from Africa) and native ungulates such as white-tailed deer (*Odocoileus virginianus*) and collared peccaries (*Pecari tajacu*). There is also a sizeable nuisance population of feral pigs. Habitat management at Mason Mountain WMA takes into consideration the needs of all species (including non-ungulate species) as well as use of the WMA for recreational hunting.

It is the former that are most amenable for use in creating and managing habitat, and I discuss these below.

I intentionally do not discuss the effects of climate and weather on landscapes. I am conceiving of landscapes as being small enough (e.g., <100 km²) that the entire landscape has only one climate. Hence climate cannot induce spatial heterogeneity within a landscape, although climatic gradients within a larger region could certainly lead to spatial heterogeneity among landscapes within the region. Weather events, however, can sometimes be localized (e.g., summer thunderstorms in the Great Plains) within a given landscape and thus induce internal spatial heterogeneity of features that respond quickly to weather events (e.g., grass height or other attributes of vegetation). Annual and seasonal variation in weather can also be substantial, and both can have a direct effect on vegetation.

Although human activities can also induce spa-

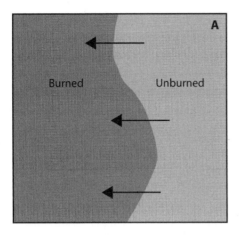
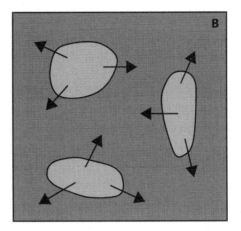

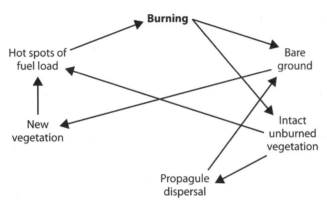

Figure 5.2. In this schematic a hypothetical prairie (perhaps 1 km²) can undergo a burning treatment in either of two scenarios. In (*A*), there is a clear burn line indicating that half the prairie has been burned, leaving the other half intact, whereas in (*B*) the burning occurred in such a way as to leave three islands of unburned intact vegetation. In each scenario, arrows show direction of propagule (seed) dispersal from unburned vegetation to the burned area, which is now essentially bare ground (alternatively, the arrows could indicate the direction of vegetative spread). The rate and spatial patterning of revegetation in each scenario will be different owing to spatial differences in propagule sources. In scenario *A* the revegetation might occur as an advancing wave, whereas in scenario *B* it might occur more as a filling-in process. In both scenarios, there likely will be regrowth of perennial species from below ground in the burned area; this regrowth adds further spatial heterogeneity to the resulting prairie in terms of species composition and grass height/thickness. The flow diagram in the bottom panel reveals how the two processes or drivers (burning and propagule dispersal) act in tandem to produce spatial heterogeneity in the vegetation. Burning results in areas of bare ground and intact vegetation, both of which could eventually become sources of fuel load for the next fire. Arrows indicate event pathways. Most importantly, the spatial heterogeneity in vegetation resulting from one burning event leads to spatial heterogeneity in fuel load that could then influence the next burning event.

tial heterogeneity, presenting an exhaustive list of anthropogenic factors on landscapes is beyond the scope of this chapter. With some exceptions, managers often have little (or no) influence over the broadscale anthropogenic trends and forces that determine habitat heterogeneity within a landscape. The same is true for climate and weather. Therefore I take the view that climate, weather, and anthropogenic factors set the context of a landscape, within which managers must attempt to control the factors that may result in meaningful positive changes to wildlife populations and habitats. Accordingly, I review several major sources of natural heterogeneity on landscapes.

Soil Heterogeneity

Within the United States, soil types are historically identified and mapped at the county level by the Soil Conservation Service (now the Natural Resources Conservation Service) of the US Department of Agriculture (Kellogg 1936, Arnold 1999, Heuvelink and Webster 2001). Even over a small distance (e.g., 1 m), soil type can change with sharp transitions between types or properties (Heuvelink and Webster 2001, Lin et al. 2005, Phillips and Marion 2005, Garten et al. 2007). Therefore in an area as small as 1 ha,

multiple soil types may be present; alternatively, the area may be homogeneous with regard to soil type consisting solely of one type. In the science of pedology, soils are classified in a hierarchical taxonomy. Within any area of land, a greater number of soil types will be represented as one goes further down the hierarchy in identifying the types. Accordingly, for an entire landscape, there are typically a few different soil types to as many as a few dozen (Kellogg 1936, Phillips 2001, Ibáñez et al. 2005, Petersen et al. 2010), and this also depends on the extent to which the soil data set is being classified for analytical purposes (Fig. 5.3; Frair and Bastille-Rousseau, Chap. 8, this volume).

Regarding wildlife management and habitat ma-

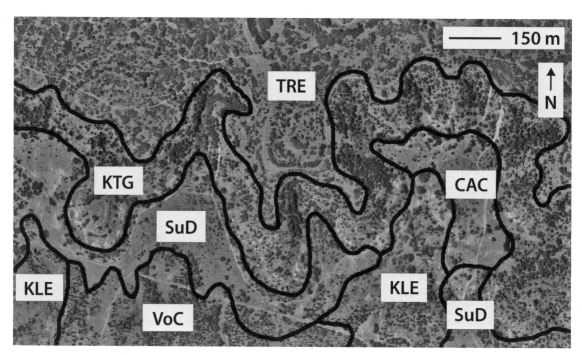

Figure 5.3. Spatial heterogeneity in soil type within a 2.3 km² portion of the Mason Mountain Wildlife Management Area (WMA) in Mason County, Texas, USA. Within the rectangular area there are 6 soil types composing the following percentages: Tarrant-rock outcrop complex (TRE; 31.9%), Kerrville/Brackett/Tarrant soils (KTG; 29.9%), Sunev clay loam (SuD; 15.8%), Keese-rock outcrop complex (KLE; 9.6%), Voca sandy loam (VoC; 8.0%), and Campwood/Sunev/Valera soils (CAC; 4.8%). Within the larger WMA, there are 20 soil types. Note that the aerial photograph appears to indicate some congruence between the soil types and vegetation, and there is greater vegetation biomass on KTG. Alphabetic codes for soil types follow the Soil Survey of Mason County, Texas (US Department of Agriculture 2011). Figure created from the Web Soil Survey (US Department of Agriculture 2019).

nipulation, spatial heterogeneity in soil type is important for two primary reasons. First, soils differ in their water-holding capacity. The extent to which precipitation can infiltrate and be retained in the substrate will partly determine the composition of plant species, and in turn the vegetation (through decomposition) increases soil fertility and nutrients (Davies et al. 1998, Clark and Clark 2000, Lechmere-Oertel et al. 2005). Some plant species prefer and grow best in mesic soil conditions, whereas others require xeric conditions promoted by well-drained soils, or species and entire vegetation types may differ in their tolerance of other soil characteristics such as acidity (Goldberg 1982). Plant species composition effectively determines the type of vegetation present, particularly if the differences in water-holding capacity and fertility between two or more soil types are great enough that one supports grassy vegetation whereas the other supports woody vegetation. In this scenario, spatial heterogeneity in soil type leads to spatial heterogeneity in surface vegetation on the landscape (Lechmere-Oertel et al. 2005, Chartier et al. 2011). Even when soils do not differ that much in water retention, minor differences in plant species composition among soil types could still occur, again producing arrangements of plant communities that are spatially heterogeneous on the landscape.

The second important physical feature of soil types is depth to a restrictive layer, or more generally the combined depth of the O, A, and B soil horizons. In pedology the restrictive layer or substratum (often horizon C or D) is defined as a relatively continuous layer of substrate or deposition material that impedes or slows the vertical penetration of water deeper (farther down) into a lower soil horizon. The restrictive layer could be a feature such as bedrock or a dense hydrophobic clay. Of importance to plant growth and aboveground vegetation, restrictive layers can constrain the depth of root growth, thus limiting a plant's zone of water acquisition to the soil horizons between the surface and the restrictive feature (Bennie 1991). When a given landscape has spatial heterogeneity in depth to restrictive layer, there may also be concordant spatial heterogeneity in surface vegetation (Phillips 2001, Molinar et al. 2002).

Of particular relevance to habitat management, soil type is not amenable to manipulation by humans. The particular soil type at any one location is the outcome of a long process involving geology and biology (Jenny 1941, 1980). The composition and arrangement of soil types over a landscape cannot typically be modified. The sole exceptions are when massive amounts of surface substrate are removed (e.g., by surface mining or creation of landfills), creating a void that is then filled in with a different type of substrate, and when the surface of a given area is overlain by some other material such as gravel, asphalt, or concrete. The former is such an extensive, extreme, and dramatic alteration of the landscape that it is best thought of as being in the purview of ecological restoration (not habitat management), and the latter is simply irreversible habitat destruction. It is also worth pointing out that anthropogenic processes of soil enrichment (e.g., adding nitrogen fertilizer to soil) do not actually change the soil type. So, for a given landscape, the wildlife manager is essentially stuck with the existing spatial heterogeneity in soil type, and habitat management must proceed within those confines. Nonetheless, it is important that the manager has knowledge of the distribution and types of soil to understand where to create and maintain certain habitat features, such as facilitating the growth of a plant species on soil types to which they are best adapted.

Hydrology

Heterogeneity in hydrologic conditions (i.e., rates and pathways of water movement) is an important factor for aquatic and wetland habitats across a landscape, particularly those habitats that depend on overland flow for water input. Hydrological heterogeneity is due to variation in topography and the spatial arrangement and connections among water bodies (Moore et al. 1991, Grayson et al. 2002). Even apart from the hydrology of water bodies, topography can interact with soil type to determine whether surface water (from direct precipitation and runoff) collects in a particular area of a landscape and subsequently the rate at which the water penetrates the soil substrate or remains in a ponding state (Western et al. 2004, Wilson et al. 2005). In turn, hydrological heterogeneity in water accumulation affects the composition, configuration, and diversity of terrestrial plant communities, thereby contributing to variation in the physiognomy of vegetation across landscapes (i.e., habitat heterogeneity). The flow of water across a landscape in the form of rivers, streams, oxbow lakes, and drainage ditches is also an example of hydrology creating heterogeneity. Whether the source is groundwater, precipitation and runoff, or flow from another water body, the movement of water creates and maintains aquatic features across a landscape. Even when flow does not result in a permanent water body, it can influence the landscape by producing areas of erosion that may have a different type of vegetative cover than non-erodible areas.

Although some components (e.g., precipitation, evapotranspiration) of an ecosystem's hydrology cannot be controlled by humans, managers can manipulate other aspects. For example, creation of new wetlands on a landscape is not done at random but rather with knowledge of the topography and soil conditions of the landscape (Moreno-Mateos et al. 2010, Darwiche-Criado et al. 2017, Uuemaa et al. 2018). Sites are selected on the basis of their capacity to collect and retain standing water and provide soils conducive to plant growth (Wall and Stevens 2015). Even the restoration of a degraded wetland may initially involve active habitat management (e.g., pumping in water or removing excess water) if the natural hydrology of the wetland is not sufficient to promote restoration. Control and manipulation of hydrology are involved in some efforts to create habitat features (e.g., wetlands); in other situations, however, the management of habitat proceeds without

the manager attempting to make radical alterations to the natural hydrological regime of a landscape.

Fire

Whether implemented naturally (wildland fires) or anthropogenically (prescribed fires), the application of fire has significant capacity to influence habitat heterogeneity on landscapes. Some ecosystems have been exposed to periodic burning for millennia, and some plant communities within the ecosystem are stimulated by fire. A classic example is the chaparral habitat of Southern California where the return interval of natural fires has been estimated to be 50 years, although estimates as low 10 years and as high as 65 years have also been suggested (Barro and Conard 1991, Steel et al. 2015). Chaparral fires are stand replacing (i.e., they burn all the vegetation). But chaparral is also fire adapted, meaning that some of the shrub species begin to die off if the fire return interval becomes too long (Keeley et al. 2012). In chaparral and similar ecosystems, fire maintains some features of wildlife habitat (e.g., vegetation).

Both natural and anthropogenic fires can influence spatial heterogeneity in vegetation within a landscape (Turner et al. 2013). Except for the most intense fires, a typical fire will burn unevenly depending on the fuel load, moisture content, wind direction, and topography. As a result, fire creates heterogeneity on the landscape because there are patches differing in the severity of the burn (extent to which the vegetation was burned away), including some patches that are relatively untouched (Turner et al. 1997). Accordingly, heterogeneity of patches on the landscape can be managed deliberately through natural fire, and in particular from an anthropogenic fire, when steps are taken ahead of time to account for the variables that influence the intensity of a burn (Fig. 5.2). The mosaic present immediately after the fire then persists as a mosaic of vegetation patches that are regrowing at different rates owing to differences in their initial conditions postburn and other sources of heterogeneity (e.g., soil type, hydrology;

Romme 1982, Shakesby 2011, Smit et al. 2013). Even the most intense fire that produces a clean, complete burn affects heterogeneity of patches on the landscape. For example, a complete burn will often leave behind spatial variation in ash residue (Kokaly et al. 2007, Caon et al. 2014, Chafer et al. 2016) and belowground vegetation that will eventually become visible as a more pronounced mosaic during regrowth of the vegetation (Kokaly et al. 2007).

Finally, the exclusion of fire can also affect spatial heterogeneity of vegetation on a landscape. Typically, this occurs as a homogenizing effect in which plant communities reach a stable climax successional state without a fire (or any other disturbance event) to perpetuate the successional cycle (Li and Waller 2015). If natural fires are suppressed throughout a landscape, then vegetation may become more spatially homogeneous, with the lack of fire leading to a decrease in heterogeneity over time (Hessburg et al. 2005, Boisramé et al. 2017). Further, such homogenization of vegetation may result in the entire landscape becoming less beneficial to some wildlife species (Doherty et al. 2015, Hovick et al. 2015, McCleery et al. 2018). Regarding fire as a management tool, fires that occur too frequently might also homogenize vegetation, which can make a landscape less suitable for some wildlife species (Doherty et al. 2015, Connell et al. 2019).

Grazing

Grazing and browsing by large indigenous mammals are natural features of many ecosystems; bison (*Bos bison*) grazing in the Great Plains region of central North America is a classic example. Many grassland and savanna ecosystems are grazing tolerant because the native plants can withstand some amount of biomass removal by consumer species. As such, grazing by livestock and wildlife can create spatial heterogeneity in vegetation structure across a landscape (Hobbs 1996, Bisigato and Bertiller 1997, Parsons and Dumont 2003, Asner et al. 2009). The spatial extent and magnitude of the heterogeneity will de-

pend somewhat on grazing intensity (Adler et al. 2001) and on how extensively the grazing animals move around on the landscape (Parsons and Dumont 2003). Further, grazing animals may prefer areas of nutrient-rich vegetation and graze those sites with greater intensity, influencing heterogeneity in vegetation across the landscape (Ben-Shahar and Coe 1992, Bailey et al. 1996, Augustine et al. 2003, Cromsigt and Olff 2006). As with fire, grazing can be thought of as an agent of disturbance that may sometimes benefit plants by stimulating regrowth and suppressing competitively dominant species, so long as the grazing is light to moderate (McNaughton 1993, Noy-Meir 1993, Hobbs 1996, Fuhlendorf and Engle 2004, Prober et al. 2013).

Grazing by livestock (primarily cattle and sheep) is often used as a habitat management tool. Compared to other grazing animals, cattle and sheep are relatively sedentary. As such, the intensity of their grazing on a site can be controlled by monitoring the amount of time that a given density of animals is allowed to graze. Fencing to control movement is typically necessary. At a landscape scale, rotational and low-intensity grazing can be used effectively to create coarse-grained heterogeneity in vegetation structure (Derner et al. 1994, Söderström et al. 2001, Sebastia et al. 2008, Fuhlendorf et al. 2012, Török et al. 2014) and thereby maintain habitat for some wildlife species (Derner et al. 2009).

Removal of Woody Biomass

Forested landscapes often seem to be relatively homogeneous when forest cover is near 100%, but typically there is natural spatial heterogeneity that can arise from events like tree falls that open up the canopy, allowing the understory plant community to more fully develop. As discussed in the previous section, fire can create substantial spatial heterogeneity on a landscape. In addition, owing to different disturbance events, different areas of a forest might be at different stages of succession, leading to differences in tree species composition, canopy height, and closure. Spatial heterogeneity in forested, and woodland-dominated landscapes may be beneficial to some wildlife species (see previous references on the effects of fire).

With regard to habitat management, removal of trees and shrubs can occur through prescribed fire (discussed above), timber harvest, and the mechanical and chemical control of brush. Further, all of these management actions can be used to create and influence heterogeneity (Franklin and Forman 1987, Mladenoff et al. 1993). Clear-cutting of timber and broadscale eradication of brush (e.g., by aerial application of herbicide) create obvious patches with no vegetation. Within these patches there is variation in the amount and type of woody biomass left behind (coarse woody debris) after the physical process of removing timber. Additionally, the landscape might contain patches of standing, intact vegetation with no biomass removal, which results in a landscape with a dichotomy of patch types. Heterogeneity on such a landscape is determined by the size and arrangement of patches of standing, intact vegetation (areas with no biomass removal) interspersed with patches of downed vegetation (areas where vegetation was intentionally killed).

Commonly referred to as coarse woody debris, researchers reported that downed wood can function as habitat features for wildlife species such as small mammals and amphibians (Sullivan and Sullivan 2001, Patrick et al. 2006, Riffell et al. 2011, Sullivan et al. 2012, Otto et al. 2013). Also, a relatively common practice in brush control is to form the dead woody biomass into scattered piles. Although a brush pile does not equate to a patch per se, brush piles can be thought of as distinct features on the landscape that add to the overall heterogeneity and possibly provide a resource to some species (Rittenhouse et al. 2008, Sperry and Weatherhead 2010, Larsen et al. 2016).

Woody biomass removal is a process controlled by managers that allows them to specifically influence the heterogeneity of landscapes. Consider that selective logging often targets particular trees in an otherwise continuous forest. This practice can create a mosaic of open-canopy patches in a matrix of dense forest. Alternatively, selective logging sometimes entails cutting all trees except specific large ones, which in effect creates a pattern of scattered solitary trees across the landscape. These residual trees (i.e., legacy trees) can provide habitat features and resources to a wide variety of wildlife species and facilitate various ecosystem processes (Mazurek and Zielinski 2004, Turner et al. 2013, Bakermans et al. 2015). Similar to fire as a process for creating spatial patchiness, the downing and removal of woody biomass lead to an initial pattern of patchiness that then persists and possibly changes as new vegetation begins to regrow in the cleared areas (Turner et al. 2013). Thus, depending on the scale of the treatment used to kill or down the vegetation, managers can target a specific spatial heterogeneity to benefit a desired wildlife species.

Appreciating Spatial Heterogeneity and Uncertainty

Managers cannot completely control spatial heterogeneity within a landscape. Even if managers could observe the pattern on process relationship precisely, management would still take place against the backdrop of significant uncertainty because there is simply too much complexity and contingency to account for. There are many ecological and environmental factors that influence heterogeneity on a landscape and in turn are affected synergistically by the heterogeneity. All the different types of heterogeneity (resulting from the different factors) are interconnected, influencing one another (Turner et al. 2013). This interconnectedness can be complex. For example, so-called large old trees create spatial heterogeneity in forest, woodland, and savanna ecosystems and have numerous ecological roles (Lindenmayer and Laurance 2017:2). The trees themselves represent heterogeneity because, as the name suggests, they are larger and older than surrounding trees. The spatial arrangement of large old trees on a landscape is determined by history (contingency) and ecology. It depends on the past events of seeds arriving at particular microsites suitable for germination and subsequent growth and survival of the adult tree. Thus heterogeneity in soil type, nutrients, moisture, canopy cover, surrounding plant species (potential competitors), and nearby herbivores can affect where the trees exist on a landscape (Lindenmayer and Laurance 2017). Even heterogeneity in human activities (e.g., tree cutting, fires) within the landscape affects the heterogeneity represented by large old trees. Clearly, many different natural, ecological, and anthropogenic factors influence the heterogeneity of large old trees on the landscape. Attempting to alter this heterogeneity has consequences for wildlife population processes. But imagine the difficulties and near impossibility of attempting to create or manage for a particular arrangement (or pattern of heterogeneity) of large old trees on a landscape. The mindset of the wildlife manager should be to appreciate ecological complexity and natural heterogeneity, and to use it to their advantage when possible (see "Managing the Landscape" below).

Along with the complexity and contingency comes uncertainty. There will always be uncertainty associated with the use of natural processes and anthropogenic management practices to create and manage habitat. For example, even with respect to anthropogenic causes of heterogeneity on the landscape, managers have only partial control. Consider a land use practice such as grazing. There is an inherent naturalness to the practice in that managers rely on some animal species (wild or domesticated) to carry out the process. The desired effect (presumably, to alter the structure and function of vegetation to enhance wildlife habitat) may not be realized because managers cannot reasonably have complete control over what and how much an animal eats (particularly wildlife species).

The lack of complete control as a result of uncertainty is magnified for natural sources of heterogeneity on landscapes. Consider a management scenario

that called for the removal of thick shrub cover to benefit a northern bobwhite population (*Colinus virginianus*). Managers might be guided by a carefully modeled relationship between bobwhite and shrub cover. As such, they develop a management plan that calls for natural (e.g., fire) and anthropogenic (e.g., mechanical, chemical) interventions to remove or thin the amount of shrub cover followed by natural or anthropogenic reseeding of the cleared areas to enhance grass growth. The managers aim to sculpt the perfect landscape for bobwhite with regard to a mix of shrubby and grassy areas. The management challenge, however, is there will always be some level uncertainty. Drought or perhaps excess precipitation during the year subsequent to the shrub clearing could affect the intended and desired amount of grass growth. Many sources of heterogeneity and their underlying processes are always present, perhaps latent, in most landscapes whether there is human intervention or not. Flux in the landscape mosaic is the natural condition if given a long enough time to play out. Thus wildlife managers should appreciate and take advantage of the factors that influence heterogeneity where they can, and also plan to be flexible in the face of uncertainty, because there is no controlling or diminishing uncertainty.

Managing the Landscape

Managing the landscape entails an explicit recognition that altering heterogeneity is desirable, and any management action at one place on the landscape can influence the overall heterogeneity on the rest of the landscape. Landscapes are dynamic mosaics (Pickett and Cadenasso 1995) with ecological processes that transcend patch boundaries. Thus wildlife managers can aim to create and maintain habitat in certain locations on the landscape, but they should also recognize that ecological processes connect the managed (or created) portions to other areas of the landscape that might not be receiving direct management. As such, it may be difficult to know completely how a particular management action will affect other areas of a landscape and vice versa, how processes emanating from the unmanaged area might be influencing the area being managed. Together the managed and unmanaged portions contribute to the overall heterogeneity of the landscape, where heterogeneity is defined in a holistic way to include distribution and composition of patches, spatial arrangement of non-patch features, and gradients of environmental properties.

The landscape-level conservation of Hermann's tortoise (*Testudo hermanni*) provides a good case study of the importance of understanding spatial heterogeneity in managing habitat. Hermann's tortoise is classified as near threatened by the International Union for the Conservation of Nature (van Dijk et al. 2004). It has a range that occupies primarily coastal and near-coastal areas of southernmost Europe from Spain to Greece, and extant populations are patchily distributed within this long linear range (van Dijk et al. 2004). Throughout its geographic range, sustainable populations occur in landscapes with substantial heterogeneity in landcover types and use, including woodland, scrubland, grassland, hayfield, grazed pasture, and abandoned or deteriorating farmland (Rozylowicz and Popescu 2013, Celse et al. 2014, Vilardell-Bartino et al. 2015). Regarding the anthropogenic landcover types (or uses), a key characteristic is the presence of a tightly knit mosaic of patches (e.g., average patch size <1 ha) with interspersed features such as hedgerows, ponds, and streams (Celse et al. 2014, Couturier et al. 2011). Telemetry studies have reported that daily movement distances are typically <100 m (Mazzotti 2004, Rozylowicz and Popescu 2013, Lecq et al. 2014, Vilardell-Bartino et al. 2015), although during a weekly to monthly time span, individuals may move greater distances, particularly when searching for mates, water sources, or hibernation sites (Rozylowicz and Popescu 2013, Celse et al. 2014). Thus the species requires a mosaic of patches and features at two (possibly more) spatial scales: the scale of the landscape wherein tortoise home ranges include a patchy mix of various landcover types needed over the course of an individual's seasonal life cycle (Berardo et al. 2015, Vilardell-Bartino et al. 2015) and the scale of the individual's thermoregulatory and foraging needs on a daily or even hourly basis (Del Vecchio et al. 2011, Rozylowicz and Popescu 2013, Couturier et al. 2011). The latter scale involves the fine-grained heterogeneity within home ranges.

Habitat management of Hermann's tortoise at the landscape scale entails ensuring that individuals (and hence the population) have access to spatial heterogeneity over the extent of the entire landscape (e.g., an area of ≥ 10 km²) and within much smaller areas of the landscape that would represent home ranges (typically 2–4 ha) of individuals (Rozylowicz and Popescu 2013, Celse et al. 2014). In most cases, habitat management for Hermann's tortoise or any other species cannot practically entail a complete rearrangement of the landscape to make it more beneficial. Nonetheless, specific and reliable knowledge of the most ideal type of landscape can be used in selecting potential sites for reintroduction (Bertolero et al. 2007).

Based on our knowledge of the ecology of Hermann's tortoises, it may be beneficial to actively manipulate and maintain patches of habitat features throughout the landscape. In some parts of the species geographic range, local people have traditionally and currently engage in land use practices (e.g., tree and shrub clearing, manual cutting of hayfields, light grazing by livestock) that produce the type of open habitat preferred by Hermann's tortoises (Rozylowicz and Popescu 2013). In other parts of the range, agricultural heterogeneity might negatively affect this species (Badiane et al. 2017). Within any landscape or region, thick ground-level vegetation can impede the movement (dispersal) of tortoises, but at the same time some amount (scattered small patches) of thick shrub cover is needed for thermoregulation, hiding from predators, and hibernation (Berardo et al. 2015, Vilardell-Bartino et al. 2015,

Golubović et al. 2017). Again, the species requires a fine-grained mosaic of patches that is linked with relatively obstacle-free dispersal or movement corridors. Given that this species requires substantial patch heterogeneity on relatively small scales of 1 ha or less, any process or activity that leads to homogenization of the landscape (e.g., large crop fields or hayfields of tens to hundreds of hectares that displace a mosaic of landcover types and features) is likely to be detrimental. In a given landscape, emphasis should be on increasing the number of patches with habitat features rather than the size of such patches (Del Vecchio et al. 2011). The best habitat management may be to simply retain the current mosaic of heterogeneous rural landscapes.

Summary

To summarize, accept (to some extent) what nature gives you. It is worth noting that even anthropogenic heterogeneity (e.g., land use patterns) is underlain by natural heterogeneity (e.g., spatial variation in soil type, topography, natural disturbance events). The latter cannot be controlled or radically altered. Nonetheless, the land manager does have control over some processes that produce and maintain heterogeneity. Further, there is no certainty with regard to the outcome of a particular management action within a landscape. This is due in part to the action of natural (and sometime unpredictable) ecosystem processes and characteristics of the managed species. The managed landscape has no real ecological boundaries and often lacks hard boundaries to dispersing (emigrating and immigrating) individuals that may or may not be retained in the landscape despite the best management actions to provide appropriate habitat.

LITERATURE CITED

Adler, P. B., D. A. Raff, and W. K. Lauenroth. 2001. The effect of grazing on the spatial heterogeneity of vegetation. Oecologia 128:465–479.

Arnold, R. W. 1999. The soil survey, past, present, and future. Natural Resources Conservation Service, US Department of Agriculture. https://www.nrcs.usda.gov/wps/portal/nrcs/detailfull/soils/survey/?cid=nrcs142p2_053369. Accessed 5 Jan 2017.

Asner, G. P., S. R. Levick, T. Kennedy-Bowdoin, D. E. Knapp, R. Emerson, J. Jacobson, M. S. Colgan, R. E. Martin, and W. G. Ernst. 2009. Large-scale impacts of herbivores on the structural diversity of African savannas. Proceedings of the National Academy of Sciences of the United States of America 106:4947–4952.

Augustine, D. J., S. J. McNaughton, and D. A. Frank. 2003. Feedbacks between soil nutrients and large herbivores in a managed savanna ecosystem. Ecological Applications 13:1325–1337.

Badiane, A., C. Matos, and X. Santos. 2017. Uncovering environmental, land-use, and fire effects on the distribution of a low-dispersal species, the Herman's tortoise *Testudo hermanni*. Amphibia-Reptilia 38:67–77.

Bailey, D. W., J. E. Gross, E. A. Laca, L. R. Rittenhouse, M. B. Coughenour, D. M. Swift, and P. L. Sims. 1996. Mechanisms that result in large herbivore grazing distribution patterns. Journal of Range Management 49:386–400.

Bakermans, M. H., C. L. Ziegler, and J. L. Larkin. 2015. American woodcock and golden-winged warbler abundance and associated vegetation in managed habitats. Northeastern Naturalist 22:690–703.

Barro, S. C., and S. G. Conard. 1991. Fire effects on California chaparral systems: an overview. Environment International 17:135–149.

Bennie, A. T. P. 1991. Growth and mechanical impedance. Pages 393–414 *in* Y. Waisel, A. Eshel, and U. Kafkafi, editors. Plant roots, the hidden half. First edition. Marcel Dekker, New York, New York, USA.

Ben-Shahar, R., and M. J. Coe. 1992. The relationships between soil factors, grass nutrients, and the foraging behavior of wildebeest and zebra. Oecologia 90:422–428.

Berardo, F., M. L. Carranza, L. Frate, A. Stanisci, and A. Loy. 2015. Seasonal habitat preference by the flagship species *Testudo hermanii*: implications for the conservation of coastal dunes. Comptes Rendus Biologies 338:343–350.

Bertolero, A., D. Oro, and A. Besnard. 2007. Assessing the efficacy of reintroduction programmes by modelling adult survival: the example of Hermann's tortoise. Animal Conservation 10:360–368.

Bisigato, A. J., and M. B. Bertiller. 1997. Grazing effects on patchy dryland vegetation in northern Patagonia. Journal of Arid Environments 36:639–653.

Boisramé, G. F. S., S. E. Thompson, M. Kelly, J. Cavalli,

K. M. Wilkin, and S. L. Stephens. 2017. Vegetation change during 40 years of repeated managed wildfires in the Sierra Nevada, California. Forest Ecology and Management 402:241–252.

Caon, L., V. R. Vallejo, C. J. Ritsema, and V. Geissen. 2014. Effects of wildfire on soil nutrients in Mediterranean ecosystems. Earth-Science Reviews 139:47–58.

Celse, J., A. Catard, S. Caron, J. M. Ballouard, S. Gagno, N. Jarde, M. Cheylan, G. Astruc, V. Croquet, M. Bosc, et al. 2014. Management guide of populations and habitats of the Hermann's tortoise. European Union LIFE+ Nature, LIFE 08 NAT/F/000475, Provence-Alpes-Côte d'Azur, France.

Chafer, C. J., C. Santin, and S. H. Doerr. 2016. Modelling and quantifying the spatial distribution of post-wildfire ash loads. International Journal of Wildland Fire 25:249–255.

Chartier, M. P., C. M. Rostagno, and G. E. Pazos. 2011. Effects of soil degradation on infiltration rates in grazed semiarid rangelands of northeastern Patagonia, Argentina. Journal of Arid Environments 75:656–661.

Clark, D. B., and D. A. Clark. 2000. Landscape-scale variation in forest structure and biomass in a tropical rainforest. Forest Ecology and Management 137:185–198.

Connell, J., S. J. Watson, R. S. Taylor, S. C. Avitabile, N. Schedvin, K. Schneider, and M. F. Clarke. 2019. Future fire scenarios: predicting the effect of fire management strategies on the trajectory of high-quality habitat for threatened species. Biological Conservation 232:131–141.

Couturier, T., M. Cheylan, E. Guérette, and A. Besnard. 2011. Impacts of a wildfire on the mortality rate and small-scale movements of a Hermann's tortoise Testudo hermanni hermanni population in southeastern France. Amphibia-Reptilia 32:541–545.

Cromsigt, J. P. G. M., and H. Olff. 2006. Resource partitioning among savanna grazers mediated by local heterogeneity: an experimental approach. Ecology 87:1532–1541.

Darwiche-Criado, N., R. Sorando, S. G. Eismann, and F. A. Comín. 2017. Comparing two multi-criteria methods for prioritizing wetland restoration and creation sites based on ecological, biophysical, and socio-economic factors. Water Resources Management 31:1227–1241.

Davies, S. J., P. A. Palmiotto, P. S. Ashton, H. S. Lee, and J. V. LaFrankie. 1998. Comparative ecology of 11 sympatric species of Macaranga in Borneo: tree distribution in relation to horizontal and vertical resource heterogeneity. Journal of Ecology 86:662–673.

Del Vecchio, S., R. L. Burke, L. Rugiero, M. Capula, and L. Luiselli. 2011. The turtle is in the details: micro-habitat choice by Testudo hermanni is based on microscale plant distribution. Animal Biology 61:249–261.

Derner, J. D., R. L. Gillen, F. T. McCollum, and K. W. Tate. 1994. Little bluestem tiller defoliation patterns under continuous and rotational grazing. Journal of Range Management 47:220–225.

Derner, J. D., W. K. Lauenroth, P. Stapp, and D. J. Augustine. 2009. Livestock as ecosystem engineers for grassland bird habitat in the western Great Plains of North America. Rangeland Ecology and Management 62:111–118.

Doherty, T. S., R. A. Davis, E. J. B. van Etten, N. Collier, and J. Krawiec. 2015. Response of a shrubland mammal and reptile community to a history of landscape-scale wildfire. International Journal of Wildland Fire 24:534–543.

Forman, R. T. T. 1995. Some general principles of landscape and regional ecology. Landscape Ecology 10:133–142.

Franklin, J. F., and R. T. T. Forman. 1987. Creating landscape patterns by forest cutting: ecological consequences and principles. Landscape Ecology 1:5–18.

Fuhlendorf, S. D., and D. M. Engle. 2004. Application of the fire: grazing interaction to restore a shifting mosaic on tallgrass prairie. Journal of Applied Ecology 41:604–614.

Fuhlendorf, S. D., D. M. Engle, R. D. Elmore, R. F. Limb, and T. G. Bidwell. 2012. Conservation of pattern and process: developing an alternative paradigm of rangeland management. Rangeland Ecology and Management 65:579–589.

Garten, C. T., S. Kang, D. J. Brice, C. W. Schadt, and J. Zhou. 2007. Variability in soil properties at different spatial scales (1 m–1 km) in a deciduous forest ecosystem. Soil Biology and Biochemistry 39:2621–2627.

Goldberg, D. E. 1982. The distribution of evergreen and deciduous trees relative to soil type: an example from the Sierra Madre, Mexico, and a general model. Ecology 63:942–951.

Golubović, A., M. Anđelković, D. Arsovski, X. Bonnet, and L. Tomović. 2017. Locomotor performances reflect habitat constraints in an armored species. Behavioral Ecology and Sociobiology 71:93–100.

Grayson, R. B., G. Blöschl, A. W. Western, and T. A. McMahon. 2002. Advances in the use of observed spatial patterns of catchment hydrological response. Advances in Water Resources 25:1313–1334.

Hessburg, P. F., J. K. Agee, and J. F. Franklin. 2005. Dry forests and wildland fires of the inland Northwest USA: contrasting the landscape ecology of the pre-settlement and modern eras. Forest Ecology and Management 211:117–139.

Heuvelink, G. B. H., and R. Webster. 2001. Modelling soil variation: past, present, and future. Geoderma 100:269–301.

Hobbs, N. T. 1996. Modification of ecosystems by ungulates. Journal of Wildlife Management 60:695–713.

Hovick, T. J., R. D. Elmore, S. D. Fuhlendorf, D. M. Engle, and R. G. Hamilton. 2015. Spatial heterogeneity increases diversity and stability in grassland bird communities. Ecological Applications 25:662–672.

Ibáñez, J. J., J. Caniego, F. San Jose, and C. Carrera. 2005. Pedodiversity-area relationships for islands. Ecological Modelling 182:257–269.

Jenny, H. 1941. Factors of soil formation: a system of quantitative pedology. Dover, New York, New York, USA.

Jenny, H. 1980. The soil resource, origin and behavior. Springer-Verlag, New York, New York, USA.

Johnson, A. R., J. A. Wiens, B. T. Milne, and T. O. Crist. 1992. Animal movements and population dynamics in heterogeneous landscapes. Landscape Ecology 7:63–75.

Keeley, J. E., W. J. Bond, R. A. Bradstock, J. G. Pausas, and P. W. Rundel. 2012. Fire in Mediterranean ecosystems. Cambridge University Press, Cambridge, UK.

Kellogg, C. E. 1936. Development and significance of the great soil groups of the United States. US Department of Agriculture Miscellaneous Publication 229, Washington, DC, USA.

Kokaly, R. F., B. W. Rockwell, S. L. Haire, and T. V. V. King. 2007. Characterization of post-fire surface cover, soils, and burn severity at the Cerro Grande Fire, New Mexico, using hyperspectral and multispectral remote sensing. Remote Sensing of Environment 106:305–325.

Larsen, A. L., J. J. Jacquot, P. W. Keenlance, and H. L. Keough. 2016. Effects of an ongoing oak savanna restoration on small mammals in lower Michigan. Forest Ecology and Management 367:120–127.

Lechmere-Oertel, R. G., R. M. Cowling, and G. I. H. Kerley. 2005. Landscape dysfunction and reduced spatial heterogeneity in soil resources and fertility in semi-arid succulent thicket, South Africa. Austral Ecology 30:615–624.

Lecq, S., J. M. Ballouard, S. Caron, B. Livoreil, V. Seynaeve, L. A. Matthieu, and X. Bonnet. 2014. Body condition and habitat use by Hermann's tortoises in burnt and intact habitats. Conservation Physiology 2:1–11.

Leyequién, E., W. F. de Boer, and V. M. Toledo. 2010. Bird community composition in a shaded coffee agroecological matrix in Puebla, Mexico: the effects of landscape heterogeneity at multiple spatial scales. Biotropica 42:236–245.

Li, D., and D. Waller. 2015. Drivers of observed biotic homogenization in pine barrens of central Wisconsin. Ecology 96:1030–1041.

Lin, H., D. Wheeler, J. Bell, and L. Wildling. 2005. Assessment of soil spatial variability at multiple scales. Ecological Modelling 182:271–290.

Lindenmayer, D., and R. J. Hobbs. 2004. Fauna conservation in Australian plantation forests—a review. Biological Conservation 119:151–168.

Lindenmayer, D., and W. F. Laurance. 2017. The ecology, distribution, conservation, and management of large old trees. Biological Reviews 92:1434–1458.

Mazurek, M. J., and W. J. Zielinski. 2004. Individual legacy trees influence vertebrate wildlife diversity in commercial forests. Forest Ecology and Management 193:321–334.

Mazzotti, S. 2004. Hermann's tortoise (Testudo hermanni): current distribution in Italy and ecological data on a population from the north Adriatic coast (Reptilia, Testudinidae). Italian Journal of Zoology 71:97–102.

McCleery, R., A. Monadjem, B. Baiser, R. Fletcher, K. Vickers, and L. Kruger. 2018. Animal diversity declines with broad-scale homogenization of canopy cover in African savannas. Biological Conservation 226:54–62.

McNaughton, S. J. 1993. Grasses and grazers, science and management. Ecological Applications 3:17–20.

Mladenoff, D. J., M. A. White, J. Pastor, and T. R. Crow. 1993. Comparing spatial pattern in unaltered old-growth and disturbed forest landscapes. Ecological Applications 3:294–306.

Molinar, F., J. Holechek, D. Galt, and M. Thomas. 2002. Soil depth effects on Chihuahuan Desert vegetation. Western North American Naturalist 62:300–306.

Moore, I. D., R. B. Grayson, and A. R. Ladson. 1991. Digital terrain modelling: a review of hydrological, geomorphological, and biological applications. Hydrological Processes 5:3–30.

Moreno-Mateos, D., U. Mander, and C. Pedrocchi. 2010. Optimal location of created and restored wetlands in Mediterranean agricultural catchments. Water Resources Management 24:2485–2499.

Morrison, M. L., B. G. Marcot, and R. W. Mannan. 2006. Wildlife–habitat relationships: concepts and applications. Third edition. Island Press, Washington, DC, USA.

Murwira, A., and A. K. Skidmore. 2005. The response of elephants to the spatial heterogeneity of vegetation in a southern African agricultural landscape. Landscape Ecology 20:217–234.

Noy-Meir, I. 1993. Compensating growth of grazed plants and its relevance to the use of rangelands. Ecological Applications 3:32–34.

Otto, C. R. V., A. J. Kroll, and H. C. McKenny. 2013.

Amphibian response to downed wood retention in managed forests: a prospectus for future biomass harvest in North America. Forest Ecology and Management 304:275–285.

Parsons, A. J., and B. Dumont. 2003. Spatial heterogeneity and grazing processes. Animal Research 52:161–179.

Patrick, D. A., M. L. Hunter, and A. J. K. Calhoun. 2006. Effects of experimental forestry treatments on a Maine amphibian community. Forest Ecology and Management 234:323–332.

Petersen, A., A. Gröngröft, and G. Miehlich. 2010. Methods to quantify the pedodiversity of 1 km² areas—results from southern African drylands. Geoderma 155:140–146.

Phillips, J. D. 2001. The relative importance of intrinsic and extrinsic factors in pedodiversity. Annals of the Association of American Geographers 91:609–621.

Phillips, J. D., and D. A. Marion. 2005. Biomechanical effects, lithological variations, and local pedodiversity in some forest soils of Arkansas. Geoderma 124:73–89.

Pickett, S. T. A., and M. L. Cadenasso. 1995. Landscape ecology: spatial heterogeneity in ecological systems. Science 269:331–334.

Prober, S. M., K. R. Thiele, and J. Speijers. 2013. Management legacies shape decadal-scale responses of plant diversity to experimental disturbance regimes in fragmented grassy woodlands. Journal of Applied Ecology 50:376–386.

Riffell, S., J. Verschuyl, D. Miller, and T. B. Wigley. 2011. Biofuel harvests, coarse woody debris, and biodiversity—a meta-analysis. Forest Ecology and Management 261:878–887.

Rittenhouse, T. A. G., E. B. Harper, L. R. Rehard, and R. D. Semlitsch. 2008. The role of microhabitats in the desiccation and survival of anurans in recently harvested oak-hickory forest. Copeia 2008:807–814.

Romme, W. H. 1982. Fire and landscape diversity in subalpine forests of Yellowstone National Park. Ecological Monographs 52:199–221.

Rozylowicz, L., and V. D. Popescu. 2013. Habitat selection and movement ecology of eastern Hermann's tortoises in a rural Romanian landscape. European Journal of Wildlife Research 59:47–55.

Sebastia, M. T., F. de Bello, L. Puig, and M. Taull. 2008. Grazing as a factor structuring grasslands in the Pyrenees. Applied Vegetation Science 11:215–222.

Shakesby, R. A. 2011. Post-wildfire soil erosion in the Mediterranean: review and future research directions. Earth-Science Reviews 105:71–100.

Shugart, H. H. 2005. Equilibrium versus non-equilibrium landscapes. Pages 36–41 in J. A. Wiens and M. R. Moss,

editors. Issues and perspectives in landscape ecology. Cambridge University Press, Cambridge, UK.

Smit, I. P. J., C. F. Smit, N. Govender, M. van der Linde, and S. MacFadyen. 2013. Rainfall, geology and landscape position generate large-scale spatiotemporal fire pattern heterogeneity in an African savanna. Ecography 36:447–459.

Söderström, B., T. Pärt, and E. Linnarsson. 2001. Grazing effects on between-year variation of farmland bird communities. Ecological Applications 11:1141–1150.

Sperry, J. H., and P. J. Weatherhead. 2010. Ratsnakes and brush piles: intended and unintended consequences of improving habitat for wildlife? American Midland Naturalist 163:311–317.

Steel, Z. L., H. D. Safford, and J. H. Viers. 2015. The fire-frequency-severity relationship and the legacy of fire suppression in California forests. Ecosphere 6(1):1–23.

Sullivan, T. P., and D. S. Sullivan. 2001. Influence of variable retention harvests on forest ecosystems. II. Diversity and population dynamics of small mammals. Journal of Applied Ecology 38:1234–1252.

Sullivan, T. P., D. S. Sullivan, P. M. F. Lindgren, and D. B. Ransome. 2012. If we build habitat, will they come? woody debris structures and conservation of forest mammals. Journal of Mammalogy 93:1456–1468.

Török, P., O. Valkó, B. Deák, A. Kelemen, and B. Tóth-mérész. 2014. Traditional cattle grazing in a mosaic alkali landscape: effects on grassland biodiversity along a moisture gradient. PLOS One 9(5):e97095.

Turner, M. G. 1989. Landscape ecology: the effect of pattern on process. Annual Review of Ecology and Systematics 20:171–197.

Turner, M. G. 2010. Disturbance and landscape dynamics in a changing world. Ecology 91:2833–2849.

Turner, M. G., D. C. Donato, and W. H. Romme. 2013. Consequences of spatial heterogeneity for ecosystem services in changing forest landscapes: priorities for future research. Landscape Ecology 28:1081–1097.

Turner, M. G., W. H. Romme, R. H. Gardner, and W. W. Hargrove. 1997. Effects of fire size and pattern on early succession in Yellowstone National Park. Ecological Monographs 67:411–433.

US Department of Agriculture. 2011. Soil survey of Mason County, Texas. Natural Resources Conservation Service. https://www.nrcs.usda.gov/Internet/FSE_MANU SCRIPTS/texas/masonTX2011/Mason.pdf. Accessed 31 Aug 2019.

US Department of Agriculture. 2019. Web Soil Survey homepage. https://websoilsurvey.sc.egov.usda.gov. Accessed 31 Aug 2019.

Uuemaa, E., A. O. Hughes, and C. C. Tanner. 2018.

Identifying feasible locations for wetland creation or restoration in catchments by suitability modelling using light detection and ranging (LiDAR) digital elevation model (DEM). Water 10:464.

van Dijk, P. P., C. Corti, V. P. Mellado, and M. Cheylan. 2004. *Testudo hermanni*. IUCN Red List of Threatened Species. http://www.iucnredlist.org/details/21648/0. Accessed 15 Jan 2017.

Vilardell-Bartino, A., X. Capalleras, J. Budó, R. Bosch, and P. Pons. 2015. Knowledge of habitat preferences applied to habitat management: the case of an endangered tortoise population. Amphibia-Reptilia 36:13–25.

Wall, C. B., and K. J. Stevens. 2015. Assessing wetland mitigation efforts using standing vegetation and seed bank community structure in neighboring natural and compensatory wetlands in north-central Texas. Wetlands Ecology and Management 23:149–166.

Western, A. W., S. L. Zhou, R. B. Grayson, T. A. McMahon, G. Blöschl, and D. J. Wilson. 2004. Spatial correlation of soil moisture in small catchments and its relationship to dominant spatial hydrological processes. Journal of Hydrology 286:113–134.

Whitehead, P. J., J. Russell-Smith, and J. C. Z. Woinarski. 2005. Fire, landscape heterogeneity, and wildlife management in Australia's tropical savannas: introduction and overview. Wildlife Research 32:369–375.

Wilson, D. J., A. W. Western, and R. B. Grayson. 2005. A terrain and data-based method for generating the spatial distribution of soil moisture. Advances in Water Resources 28:43–54.

6 — Translating Landcover Data Sets into Habitat Features

David D. Diamond and
Lee F. Elliott

Introduction

Classified landcover data sets can provide a "bird's-eye view" of planning regions. The desired region of analysis may be smaller, such as a land ownership or small stream watershed, or larger, including entire continents or all terrestrial habitats on earth. Different types of input imagery and classifications are appropriate for different-sized planning regions. Regardless of the size of the planning region, landcover maps offer a means to quantify the amount and configuration of habitats that cannot be gained from ground-level analysis.

Landcover classifications and maps are considerably more useful for habitat analyses than images. Mapped landcover types can be quantified in terms of area and pattern, not simply viewed. The mapped types can be combined and manipulated in a variety of ways using geographic information system techniques. For example, a viewer might be able to distinguish different vegetation types from an unclassified aerial photo or satellite image by inspection but cannot determine the area or patch size of each type. Moreover, the number of landcover types may not be immediately apparent from viewing an aerial photo but can be defined for landcover classification and mapping. Classified landcover data sets can be analyzed, manipulated, and merged (or overlain with other digital data sets) to help define characteristics of different habitats, but aerial photos or unclassified satellite images cannot.

Our objective in this chapter is to provide an overview of landcover classification and mapping methods with an emphasis on understanding how such data sets can be used to help define wildlife habitat. We attempted to keep the content at a core level that will facilitate understanding without compromising substance, and we emphasize references to widely available data sets produced for the United States.

Landcover Classification Overview

Landcover or habitat classification target types must be selected prior to development of a map. The number of classification targets is termed thematic resolution (Anderson et al. 1976). Water almost always serves as one mapped type, and beyond that land use (e.g., urban, row crop) and landcover, usually expressed as life form (e.g., deciduous forest, evergreen forest, shrubland, herbaceous vegetation), might serve as classification targets. Several recent large-area or nationwide maps within the United States used ecological systems or subsystems as classification targets. Ecological systems are major habitats or combinations of habitats (biological communities) that are found in similar physical environments and are influenced by similar processes, such as fire or flooding (Comer et al. 2003). Those systems are recognized and named mostly by region and dominant plant species or life form. Examples include Allegheny-Cumberland Dry Oak Forest and Woodland, Central Mixed-Grass Prairie, Colorado Plateau Pinyon-Juniper Shrubland, Inter-Mountain Basins Aspen-Mixed Conifer Forest and Woodland, Mississippi River Riparian Forest, Tamaulipan Mixed Deciduous Thornscrub, and West Gulf Coastal Plain Pine-Hardwood Forest.

Imagery is composed of a characteristic pixel, or grain size. Images with smaller pixels are termed higher spatial resolution than images with larger pixels. Pixel size is typically referred to by the length of a pixel edge. So, resolution may be referred to as 1 m (representing 1 m^2), 10 m (representing 100 m^2), 30 m (representing 900 m^2), and so on. When viewed on a screen or in hard copy, smaller pixels may yield clearer images. For example, the square edges that make up 30-m pixels (a common grain size for satellite imagery) become visible at 1:24,000 resolution (US Geological Survey [USGS] 7.5′ quadrangle resolution), but smaller pixels (e.g., 10 m or 1 m) appear smooth. However, images made up of smaller pixels are larger digital files for the same area covered and require more computing power and time for processing. For landcover mapping, typical pixel size ranges from <1 m for aerial photos to 1 km for satellite imagery designed to cover continents. For use in landcover classification, imagery is orthorectified, so spatial location and pixel size are corrected for the curvature of the earth and the roughness of the land surface.

In the United States, national land use, landcover, and, more recently, ecological systems data sets have been produced using data collected from 30-m square pixels, referred to as 30-m spatial resolution (Homer et al. 2004, 2015). The National Aeronautics and Space Administration launched the Landsat 4 mission in 1982 and began collecting 30-m resolu-

tion data for landcover mapping, first using the thematic mapper (TM) sensor and later the enhanced thematic mapper (ETM+). On 11 February 2013, Landsat 8 (formerly the Landsat Data Continuity Mission) was launched with the Operation Land Imager (OLI) sensor on board, which is similar to the TM and ETM+ sensors. These three sensors collect data from seven reflectance bands designed, in part, to enhance vegetation classification. Two additional bands related to water and cloud detection were added to the OLI sensor. Important features of the Landsat missions include raw data that are free to users, national landcover maps that have typically used these sensors to provide results, data dating back >30 years that are basically compatible, and absolute limitations in terms of spatial and thematic resolution that are inherent in these data sets.

The European Space Agency launched the Sentinel-2A and 2B satellites under the umbrella of their Copernicus Programme. Data from Sentinel-2A became available in fall 2015, and data from 2B in June 2017. The sensors aboard Sentinel-2A and 2B are appropriate for use in vegetation classification. Data from three visible bands and near infrared are collected at 10-m resolution, and additional bands are collected at 20 m and 60 m. Together, these two satellites cover the continental United States about once every five days. The relatively new Sentinel-2 data may prove highly useful for landcover map production.

The National Aeronautics and Space Administration launched a moderate-resolution imaging spectroradiometer sensor on board the Terra (EOS AM) satellite in 1999 and on the Aqua (EOS PM) satellite in 2002. This sensor collects data in different reflectance bands ranging from 250-m to 1-km resolution. While this spatial resolution is coarse for local landcover mapping applications, the sensor has the advantage of high temporal resolution (quick return intervals) over the OLI sensor and is therefore useful for tracking vegetation health (green up and brown down) across seasons and years. The US Forest Service uses data from this satellite in their ForWarn change recognition and tracking system to help recognize areas of increased wildfire hazards (US Forest Service 2019).

Data sets from the 30-m-resolution OLI sensor and the 10-m Sentinel-2 sensors have high spectral resolution (enhanced usefulness) for vegetation mapping (Table 6.1). For example, data collected from a near infrared reflectance band, which is reflected by healthy green vegetation, can be compared with data from a red band, which is absorbed by green vegetation, to provide an index of vegetation greenness. This index is called the Normalized Difference Vegetation Index (NDVI). Because OLI pixels are 30-m resolution, however, they are not designed to provide highly accurate maps of small areas, such as a 100-ha property ownership. Sentinel-2 data have 9 times higher resolution (100-m² pixels versus 900-m² pixels) and are more useful for mapping smaller areas. Aerial photos, such as those from the National Agriculture Imagery Program (NAIP), which are available across most of the nation, have high spatial resolution (often 1-m pixels). Standard NAIP photos collect data in three visible reflectance bands, whereas 4-band NAIP adds information in a near infrared band. Therefore NAIP has lower spectral resolution versus OLI satellite data. National Agriculture Imagery Program photos are collected once a year during the growing season, whereas OLI data sets are collected about every 16 days over the same location, and Sentinel-2 data about every 5 days. Use of data from both leaf-on and leaf-off collection dates enhances the ability to distinguish among vegetation types, such as cropland versus grassland and evergreen versus cold-deciduous forest. Thus aerial photos are often visually clear when viewed at fine spatial resolution, but this does not translate into enhanced results when classifying vegetation. Use of multiple-date OLI and Sentinel-2 satellite data allows for relatively accurate classification of 12 to 14 landcover classes within a given region. In contrast, only 2 to 6 landcover types can be classified using free, widely available 3- or 4-band aerial photos. Even

Table 6.1. Common No-Cost Imagery Available for Landcover Mapping in the United States

Data Source	Spatial Resolution	Spectral Resolution	Comments
National Aeronautics and Space Administration (NASA), Operational Land Imager (OLI), and earlier thematic mapper and enhanced thematic mapper sensors	30 m	7–9 reflectance bands; designed primarily for vegetation mapping	Same area covered about every 16 days, allowing users to merge leaf-on and leaf-off images; used to produce nationwide landcover classifications since 1992; maximum of about 12 to 14 types can be classified directly within a given region
NASA, moderate-resolution imaging spectroradiometer (MODIS)	250 m to 1 km	36 bands; two collected at 250-m spatial resolution most relevant	Same area covered every 1 to 2 days; coarse spatial resolution but frequent return interval allows detection of vegetation health (greening or browning)
Aerial photos, National Agriculture Imagery Program (NAIP), and others	Varies; 1 m is common	Typically 3 or 4 reflectance bands	Free imagery mostly available as leaf-on products, with spotty coverage of both leaf-on and leaf-off dates; spatial resolution allows for clarity when viewed, but reduced spectral resolution and lack of both growing season and nongrowing season dates often do not allow accurate classification of more than 2 to 6 landcover classes
European Space Agency, Copernicus Programme, Sentinel-2A and 2B	10 m, 20 m, and 60 m	Four bands at 10 m, including visible and near-infrared	Sentinel-2A data available fall 2015 and 2B available summer 2017; coverage about every 4 to 5 days in USA; spectral bands good for classifications similar to NASA OLI data

then, the accuracy of results from classified aerial photos is often disappointing.

Lidar (a linguistic blend of the words light and radar) is used to accurately measure the location and elevation of surfaces by bouncing light pulses off of them. Lidar data sets are rapidly becoming available for the United States. Many light pulses are recorded by the sensors, and these pulses are then aggregated and analyzed within pixels that each contain multiple light return data points. The lowest recorded elevation within a pixel is the land surface, and the highest elevation is the top of the tree canopy or building within the pixel. Thus digital elevation models (ground surface) and canopy height can be generated from lidar. For modern lidar data collections, these values can accurately be generated for pixels at 10-m resolution or smaller. These data sets can be used to enhance landcover maps by providing an index to water regime and vegetation height (Fig. 6.1). Lidar data also can provide an indicator to habitat features such as tree height and woody species density. Lidar data sets are large, but improvements in computing power and data storage are rapidly making them practical to use across planning regions as large as counties, multiple-county regions, and states.

Landcover Classification Methods

Georeferenced landcover classifications can be produced by digitizing polygons from on-screen viewing of original satellite or aerial photo imagery. This can be an arduous process if attempted over large areas and requires a high degree of skill and knowledge of on-the-ground conditions. This method may be as prone to inaccuracies as automated methods. For these reasons, most landcover classifications that cover larger areas are based on grouping of pixels from satellite data (or aerial photographs) into different landcover types using the attributes of the pixels.

Pixels are classified on the basis of differences and similarities among attributes (often referred to as explanatory data) assigned to each pixel. These attributes can be derived from a single image or from several images (such as leaf-on and leaf-off images) and additional information derived from nonimage sources that represent the same area covered by the pixel. This set of data is often referred to as a stack. For 30-m satellite data, pixel attributes often include values for reflectance in 7 to 9 or more bands (designed in part specifically to help classify vegetation) and data generated from digital elevation models such as percentage slope, elevation, and slope exposure.

Images may be classified using unsupervised or supervised methods (Anderson et al. 1976, Lu and Weng 2007). In both methods the pixels from raw imagery are classified on the basis of the attributes of the pixels (e.g., reflectance band values). In unsupervised classification, pixels are classified into a set number of groups designated by the user. Each group is then assigned by an analyst to a landcover type. Several algorithms, often imbedded within image classification software, can be used to accomplish the classification, including iterative self-organizing data analysis techniques, and other cluster analyses. The result of such a classification may include groups that may or may not reflect landcover concepts that an analyst anticipated, and further grouping or splitting of some of the original pixel clusters may be required. Supervised classification relies on initially identifying all target landcover types to be mapped and then selecting a set of pixels that represents each of the desired classification types. For example, a training data set for supervised classification might consist of 50 to several hundred samples each of 15 target landcover types. The training data might be generated from georeferenced ground-collected data that are supplemented with additional samples generated by viewing aerial photos on-screen. The training data set is used to create a model, and then all pixels can be assigned to a target classification type based on the model. Some of the more commonly used methods for supervised classification are referred to

Figure 6.1. Lidar data sets can be used to provide fine-resolution digital elevation models (*top*) and vegetation height (*bottom*). Individual trees or groups of trees are 3 to 6 m in diameter, and the tree canopy height is about 20 m. Data from Karnes County, Texas, USA.

generally as classification and regression tree models. Algorithms such as randomForest and other support-vector machine learning algorithms have been used to accomplish supervised classification of images (Prasad et al. 2006).

Supervised and unsupervised classification algorithms may be applied to image objects created from the original pixel-based image data set. Image objects are polygons that consist of geographically co-located patches of pixels that are aggregated together into homogeneous units (Blaschke 2010). The process of delineating image objects is often called image segmentation. The characteristics of the pixels within these image objects are summarized (e.g., the mean of the pixel values from each of the sensor reflectance bands), and the summary values of these attributes are used to perform the classification. Alternatively, classified pixels can be used to label an image object by assigning the object with the class of the majority of pixels included within the object. Classified pixels or image objects, since they are georeferenced, form the map.

A typical map production process flow might begin with delineation of the planning region and determination of landcover classification targets. Next, acquisition of appropriate imagery (typically, satellite data, or aerial photos) and ancillary data (such as digital elevation models or digital soils

maps) must occur. Then, collection or generation of georeferenced samples of all classification targets is needed if the classification is supervised. Finally, a classification algorithm and successive refinement (e.g., more training data, modification of the classification targets) is applied to achieve the highest accuracy possible. Sometimes, successive segmentation and reclassification of segments of the data are needed (e.g., initial separation of wooded from non-wooded segments of the data and separate classification within these two segments). Overlay of other map layers (ancillary data) on the initial classification results may also be necessary—such as ecoregion boundaries, digital soils, surface geology, slope, or hydrography—to increase the number of types mapped (Fig. 6.2). The process flow ends with ground checking of results and modifications to the data set.

Different classifications are discussed below. For supervised classifications, georeferenced samples (either from ground-collected data or from photo interpretation) are important to the development of products with high accuracy. Compositional data of plant communities from ground-collected and georeferenced points or plots may be used to characterize the final mapped types (map class descriptions). Additionally, the development of data to perform accuracy assessments on the products can be costly but may be extremely useful.

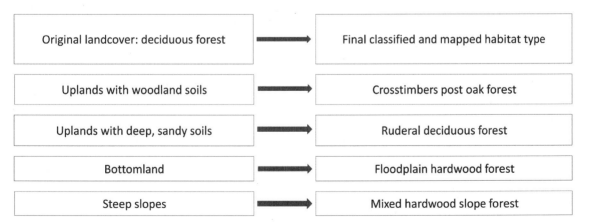

Figure 6.2. Example of use of soil type, land position, and percentage slope as map overlays (ancillary data) to increase the thematic resolution of a classification. Adapted from Diamond and Elliott (2015).

Examples of Landcover Products
National Land Cover Database

The National Land Cover Database (NLCD) is produced by the Multi-Resolution Land Characteristics Consortium, a group of 10 federal agencies led by the USGS. The initial NLCD was released in 1992, and since then 3 iterations have been produced, the latest in 2016. These data sets are free and available for download (Multi-Resolution Land Characteristics Consortium 2016). Additional products have been produced as part of the NLCD, including landcover change, percentage impervious cover, and percentage tree canopy. All products are produced at 30-m spatial resolution from the similar TM, ETM+, and, most recently, OLI sensors. The land use / landcover mapping targets have changed only slightly since 1992. They consist of a mix of land use (e.g., developed, agriculture) and landcover (e.g., deciduous forest, shrub, herbaceous) concepts with 16 to 22 major types nationwide (Anderson et al. 1976; Table 6.2). Methods used to produce the database have evolved and improved, but the basic intent has been to maintain close to the same land use / landcover mapping targets, to keep this number at a level that can be classified reasonably accurately with the available satellite data, and to keep the input data and classification methods relatively stable, simple, and focused on direct classification of pixels. The spatial and thematic resolutions of the NLCD data sets are best suited for larger planning regions, such as states or counties. These free data sets can be manipulated (i.e., types can be analyzed in terms of patch size and configuration) or overlain with other digital data (e.g., soils, geology, hydrology, slope, exposure) to help identify important habitat features. The simplicity of the original mapping targets, which are essentially plant physiognomy and life form types, facilitates user modifications. For example, the NLCD deciduous forest landcover type could be subdivided into bottomland and upland types by overlaying a digital floodplain data layer. Local knowledge might allow workers to infer the main composition of these types within a planning region.

LANDFIRE Existing Vegetation Type and Other Products

The LANDFIRE Program is a joint effort between the US Forest Service and the Department of the Interior that provides free, downloadable data sets relevant to landcover classification and habitat mapping (LANDFIRE 2019). The first data sets were released in 2001, and several updates have been made available. The existing vegetation type (EVT) data layer uses ecological systems and subsystems as mapping targets, and a total of 849 types were mapped across the nation in the 1.3.0 version (LANDFIRE 2019). Original input data are from 30-m multispectral sensors, plus variables from 30-m digital elevation models, just as for the NLCD. The increased thematic resolution of EVT makes it more appealing than the NLCD in this regard, but the EVT contains some difficult-to-explain inaccuracies owing to the methods used for production and the increased number of types mapped. Methods included successive segmentation of imagery with training data from plots assigned to target map classification types. Each of

Table 6.2. Landcover Classes from the National Land Cover Database

General Category	Class
Non-vegetated	Open water
	Perennial ice/snow
Developed	Developed, open space
	Developed, low intensity
	Developed, medium intensity
	Developed, high intensity
Agriculture	Hay/pasture
	Cultivated crops
Natural	Barren land
	Deciduous forest
	Evergreen forest
	Mixed forest
	Shrub/scrub
	Grassland/herbaceous
	Woody wetland
	Herbaceous wetland

67 map zones across the conterminous United States was completed separately. The LANDFIRE program has produced a variety of data sets that may be relevant to habitat features, including biophysical setting (prevailing pre-European vegetation), existing vegetation cover (percentage cover of the dominant plant life form), existing vegetation height, and a suite of information related to fire fuel modeling. It is an ongoing program, and, like the NLCD, updates are planned, at least into the near-term future. These data sets are best used at resolutions that range from multistate regions to states or counties. Existing vegetation type is best used as it comes, and users will find simple overlay of other map products a poor strategy for modification.

Use of Map Overlays for Enhanced Landcover Mapping

A number of regional efforts covering one or more states have used map overlays to improve the accuracy or spatial resolution of maps from 30-m-resolution satellite imagery. Generally, these efforts generate 12 to 14 landcover classes from 30-m satellite data (often Landsat) using pixel classification methods similar to those used by NLCD. Digital abiotic map layers such as soils, geology, slope, elevation, and hydrography are overlain on the landcover, and new landcover or habitat types are mapped in an expert driven system. Essentially, original landcover and abiotic data are combined to infer a new landcover type or habitat type class. For example, within a county or ecoregion, deciduous forest (original landcover) might be overlain on a digital soils data layer and reclassified as oak-hickory if on typical upland soils, sandyland sparse oak if on deep sands, and elm-ash if on bottomland soils (Fig. 6.2). This process can be used to increase the thematic resolution of landcover classifications generated from satellite data. Workers also can use this concept to map habitat relevant to a given species, if the species is associated both with landcover and abiotic variables. The USGS Gap Analysis Project produced and has made available some data sets that were generated using this concept, but these are quickly becoming outdated, are not likely to be updated, and were not produced using consistent methodologies across the nation (US Geological Survey 2019). A data set produced for Texas and Oklahoma (11% of the conterminous United States) used this map overlay methodology and added improved spatial resolution by producing image objects (visually homogeneous polygons generated by automated software programs) at 10-m resolution from NAIP imagery (Diamond and Elliott 2015). Similar map overlay concepts were used to provide a terrestrial wildlife habitat map for the northeastern United States, north of Virginia (Ferree and Anderson 2013). Regional landcover maps may or may not be updated consistently over time.

Maps with Fine Spatial Resolution

The National Park Service has implemented a vegetation inventory and mapping protocol for all National Park lands in the United States (National Park Service 2019). The national vegetation classification is used to provide mapping targets (plant associations or combinations of similar associations; US National Vegetation Classification 2019), and generally aerial photos provide input data at 1-m spatial resolution. Workers use vegetation sampling plots to sample plant communities and either digitize polygons on-screen or generate image objects for classification. The results of these National Park Service classification efforts represent the largest repository of fine-resolution habitat maps in the United States, with more than 300 Park Service properties mapped and data available for download. Although they only cover areas within the parks, these park maps and associated reports are often circumscribed within a larger planning region and are a ready source of information for planners and managers.

Use of Common Landcover Classifications to Define Habitat Features

No single classification or map will serve all needs because goals and spatial planning regions vary. For example, planners may use results from 30-m-resolution landcover maps to target large patches of different landcover types or a threshold percentage of major types in development of an ecoregion conservation plan across a given region, state, or number of states (Diamond et al. 2003, Groves et al. 2000). Others may desire accurate mapping of specific habitat features in order to produce habitat suitability models for a target species or a group of species. For example, the USGS Gap Analysis Project has systematically attempted to model species distribution patterns across the United States over the past decade, and 30-m-resolution landcover data are a key input for these models (US Geological Survey 2019). The NLCD has recently emphasized detection of change over time, which affects habitat and species distribution patterns (Jin et al. 2013, Sleeter et al. 2013, Homer et al. 2015). Change detection may help explain observed alterations in species distribution patterns across time and may prove especially important as global heating shapes new communities over time. The US Forest Service has used 250-m moderate-resolution imaging spectroradiometer satellite data to assess forest stress across the United States every 8 to 16 days to help predict forest fire risk (US Forest Service 2019).

Users should understand the limitations of spatial and thematic accuracy when using remote sensing. For example, past and current versions of the NLCD, which map about 22 landcover types, have a reported accuracy in the low 80% range (Wickham et al. 2017). Likewise, specific landcover types and habitats may be difficult or impossible to map. For example, distinguishing among local grassland types or conditions or determining species-specific forest composition using remote sensing at any resolution is fraught with difficulty and may be costly, impossible, or only possible for relatively small areas. The

usefulness of freely available landcover classifications versus the cost of improved classifications that better meet specific habitat mapping needs is often a consideration.

Remote sensing landcover classifications are limited by the spatial and thematic resolution that can be achieved depending on the original input data (Fig. 6.3; Table 6.3). National data sets in the United States, such as the NLCD and LANDFIRE EVT, are produced from 30-m-resolution imagery. Small features, less than 3 or 4 pixels (2,700–3,600 m²), cannot be reliably mapped. Types with similar spectral signals, including similar biomass or phenology of growth, are difficult to separate accurately. These similar types include cropland versus grassland and grassland versus shrubland. Plant community composition usually cannot be directly mapped. This includes separation of grassland types that are based, for example, on native versus non-native species composition and separation of deciduous forest types based on species composition. Workers may infer composition based on direct use of abiotic variables in the classification (e.g., LANDFIRE EVT) or map overlays (Diamond and Elliott 2015). For example, different woodland or forest composition may be inferred from location (ecoregion), elevation, slope, or soil texture simply by understanding that different plant communities prevail in different locations and on different geophysical settings.

Cross-walking is the process of developing a dictionary that translates one classification to another. In some instances, this is merely an exercise of providing a different naming convention to existing data. In other situations, a certain amount of lumping (aggregation) may be required. For example, if a researcher or manager is using LANDFIRE data and is interested in developing a habitat model for a particular species that uses almost any kind of shrubland, it may be beneficial to combine all the shrubland types into a single map class. Splitting of map classes in a cross-walk is more problematic and will require focused additional mapping if spatial representation of the split classes is required. For

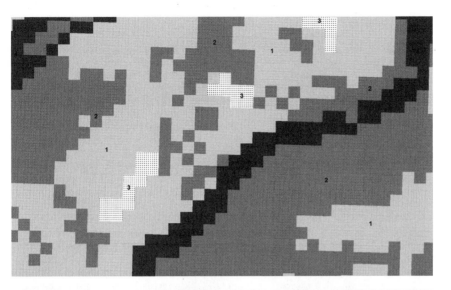

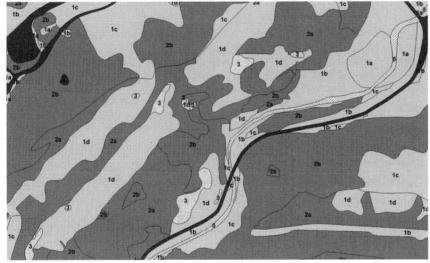

Figure 6.3. Landcover classification for a 5-ha section of Hot Springs National Monument (Arkansas, USA) from the National Land Cover Database, generated using 900-m² pixels (*top*) versus 1-m² pixels from aerial photos (*bottom*; Diamond et al. 2015). In the top panel, deciduous forest is light gray (1), evergreen and mixed forest is medium gray (2), grassland is stippled (3), and urban land is dark gray (4). In the bottom panel, light gray includes disturbance (1a), bottomland sweetgum (*Liquidambar styraciflua*) (1b), dry-mesic white oak (*Quercus alba*)-hickory (*Carya* spp.) (1c), and dry post oak (*Q. stellata*)-hickory forests (1d); medium gray includes shortleaf pine (*Pinus echinata*)-post oak dry (2a) and shortleaf pine-white oak dry-mesic forests (2b); stippling is novaculite glade (3); dark gray is urban (4); and striping is water (5).

this reason, it is usually beneficial to have products with high thematic resolution because combining map classes is relatively simple, but splitting them is difficult.

Fine-resolution input data such as orthorecti-fied photos (collected regularly across much of the United States) and lidar elevation and plant height data can be used to provide landcover information at resolutions <10 m. These landcover data sets generally must be produced at user cost and require

Table 6.3. Pros and Cons of Common Landcover Maps

Mapping Effort	Spatial Resolution	Thematic Resolution	Pros and Cons for Mapping Habitat Features
National Land Cover Database (NLCD)	30-m pixels; best used at 1:50,000 resolution or greater	Life forms and land use are mapping targets; typically 12 to 14 classes within an ecoregion	Free; methods and issues are well known and program is ongoing; difficulty separating cropland/grassland and grassland/shrubland; plant composition or quality of types such as grassland and deciduous forest not mapped; overlay of abiotic features such as soils, geology, elevation, and slope can be used to infer additional information
LANDFIRE Existing Vegetation Type	30-m pixels; best used at 1:50,000 resolution or greater	Ecological systems and subsystems mapped; typically >20 major types per ecoregion	Free; program is ongoing; more types are mapped versus NLCD, and types have more formal descriptions of composition and abiotic characteristics; plant composition or quality of habitats may be better inferred but still imperfect; mapping issues are varied and difficult to understand versus NLCD; overlay of abiotic features to infer additional information may be problematic
State and ecoregional efforts	30-m pixels or sometimes polygons generated at 10-m resolution	Varied mapping targets; usually ecological systems and subsystems	Local users pay; quality varies but can be better than NLCD or LANDFIRE; plant composition or quality of habitats may be better inferred but still imperfect; updates are not ensured; local map producers may be better able to explain pros and cons to users and make modifications
Fine-resolution efforts	Varies (1 m to 5 m); appropriate for use at 1:12,000 or better resolution	Varied mapping targets; often centers on plant associations or combinations of associations	The National Park Service has completed many parks using standardized methods, but for other lands, local users pay; may include the opportunity for close collaboration between producers and users so that results can be customized to suit user needs; updates are not ensured

local knowledge of on-the-ground conditions (Fig. 6.3). Aerial photos alone are most often no more useful than satellite imagery in mapping specific grassland or forest plant community composition unless map production is informed by georectified ground-collected information. Lidar vegetation height information has a limited shelf life, as vegetation patterns change over time. Large-area (multicounty) lidar data acquisitions are available across the United States but, thus far, data collections represent a single point in time and are not usually repeated. Fine-resolution landcover data sets will become more available in the future but will be mostly produced and applicable across relatively small areas (e.g., an ecoregion, or several counties) because thematic resolution (e.g., specific plant community composition) varies across the landscape.

Summary

The strength of landcover classification for defining habitat features resides in the ability to quantify and manipulate data across a study area that are often at a scale of tens to thousands to hundreds of thousands of hectares, and in the ability to integrate the information with other mapped georeferenced data to quantify habitat features. Habitat for a given species or group of species may depend not only on landcover, but also on the configuration of different landcover types or their proximity to other elements of the landscape. Mapped landcover types can be aggregated into patches of desired types and analyzed for size, configuration, and context with respect to each other (e.g., proximity of natural habitats to urban, row crop, water landcover types) or other elements on the landscape (e.g., roads, streams). Known species locations can be overlain on a landcover classification map to help define or model the amount and location of habitat for that species within a study area (Drew et al. 2011). Elements of habitat such as geophysical setting (geology, landform, soils, elevation) can be analyzed together with mapped landcover to refine habitat mapping or species distribution models. For example, a modeled species distribution pattern might be improved by including elevation or percentage slope, with landcover, to sharpen models of the amount and location of habitat.

LITERATURE CITED

Anderson, J. A, E. E. Hardy, J. T. Roach, and R. E. Witmer. 1976. A land use and land cover classification system for use with remote sensor data. US Geological Survey Professional Paper 964, Washington, DC, USA.

Blaschke, T. 2010. Object based image analysis for remote sensing. Journal of Photogrammetry and Remote Sensing 65:2–16.

Comer, P., D. Faber-Langendoen, R. Evans, S. Gawler, C. Josse, G. Kittel, S. Menard, M. Pyne, M. Reid, K. Schultz, et al. 2003. Ecological systems of the United States: a working classification of U.S. terrestrial systems. NatureServe, Arlington, Virginia, USA.

Diamond, D. D., and L. F. Elliott. 2015. Oklahoma ecological systems mapping interpretive booklet: methods, short type descriptions, and summary results. Oklahoma Department of Wildlife Conservation, Norman, Oklahoma, USA.

Diamond, D. D., L. F. Elliott, M. D. Debacker, K. M. James, D. L. Pursell, and A. Struckhoff. 2015. Vegetation mapping and classification of Hot Springs National Park, Arkansas: Project report. Natural Resource Report NPS/HTLN/NRR—2015/1075, National Park Service, Fort Collins, Colorado, USA.

Diamond, D. D., T. M. Gordon, C. D. True, and R. D. Lea. 2003. An ecoregion-based conservation assessment and conservation opportunity area inventory for the lower midwestern USA. Natural Areas Journal 23:129–140.

Drew, C. A., Y. F. Wiersma, and F. Huettmann, editors. 2011. Predictive species and habitat modeling in landscape ecology: concepts and applications. Springer, New York, New York, USA.

Ferree, C., and M. G. Anderson. 2013. A map of terrestrial habitats of the northeastern United States: methods and approach. Eastern Conservation Science, Eastern Regional Office, The Nature Conservancy, Boston, Massachusetts, USA.

Groves, C., L. Valutis, D. Vosick, B. Neely, K. Wheaton, J. Touval, and B. Runnels. 2000. Designing a geography of hope: a practitioner's handbook for ecoregional conservation planning. Volume 1. The Nature Conservancy, Arlington, Virginia, USA.

Homer, C., J. A. Dewitz, L. Yang, S. Jin, P. Danielson, G. Xian, J. Coulston, N. D. Herold, J. Wickham, and K. Megown. 2015. Completion of the 2011 National Land Cover Database for the conterminous United States—representing a decade of land cover change information. Photogrammetric Engineering and Remote Sensing 81:345–354.

Homer, C., C. Huang, L. Yang, B. Wylie, and M. Coan. 2004. Development of a 2001 National Land Cover Database for the United States. Photogrammetric Engineering and Remote Sensing 70:829–840.

Jin, S., L. Yang, P. Danielson, C. G. Homer, J. A. Fry, and G. Xian. 2013. A comprehensive change detection method for updating the National Land Cover Database to circa 2011. Remote Sensing of Environment 132:159–175.

LANDFIRE. 2019. LANDFIRE homepage. https://www .landfire.gov/. Accessed 16 Aug 2019.

Lu, D., and Q. Weng. 2007. A survey of image classification methods and techniques for improving classification performance. International Journal of Remote Sensing 28:823–870.

Multi-Resolution Land Characteristics Consortium. 2016. National Land Cover Dataset (NLCD) 2016. https:// www.mrlc.gov/national-land-cover-database-nlcd-2016. Accessed 16 Aug 2019.

National Park Service. 2019. Vegetation mapping inventory. https://www.nps.gov/im/vegetation-inventory.htm. Accessed 16 Aug 2019.

Prasad, A. M., L. R. Iverson, and A. Liaw. 2006. Newer classification and regression tree techniques: bagging and random forests for ecological prediction. Ecosystems 2:181–199.

Sleeter, B. L, T. L. Sohl, T. R. Loveland, R. F. Auch, W. A. Acevedo, M. A. Drummond, K. L. Sayler, and S. V. Stehman. 2013. Land-cover change. I. the conterminous United States from 1973 to 2000. Global Environmental Change 23:733–748.

US Forest Service. 2019. ForWarn II homepage. https:// forwarn.forestthreats.org/. Accessed 16 Aug 2019.

US Geological Survey. 2019. Gap Analysis Project homepage. https://www.usgs.gov/core-science-systems /science-analytics-and-synthesis/gap/science/. Accessed 16 Aug 2019.

US National Vegetation Classification. 2019. USNVC homepage. http://usnvc.org/. Accessed 16 Aug 2019.

Wickham, J., S. V. Stehman, L. Gass, J. A. Dewitz, D. G. Sorenson, B. J. Granneman, R. V. Poss, and L. A. Baer. 2017. Thematic accuracy assessment of the 2011 National Land Cover Dataset (NLCD). Remote Sensing of Environment 191:328–341.

7

Influence of Habitat Loss and Fragmentation on Wildlife Populations

Amanda E. Martin,
Joseph R. Bennett, and
Lenore Fahrig

Introduction

Evidence of wildlife population declines and extinctions suggests that we are in the midst of a biodiversity crisis (Dirzo et al. 2014, Pimm et al. 2014, Ceballos et al. 2015). Habitat fragmentation is frequently implicated as a causal factor in these wildlife declines (Novacek and Cleland 2001, Brook et al. 2008, Haddad et al. 2015). In this chapter we introduce a landscape ecology perspective on habitat fragmentation and explain how this perspective can be used to guide wildlife management. Specifically, after reading this chapter, the reader should understand that the term *habitat fragmentation* is used in two primary ways. First, habitat fragmentation has been used to encompass habitat loss and how that loss affects the spatial pattern of remnant habitat (through the breaking apart of habitat into smaller patches). Second, it has been used to refer only to the spatial pattern of remnant habitat, explicitly separating habitat fragmentation from habitat loss. From a landscape ecology perspective, habitat fragmentation can be separated from loss. This chapter goes on to discuss the theoretical and empirical evidence that shows that habitat loss has consistently strong, negative effects on wildlife, whereas effects of habitat fragmentation (independent of habitat loss) on wildlife can

be neutral, positive, or negative. Then we introduce the major contemporary sources of habitat loss and habitat fragmentation as well as evidence that suggests that habitat loss and habitat fragmentation will continue without major changes in human behavior. The chapter concludes with a discussion of the implications of habitat loss and habitat fragmentation for management of individual wildlife species and biodiversity.

What Is Habitat Fragmentation?

The answer to this question is more complex than one might think. Although habitat fragmentation is an important topic in conservation biology (Fazey et al. 2005), what this term actually means (and how it should be used) is still intensely debated (Didham et al. 2012, Hadley and Betts 2016, Fletcher et al. 2018, Fahrig et al. 2019). Indeed, the wide-ranging ways in which the term *habitat fragmentation* has been used has led some to suggest this term has become a panchreston (i.e., a term used in so many ways that it has become almost meaningless; Lindenmayer and Fischer 2007).

This debate is centered on whether habitat fragmentation should explicitly encompass habitat loss and how that loss affects the spatial pattern of rem-

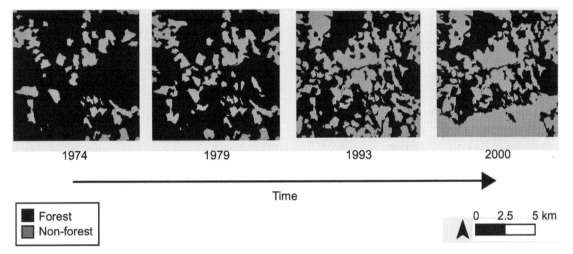

Figure 7.1. Example showing conversion of forest cover to non-forest cover over time in a landscape near Whitefish, Montana, USA. Landcover data were taken from the US Geological Survey Land Cover Trends Dataset (Soulard et al. 2014).

nant habitat (through the breaking apart of habitat into smaller patches) or if it should be used to refer only to the spatial pattern of remnant habitat, explicitly separating habitat fragmentation from habitat loss. In the following subsections we discuss each definition of habitat fragmentation and argue that adopting the latter view of fragmentation, which separates habitat fragmentation from habitat loss, can have important implications for conservation and landscape management.

Habitat Fragmentation as a Combination of Habitat Loss and the Breaking Apart of Habitat

Habitat fragmentation has typically been defined as the process through which "a large expanse of habitat is transformed into a number of smaller patches of smaller total area, isolated from each other by a matrix of habitats unlike the original" (Wilcove et al. 1986:237). In most studies, habitat (i.e., an area containing the features needed for a given wildlife species to persist) is represented by one or several landcover types (e.g., forest or wetland; Trzcinski et al. 1999, Quesnelle et al. 2015). Thus habitat

fragmentation can be viewed as the change in the amounts and distributions of landcover types over time, where a landcover type that a wildlife species uses as habitat is replaced by other landcover types (termed the matrix) that are not part of the habitat for the species. For example, habitat fragmentation for forest specialist species occurs when forested areas of a landscape are converted to other landcover types (Fig. 7.1).

A Landscape Ecology Perspective: Habitat Fragmentation as Separate from Habitat Loss

In landscape ecology we describe two components of landscape structure: landscape composition and landscape configuration (McGarigal and Marks 1995, Perotto-Baldivieso, Chap. 4, this volume). Landscape composition is what landcover types are in the landscape, and how much of each landcover type is present. For example, the composition of one landscape may be 50% native grassland and 50% crop field, which represents a different composition than a landscape composed of 25% each of four landcover types: grassland, crop field, forest, and urban (Fig. 7.2). Landscape configuration refers to

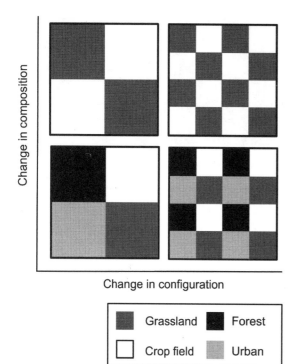

Figure 7.2. Illustration of the differences between landscape composition (what landcover types are in the landscape and how much of each landcover type is present) and landscape configuration (the spatial arrangement of those landcover types).

the spatial arrangement of those landcover types. For example, two landscapes, each composed of 50% grassland and 50% crop field, may have those land-cover types arranged in a few large patches or many small patches (Fig. 7.2).

The process of habitat fragmentation, as defined above (see "Habitat Fragmentation as a Combination of Habitat Loss and the Breaking Apart of Habitat"), affects landscape composition and configuration. It changes landscape composition because it reduces the amount of the landcover type(s) a wildlife species uses as habitat (i.e., habitat loss). It also changes the configuration of the landscape. This change in configuration can be considered independently of the change in composition (habitat loss) because, although the change in configuration occurs because of habitat loss, removal of a given amount of habitat can result in different configurations. That is, habitat

removal can result in a more fragmented distribution of remnant habitat (e.g., more, smaller patches and greater habitat-matrix edge density) or a less frag-mented one (fewer, larger patches and lower edge density; Fig. 7.3). Thus, from this landscape ecology perspective, habitat fragmentation is the breaking apart of habitat (i.e., a configurational change) that occurs for a given loss of habitat. This definition of habitat fragmentation will be used for the remainder of the chapter. We note that habitat fragmentation independent of habitat amount is also sometimes re-ferred to as habitat fragmentation per se (Haila and Hanski 1984, Fahrig 2003).

Habitat loss and measures of habitat fragmenta-tion are often highly correlated with one another. For example, landscapes with less forest tend to have smaller forest patches, while the number of forest patches and the total forest edge density (length of forest-matrix edge divided by landscape area) tend to peak at intermediate forest amounts (Fahrig 2003; Fig. 7.4). However, habitat loss and measures of hab-itat fragmentation are not perfectly correlated: for a given habitat amount, there is a range of potential levels of habitat fragmentation (Fig. 7.4). This means that, although landscapes with low habitat amounts typically have high levels of habitat fragmentation, there are low-habitat landscapes with low levels of habitat fragmentation (Fig. 7.5). And, although landscapes with high habitat amounts typically have low levels of habitat fragmentation, there are high-habitat landscapes with high levels of habitat frag-mentation (Fig. 7.5).

Benefits of Considering Habitat Fragmentation in Conservation and Landscape Management

There are three main reasons why managers should consider habitat fragmentation separate from habi-tat loss. First, failure to discriminate between habi-tat loss and fragmentation implies that these must occur together. This is not always the case, as dis-cussed above (see "A Landscape Ecology Perspective:

Habitat Fragmentation as Separate from Habitat Loss"). Thus consideration of habitat fragmentation separate from habitat loss allows for a better understanding of the structure of landscapes under management.

Second, failure to discriminate between habitat loss and fragmentation leads to the assumption that both habitat loss and fragmentation have consistently negative effects on wildlife populations. Empirical evidence suggests that this is not the case: fragmentation can have neutral, positive, or negative effects on wildlife (see "Effects of Habitat Fragmentation," below). The assumption that habitat fragmentation has negative effects on wildlife can have negative consequences for conservation when the true effects of fragmentation are positive because it incorrectly implies that reducing fragmentation can offset habitat loss. For example, Ethier and Fahrig (2011) found that the abundance of northern long-eared bats (*Myotis septentrionalis*) and little brown bats (*M. lucifugus*) increased as habitat fragmentation increased (i.e., positive effects of habitat fragmentation). Both species are endangered within the study region (Ontario, Canada; Government of Canada 2016). In such cases the assumption that reducing fragmentation when habitat is removed (e.g., by selectively removing the smallest habitat fragments) can offset the effects of habitat loss could lead to policies that would exacerbate population declines caused by habitat loss.

Finally, a definition of habitat fragmentation that does not discriminate between habitat loss and fragmentation implies that one cannot make management recommendations or policies to affect habitat fragmentation independent of habitat loss. This is not the case. Although habitat fragmentation occurs as a result of habitat loss, landscape planning and land use practices can be altered to either increase or decrease fragmentation for a given loss of habitat (Fig. 7.3). For example, an objective of forestry policies can be to increase or decrease the number of forest patches, for the same total forest removal.

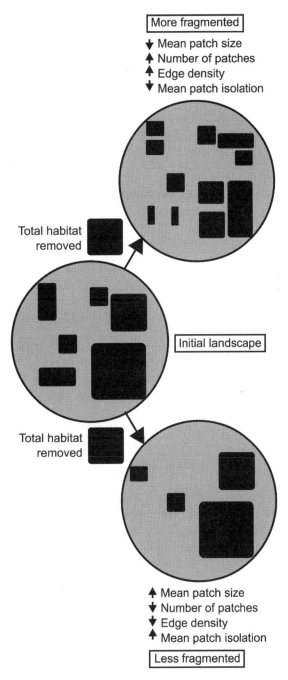

Figure 7.3. Illustration of how habitat loss can result in different levels of habitat fragmentation, depending on where habitat is removed. Changes in the mean patch sizes, number of patches, edge density (length of edge between habitat and matrix divided by landscape area), and mean patch isolation (mean distance between neighboring habitat patches) relative to the initial landscape are indicated by arrows.

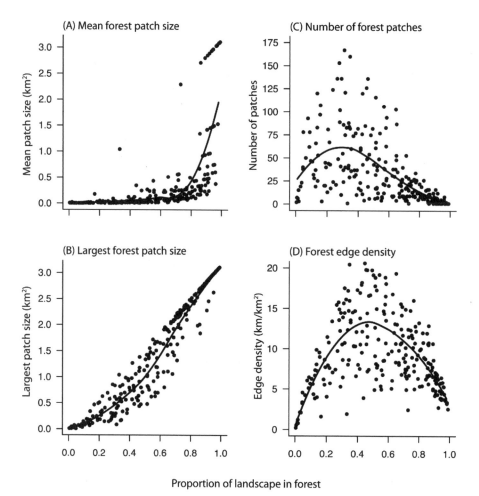

Figure 7.4. Empirical example showing the typical relationships between habitat loss, measured as the proportion of the landscape in forest, and four measures of habitat fragmentation: (*A*) mean forest patch size, (*B*) largest forest patch size, (*C*) number of forest patches, and (*D*) forest edge density (length of edge between forest and non-forest divided by landscape area). Empirical relationships were derived from a random sample of 250 circular landscapes (1-km radius) in Ontario, Canada. Each point represents a landscape. Forest cover from Agriculture and Agri-Food Canada (2015).

How Do Habitat Loss and Habitat Fragmentation Affect Wildlife?
Effects of Habitat Loss

Habitat loss has consistently strong, negative effects on wildlife. Empirical studies have reported that population abundances and wildlife species richness are lower in landscapes with less habitat (Fahrig 2003). Much of this negative effect is likely explained by the fact that when there is less habitat in the landscape, there are fewer resources available for the species, and thus the landscape supports a smaller number of individuals. For example, there are typically fewer home ranges or territories in landscapes with less habitat than in landscapes with more habitat (Kajtoch et al. 2015). Habitat loss is also associated with declines in breeding, foraging, and dispersal success (Matthysen and Currie

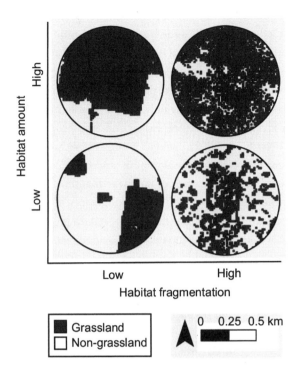

Grassland
Non-grassland

0   0.25   0.5 km

Figure 7.5. Example landscapes in Saskatchewan, Canada, showing high and low levels of grassland habitat fragmentation (i.e., the breaking apart of habitat) for a given habitat amount. Grassland cover from Agriculture and Agri-Food Canada (2015).

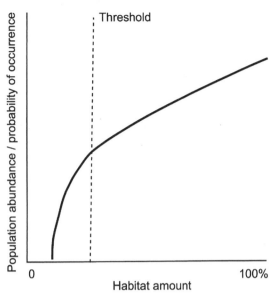

Figure 7.6. Illustration of the extinction threshold for habitat loss. A threshold relationship occurs when, below some level of habitat loss, the wildlife species' abundance or occurrence rapidly drops to zero.

1996, Kurki et al. 2000, Hinam and St. Clair 2008), which contribute to population declines when habitat is lost.

Population declines can lead to wildlife species extirpations or extinctions when enough habitat has been lost. There is evidence of a threshold habitat amount below which a population cannot sustain itself and becomes extinct (Fig. 7.6). Most support for this hypothesis comes from theoretical studies (Fahrig 2001, Swift and Hannon 2010); however, there is also empirical evidence of extinction thresholds. Researchers have reported that below some level of habitat availability, wildlife population abundance or probability of occurrence rapidly drops to zero (Homan et al. 2004, Holland et al. 2005, Betts et al. 2007, Zuckerberg and Porter 2010).

Effects of Habitat Fragmentation

Empirical studies generally suggest that habitat fragmentation has weaker effects on wildlife than habitat loss (Fahrig 2003). Perhaps even more surprising is the evidence that habitat fragmentation is more likely to have positive effects on wildlife than negative effects. Fahrig (2017) reviewed the results of 118 studies that reported 381 significant responses to habitat fragmentation; 76% of these effects were positive (e.g., higher abundance or species richness in more fragmented landscapes, when controlling for the habitat amount). This phenomenon is found across taxonomic groups: more positive than negative effects of fragmentation were reported for all studied taxa (plants, microorganisms, invertebrates, fish and herptiles, birds, mammals; Fahrig 2017). Positive effects of fragmentation are also not limited to the most abundant wildlife species or habitat generalists. For example, 29 of 30 of the significant responses of specialist, rare, or threatened species richness to fragmentation were positive, as were 49 of 79 of the responses of globally threatened or declining species (Fahrig 2017).

The remainder of this section focuses on explanations for why effects of habitat fragmentation are weaker than the effects of habitat loss, and why habitat fragmentation has positive effects on some wildlife populations and communities but negative effects on others.

WHY ARE OBSERVED EFFECTS OF HABITAT FRAGMENTATION ON WILDLIFE WEAKER THAN THE EFFECTS OF HABITAT LOSS?

The simplest, and most likely, explanation for why effects of habitat fragmentation on wildlife are weaker than those of habitat loss is that the amount of habitat in the landscape is much more important for wildlife species abundance and persistence than how that habitat is arranged. This explanation is consistent with findings from theoretical models (Fahrig 1997, Jackson and Fahrig 2014, 2016) and the bulk of the empirical literature (see "Effects of Habitat Loss," above).

Observed effects of habitat fragmentation on wildlife could also be weaker than those of habitat loss because habitat fragmentation is not typically studied in the contexts where it is most important for wildlife. Such a situation could occur if effects of habitat fragmentation are only important when habitat is limited. If true, studies that include landscapes with moderate to high habitat amounts would underestimate the effects of habitat fragmentation that occur at low habitat amounts. This explanation is based on theoretical models that suggest the spatial arrangement of habitat patches only affects wildlife abundance and occurrence when habitat is limited (i.e., less than ~20% to 25% of the landscape in habitat; Andrén 1996, Fahrig 1998). There are two problems with this explanation. First, empirical evidence of this fragmentation threshold is weak. Researchers have reported that the effects of fragmentation on wildlife are strongest at both low (below 10% to 30%

of the landscape in habitat; Andrén 1994) and intermediate (at 40% of the landscape in habitat but not below 30%; With and Pavuk 2012) habitat amounts. Others found no effect of habitat amount on the wildlife species and fragmentation relationship, or that this occurred in only a small number of cases (Trzcinski et al. 1999, Parker and Mac Nally 2002, Betts et al. 2007). Second, even if there are stronger effects of fragmentation in landscapes with low habitat amounts, it does not mean that its effects on wildlife are as strong as that of habitat loss. Therefore it is unlikely that the effects of habitat fragmentation have been underestimated because its effects are not typically studied in landscapes where it should be most important.

Effects of habitat fragmentation may also be underestimated because studies do not typically measure the biological responses most affected by fragmentation. For example, simulation modeling has suggested that habitat fragmentation has stronger effects on genetic structure than population size (Bruggeman et al. 2010). If true, it suggests that long-term effects of fragmentation on wildlife may be underestimated by studies that look at only estimates of abundance or occurrence. More recent simulations, however, suggest that habitat amount is a better predictor of genetic structure than habitat fragmentation when correlations between habitat amount and fragmentation are minimized (Cushman et al. 2012, Jackson and Fahrig 2016). Stronger effects of habitat amount on genetic structure may occur because population size, which is mainly determined by habitat amount, has stronger effects on genetic structure than does population subdivision (Jackson and Fahrig 2016). Empirical study is needed to determine which of these predictions is correct. To our knowledge, however, there is currently no empirical evidence to support the suggestion that effects of habitat fragmentation have been underestimated because studies do not typically measure the biological responses most affected by fragmentation.

Wildlife may be at risk from fragmentation because more fragmented landscapes have smaller patches, albeit more of them (Fig. 7.7A). Small patches contain small populations, which will have a high probability of local extinction (Matthies et al. 2004). At extreme levels of fragmentation, patch sizes may become too small to support even an individual's or breeding pair's territory. But this will only be the case if individuals cannot include multiple, small patches within their territories.

The smaller patch sizes in more fragmented landscapes may also negatively affect wildlife because rates of dispersal mortality are higher in landscapes with smaller patches (Fig. 7.7A). The probability of a disperser encountering a habitat boundary, and entering the matrix, is higher when a habitat patch is smaller (Fahrig 2007). Thus, for a given amount of habitat, rates of movement through the matrix should be higher in a landscape with more, smaller patches. This should increase the per capita mortality rates because the probability of dispersal mortality is generally higher in the matrix than in habitat.

Fragmentation may also have negative effects on wildlife because of negative edge effects. Negative edge effects occur when some measure of the wildlife species or community viability (e.g., nest predation rates, abundance, species richness) declines from the interior to the edge of a habitat patch. Negative edge effects on wildlife may result in negative effects of fragmentation on wildlife because a more frag-

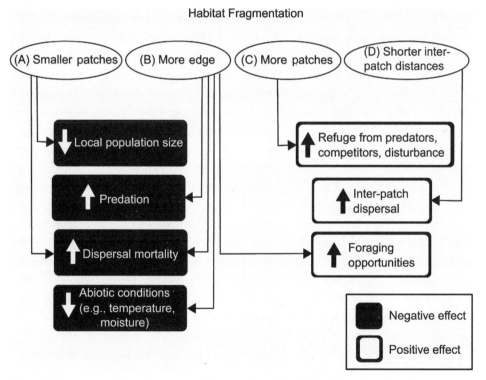

Figure 7.7. Summary of hypotheses to explain negative and positive effects of habitat fragmentation (independent of habitat amount) on wildlife, showing the link between each hypothesized driver of population size and one of four characteristic changes in landscape configuration that occur with fragmentation. More fragmented landscapes have (A) smaller habitat patches, (B) more edge, (C) more patches, and (D) shorter inter-patch distances than less fragmented landscapes.

mented landscape has more edge for a given habitat amount than a less fragmented landscape (Fig. 7.7B). Increased edge can have negative effects on wildlife through several mechanisms. Edge effects may occur because habitat quality declines near edges (Watson et al. 2004). For example, studies have shown that there can be differences in temperature, moisture, and wind speed from forest edge to interior (Davies-Colley et al. 2000, Arroyo-Rodríguez et al. 2017). Thus conditions at edges may be outside the optimal temperature, moisture, or wind speed range for a forest species. Second, there can be increased rates of predation near habitat edges because individuals near edges are at risk from predators that use the matrix landcover types and forage along boundaries between the two landcover types. There is, however, mixed support for this mechanism, which has been most extensively studied for nest predation of forest breeding birds. Although some studies have reported higher rates of nest predation near edges relative to the forest interior, most have not (Lahti 2001). This finding suggests that increased predation along edges may have less influence on wildlife responses to edge than declines in habitat quality in edge versus interior habitat.

REASONS FOR POSITIVE EFFECTS OF

HABITAT FRAGMENTATION

If habitat patches function as temporary refugia from predators, competitors, or disturbance, then having a larger number of (albeit smaller) patches should benefit the wildlife community (Fig. 7.7C). In his seminal experiments, Huffaker (1958) reported that subdivision of a given amount of habitat into more, smaller patches increased the persistence of a predator-prey system, and this result has been replicated in later theoretical and empirical studies (Ellner et al. 2001, Karsai and Kampis 2011). Such positive effects of subdivision on a predator-prey system are likely to occur when prey are able to colonize a habitat patch after predator extirpation, increase in numbers, and then disperse before the predator reaches the patch. Similarly, habitat fragmentation

may allow for coexistence of competitors, where the inferior competitor(s) can only coexist with the superior competitor(s) when they are able to move among habitat patches and exploit patches that are temporarily free of the superior competitor(s). Finally, theory suggests that having more habitat patches can reduce the risk of an entire population going extinct from some disturbance event (den Boer 1968, Bascompte et al. 2002).

Habitat fragmentation may also positively affect wildlife because, for a given habitat amount, mean distances among habitat patches are typically shorter in a more fragmented landscape (Figs. 7.7D, 7.8). Shorter inter-patch distances should translate into more successful inter-patch dispersal events (Johnson et al. 2009). Theory suggests that wildlife species with higher rates of successful inter-patch dispersal should be less likely to be extirpated or to go extinct than less dispersive species when habitat is lost and fragmented (Martin and Fahrig 2016). Thus, for some wildlife species, the benefits of frequent inter-patch dispersal may result in positive effects of fragmentation. Such positive effects likely depend on the relative importance of inter-patch dispersal versus dispersal mortality for wildlife species persistence because we expect that dispersal mortality is higher in more fragmented landscapes.

Finally, while some wildlife species show negative edge effects, others show the opposite. These species should increase with fragmentation because more fragmented landscapes have more edge (Fig. 7.7B). Positive responses to edges between different landcover types may occur if food resources are more abundant in edges (Harding and Gomez 2006, Macreadie et al. 2010).

WHY ARE EFFECTS OF FRAGMENTATION POSITIVE

IN SOME CASES BUT NEGATIVE IN OTHERS?

Above we provided explanations for the positive and negative effects of habitat fragmentation. We lack, however, a way to predict when an effect should be positive and when an effect should be negative. The ability to predict how a wildlife species responds to fragmentation could be beneficial for managers in cases where detailed studies of responses to fragmentation are lacking for many of the managed wildlife species. Below we discuss two potential predictors of a wildlife species' response to fragmentation—its traits and the matrix quality—although we note that both require further empirical validation before they can be reliably used in landscape management.

The response of a wildlife species to fragmentation may depend on its traits because a species' traits influence its susceptibility to the potential positive and negative effects of fragmentation discussed above (see "Reasons for Negative Effects of Habitat Fragmentation" and "Reasons for Positive Effects of Habitat Fragmentation"). If a wildlife species is more responsive to the mechanisms causing positive effects of fragmentation than negative effects, then the benefits of fragmentation should outweigh its costs, resulting in an overall positive effect on the species (and vice versa). For example, a wildlife species' behavior at habitat boundaries (e.g., whether a disperser crosses, or avoids crossing, from habitat to matrix) may affect its response to fragmentation. A species that does not readily disperse among habitat patches is likely to be most sensitive to the negative fragmentation effects associated with having smaller local population sizes, such as inbreeding depression (Fig. 7.7A). Researchers conducting empirical studies have reported that a wildlife species' traits can influence its response to the process of fragmentation (i.e., the combination of habitat loss and fragmentation; Vetter et al. 2011, Carrié et al. 2017). To our knowledge, however, there are no empirical tests of the influence of wildlife species traits on the response to habitat fragmentation independent of habitat amount.

Another factor that may influence whether effects of fragmentation are positive versus negative is the quality of the matrix surrounding the remnant habitat patches. Although an increase in any matrix landcover type has a negative effect on wildlife (because matrix gain equals habitat loss), there is a growing body of literature that supports the idea that

(A) Low fragmentation

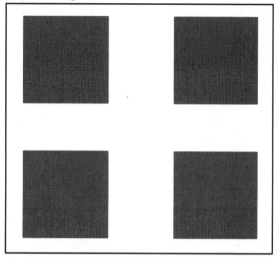

(B) High fragmentation

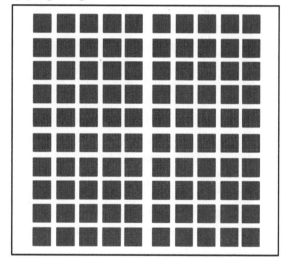

Figure 7.8. Illustration showing why landscapes with lower habitat fragmentation (*A*) typically have longer distances among neighboring habitat patches than landscapes with higher fragmentation (*B*). Both landscapes have the same total amount of habitat. In (*B*), however, the distances between habitat patches are shorter because the remaining habitat is subdivided into more habitat patches.

matrix landcover types are not all equal. Wildlife species richness, occurrence, and abundance in habitat patches can be significantly affected by the surrounding matrix type (Prevedello and Vieira 2010). The relative effects of different matrix types on a given wildlife species are often discussed in terms of matrix quality, where a high-quality matrix is one that has a less negative effect on the species of interest than a low-quality matrix.

We hypothesize that when matrix quality is low, fragmentation effects are more likely to be negative than positive, primarily because the probability of dispersal mortality should be higher in a low-quality matrix and populations in fragmented landscapes have higher rates of movement into the matrix than populations in non-fragmented landscapes. The resulting higher rates of dispersal mortality should put wildlife in more fragmented landscapes at greater risk. When matrix quality is high, however, such that there is less risk of mortality during movement through matrix, fragmentation effects are more likely to be positive because wildlife in fragmented landscapes will have higher rates of inter-patch dispersal and thus greater access to potential refugia from predators, competitors, or disturbance. Matrix quality may also affect wildlife species responses to fragmentation through its role in determining whether edge effects are positive or negative. For example, Keyser (2002) reported that predation of bird nests was higher in the patch edge versus interior when the edge was next to residential landcover, but predation was lower in the edge versus interior when the edge was next to forest. While there is ample evidence that matrix quality affects wildlife species abundance and habitat occupancy (Prevedello and Vieira 2010), we know of no direct evidence that matrix quality influences wildlife responses to habitat fragmentation. In one study, Öckinger et al. (2012) reported that matrix quality affected the strength of relationship between habitat patch size and the richness of grassland butterflies, with weaker effects of patch size on species richness when matrix quality was higher. Because habitat patch size is frequently correlated with habitat amount, however, it is unclear whether in this case matrix quality was moderating the effects of habitat fragmentation or habitat loss.

Alternatively, matrix quality could cause apparently opposite effects of habitat fragmentation for different species and geographic regions, if effects of matrix quality on wildlife are stronger than that of habitat fragmentation, and the cross-landscape correlations of matrix quality and fragmentation are in opposite directions for those species or regions. For example, in a region where landscapes with greater habitat fragmentation have more roads than landscapes with less fragmentation, fragmentation may appear to have a negative effect on wildlife. This negative effect could occur because road mortality, which can have negative effects on wildlife abundances (Fahrig et al. 1995, Jack et al. 2015, Martin et al. 2018), is higher in landscapes with greater fragmentation. In contrast, in an agricultural region where organic farms tend to have higher forest fragmentation than conventional farms (e.g., because of greater hedgerow densities on organic farms; Fuller et al. 2005), fragmentation may appear to have a positive effect on forest species because organic crop fields have less negative effects on wildlife than conventional fields (Tuck et al. 2014).

What Are the Contemporary Sources of Habitat Loss and Habitat Fragmentation?
Anthropogenic Sources of Habitat Loss

There has been widespread conversion of native landcover types (e.g., forest, wetland, grassland) to human-dominated ones (e.g., agriculture), resulting in habitat loss for wildlife species that cannot use the human-dominated landcover types. It is estimated that the percentage of ice-free area of the earth that is used by humans went from 5% to 55% in just 300 years (1700–2000), replacing native landcover types with agricultural lands and human settlements (Ellis et al. 2010). Habitat losses for wildlife species in some regions have been extreme. For example,

>95% of the native tallgrass cover has been lost in North America's Great Plains region (Samson and Knopf 1994), and >88% of Brazil's original Atlantic Forest, a biodiversity hotspot, has been lost (Ribeiro et al. 2009).

Conversion to agricultural lands is likely to continue, although there is evidence that the rate of conversion has slowed (Food and Agriculture Organization 2003). Projections of future rates of land conversion vary. For example, Wirsenius et al. (2010) predicted a 13% increase in the area of cropland and 4% increase in the area of pastureland between 1992 and 2030, while Tilman et al. (2001) predicted a 23% increase in the area of cropland and 16% increase in the area of pastureland between 2000 and 2050. These studies generally agree, however, that if farming practices, food consumption, and human population growth follow current trends, conversion of native landcovers to crop and pasture fields will continue.

While rates of agricultural land clearing may be slowing over time, rates of urban expansion have remained high. For example, a meta-analysis of 326 studies of urban expansion reported increases in urban land area in all geographic regions, with a global increase of 58,000 km^2 between 1970 and 2000 (Seto et al. 2011). This conversion is likely to continue as the human population grows, with estimates of a 40% to 257% increase in urban area between 2000 and 2030 (Seto et al. 2011).

In addition to land clearing for agriculture and urban development, expansion of the road network is an important source of habitat loss. For example, in the United States, there was a 9% increase in road lanes over 30 years, increasing the length of road lanes by 1.2 million km (1980–2013; Federal Highway Administration 2016). This growth has resulted in a pervasive influence of roads on wildlife. For example, Riitters and Wickham (2003) reported that only ~18% of land was >1 km from a road in the conterminous United States. Such rapid expansion of the road network is not limited to the United States; globally, the length of paved roads is expected

to increase by approximately 60% between 2010 and 2050 (Dulac 2013).

Conversion of native landcovers to human-dominated ones results in habitat loss for wildlife species that do not use human-dominated landcover types as habitat. Wildlife species that have habituated to farmland and urban areas are also experiencing habitat loss as intensification of those land uses degrades habitat quality to the point that it is uninhabitable. For example, widespread declines in European farmland bird populations have been linked to agricultural intensification practices aimed at increasing crop yield per unit area (e.g., increasing pesticide and fertilizer inputs; Donald et al. 2001, Wretenberg et al. 2007). Urban densification, which increases building density at the expense of greenspace, has also been related to declines in richness and abundance of wildlife species typically found within cities (Ikin et al. 2013).

Climate change has the potential to exacerbate habitat loss. For example, climate change can reduce habitat amounts directly for high-altitude and northern latitude wildlife species. As climate warms, the habitat for these cold-adapted species shrinks toward the mountain peaks or northernmost coast, respectively (Virkkala et al. 2008, Dirnböck et al. 2011). Island and coastal wildlife species are also likely to lose habitat owing to rising sea levels (Rahmstorf 2007). Even if the location of a wildlife species' habitat can shift it may be inaccessible, or only partially accessible, if species cannot move to keep pace with changes in the habitat location (Schloss et al. 2012). Finally, habitat loss may occur because of the increasing frequency of extreme weather events, including heat waves, droughts, floods, tornados, and fires (Intergovernmental Panel on Climate Change 2014).

Anthropogenic Sources of Habitat Fragmentation

The level of habitat fragmentation for a given loss of habitat depends on how habitat is removed (Fig. 7.3). For example, clearing long, narrow areas (e.g.,

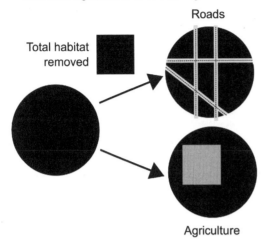

(A) Greater fragmentation from road construction than from agricultural land clearing

Roads

Total habitat removed

Agriculture

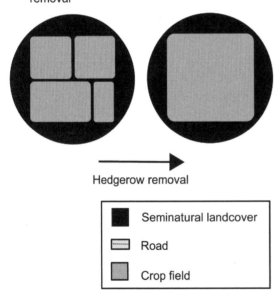

(B) Reduced fragmentation with hedgerow removal

Hedgerow removal

■	Seminatural landcover
▦	Road
▨	Crop field

Figure 7.9. Examples of the sources of habitat fragmentation. (*A*) Road construction can result in higher levels of fragmentation than land clearing for agriculture because clearing long, narrow areas for roads results in greater edge densities relative to removing the same area in a large, square plot. (*B*) Habitat loss through removal of hedgerows among crop fields reduces fragmentation for wildlife species using seminatural landcover types, because removal of hedgerows reduces the density of edges between crop and seminatural landcover types.

for road construction) will result in higher levels of fragmentation relative to removing the same area in a large, square plot (e.g., for a crop field) because the former pattern of habitat removal results in greater edge densities (Fig. 7.9A). In some cases, habitat may even become less fragmented with habitat removal. For example, removal of hedgerows to create larger crop fields reduces the density of edges between crop and seminatural landcover types, thereby reducing the level of fragmentation for wildlife species using seminatural landcover types (Fig. 7.9B). Land clearing for agriculture, urbanization, and roads can all be planned to result in more or less fragmentation for the same loss of habitat.

Non-anthropogenic Sources of Habitat Loss and Fragmentation

Anthropogenic activities are not the only sources of habitat loss and fragmentation. Historically, disturbances including wildfires, floods, landslides, tornadoes, and insect outbreaks caused habitat loss. Such disturbances could maintain a fragmented distribution of habitat, particularly in regions where disturbances occurred frequently but were small. Foraging and migrations of large herbivore herds could also maintain a fragmented distribution of habitat. For example, American bison (*Bison bison*) played an important role in the maintenance of a patchy mosaic of grassland landcover types in North America's Great Plains because their foraging locations and intensities varied spatially (Truett et al. 2001). Such ecological systems are naturally fragmented, and wildlife in these systems are adapted to the sources of fragmentation that maintain those systems. Thus non-anthropogenic sources of fragmentation may be less likely to cause long-term population declines and wildlife species extirpations or extinctions than the anthropogenic sources discussed above. We note, however, that the frequency and size of such events may be altered by human activities, putting wildlife species in these systems at risk (e.g., climate change increasing the frequency and severity of extreme weather events).

What Are the Implications of Habitat Loss and Fragmentation for Management of Individual Wildlife Species and Biodiversity?

Our understanding of the independent effects of habitat loss and fragmentation has implications for management of individual wildlife species and biodiversity. First and foremost, the evidence of strong, negative effects of habitat loss on wildlife suggests that wildlife management should focus on reducing habitat loss and on habitat restoration.

Some level of habitat loss, however, is often inevitable (see "What Are the Contemporary Sources of Habitat Loss and Habitat Fragmentation?" above). When habitat loss is inevitable, one may want to recommend where that habitat should be removed. Management for an optimal level of fragmentation will depend on the management objective because fragmentation has positive effects on some wildlife species but negative effects on others. If management is species based (e.g., planning for an at-risk species) then research should focus on the species-specific response to fragmentation to determine whether the wildlife species will benefit from an increase or decrease in the level of fragmentation. If the management objective is to increase or maintain biodiversity, then empirical evidence thus far suggests that, for a given level of habitat loss, managing for a higher level of habitat fragmentation may be more likely to meet that objective than managing for less fragmentation. Positive effects of fragmentation have been observed more often than negative effects (Fahrig 2017), and thus a more fragmented landscape, for a given habitat amount, will likely produce positive outcomes for more wildlife species than a less fragmented landscape.

A recommendation to increase the level of habitat fragmentation may seem controversial, mainly because many people believe that habitat fragmentation equates to habitat loss. Thus they interpret a recommendation to increase habitat fragmentation as giving license to remove habitat. This is not the case. On the contrary, maintaining a more fragmented landscape may increase the impetus to save habitat because it implies that small habitat patches are important for reaching conservation and management objectives, and thus are worth protecting. Interestingly, there is empirical support for the idea that small habitat patches can be important for biodiversity conservation. For example, Fischer and Lindenmayer (2002) demonstrated that small forest patches can contribute to conservation of bird species richness, showing that the number of bird species conserved per unit area would accumulate faster when small habitat patches were conserved first than when large patches were conserved first. Scheffer et al. (2006) similarly demonstrated the benefits of conserving small lakes and ponds, finding that the richness of aquatic birds, plants, amphibians, and invertebrates was often higher in smaller waterbodies.

Summary

In this chapter we introduced readers to a landscape ecology perspective on habitat fragmentation. From this perspective the process of habitat fragmentation has two components: habitat loss and habitat fragmentation, the latter referring to the breaking apart of habitat, for a given loss of habitat. Although habitat loss and fragmentation tend to occur together, it is important to understand the independent effects of each component. This understanding is ultimately important because failure to discriminate between habitat loss and fragmentation leads to the assumption that both habitat loss and fragmentation have consistently negative effects on wildlife populations. In contrast, empirical evidence shows that habitat loss and fragmentation do not have the same effects on wildlife: habitat loss has strong, negative effects on wildlife, but fragmentation has weaker effects that can be either positive or negative. There are various mechanisms that should cause negative effects of habitat fragmentation on wildlife and mechanisms that should cause the opposite. This may explain, at least in part, why fragmentation effects on wildlife are typically weak and positive. If multiple mechanisms act to affect their responses to habitat fragmentation, and mechanisms causing positive effects of habitat fragmentation only slightly outweigh those driving negative effects of habitat fragmentation, then the overall effects of habitat fragmentation on wildlife would be weak (Fahrig et al. 2019).

Conversion of native landcover types to agricultural lands, urban areas, and roads are major contemporary sources of habitat loss and habitat fragmentation, and evidence suggests that this conversion of native to human-dominated landcover types will most likely continue. This habitat loss may be exacerbated by climate change. Climate change can reduce habitat amounts directly, particularly for high-altitude and high-latitude wildlife species. It can also reduce availability of remaining habitat, if a species cannot move to keep pace with changes in the location(s) of its habitat.

A major implication of this landscape ecology perspective on habitat loss and fragmentation for management of individual wildlife species and biodiversity is that protecting and restoring habitat for at-risk wildlife species are paramount. Just because a given habitat patch is small does not mean it is worthless. On the contrary, conserving large numbers of small habitat patches can benefit many wildlife species and help land managers reach their conservation goals.

LITERATURE CITED

Agriculture and Agri-Food Canada. 2015. 2015 Annual Crop Inventory [data set]. Agriculture and Agri-Food Canada, Ontario, Canada.

Andrén, H. 1994. Effects of habitat fragmentation on birds and mammals in landscapes with different proportions of suitable habitat: a review. Oikos 71:355–366.

Andrén, H. 1996. Population responses to habitat fragmentation: statistical power and the random sample hypothesis. Oikos 76:235–242.

Arroyo-Rodríguez, V., R. A. Saldaña-Vázquez, L. Fahrig, and B. A. Santos. 2017. Does forest fragmentation cause an increase in forest temperature? Ecological Research 32:81–88.

Bascompte, J., H. Possingham, and J. Roughgarden. 2002. Patchy populations in stochastic environments: critical number of patches for persistence. American Naturalist 159:128–137.

Betts, M. G., G. J. Forbes, and A. W. Diamond. 2007. Thresholds in songbird occurrence in relation to landscape structure. Conservation Biology 21:1046–1058.

Brook, B. W., N. S. Sodhi, and C. J. A. Bradshaw. 2008. Synergies among extinction drivers under global change. Trends in Ecology and Evolution 23:453–460.

Bruggeman, D. J., T. Wiegand, and N. Fernández. 2010. The relative effects of habitat loss and fragmentation on population genetic variation in the red-cockaded woodpecker (*Picoides borealis*). Molecular Ecology 19:3679–3691.

Carrié, R., E. Andrieu, S. A. Cunningham, P. E. Lentini, M. Loreau, and A. Ouin. 2017. Relationships among ecological traits of wild bee communities along gradients of habitat amount and fragmentation. Ecography 40:85–97.

Ceballos, G., P. R. Ehrlich, A. D. Barnosky, A. García, R. M. Pringle, and T. M. Palmer. 2015. Accelerated modern human-induced species losses: entering the sixth mass extinction. Scientific Advances 1:e1400253.

Cushman, S. A., A. Shirk, and E. L. Landguth. 2012. Separating the effects of habitat area, fragmentation and matrix resistance on genetic differentiation in complex landscapes. Landscape Ecology 27:369–380.

Davies-Colley, R. J., G. W. Payne, and M. van Elswijk. 2000. Microclimate gradients across a forest edge. New Zealand Journal of Ecology 24:111–121.

den Boer, P. J. 1968. Spreading of risk and stabilization of animal numbers. Acta Biotheoretica 18:165–194.

Didham, R. K., V. Kapos, and R. M. Ewers. 2012. Rethinking the conceptual foundations of habitat fragmentation research. Oikos 121:161–170.

Dirnböck, T., F. Essl, and W. Rabitsch. 2011. Disproportional risk for habitat loss of high-altitude endemic species under climate change. Global Change Biology 17:990–996.

Dirzo, R., H. S. Young, M. Galetti, G. Ceballos, N. J. B. Isaac, and B. Collen. 2014. Defaunation in the Anthropocene. Science 345:401–406.

Donald, P. F., R. E. Green, and M. F. Heath. 2001. Agricultural intensification and the collapse of Europe's farmland bird populations. Proceedings of the Royal Society B: Biological Sciences 268:25–29.

Dulac, J. 2013. Global land transport infrastructure requirements: estimating road and railway infrastructure capacity and costs to 2050. International Energy Agency, Paris, France.

Ellis, E. C., K. K. Goldewijk, S. Siebert, D. Lightman, and N. Ramankutty. 2010. Anthropogenic transformation of the biomes, 1700 to 2000. Global Ecology and Biogeography 19:589–606.

Ellner, S. P., E. McCauley, B. E. Kendall, C. J. Briggs, P. R. Hosseini, S. N. Wood, A. Janssen, M. W. Sabelis, P. Turchin, R. M. Nisbet, et al. 2001. Habitat structure and population persistence in an experimental community. Nature 412:538–543.

Ethier, K., and L. Fahrig. 2011. Positive effects of forest fragmentation, independent of forest amount, on bat abundance in eastern Ontario, Canada. Landscape Ecology 26:865–876.

Fahrig, L. 1997. Relative effects of habitat loss and fragmentation on population extinction. Journal of Wildlife Management 61:603–610.

Fahrig, L. 1998. When does fragmentation of breeding habitat affect population survival? Ecological Modelling 105:273–292.

Fahrig, L. 2001. How much habitat is enough? Biological Conservation 100:65–74.

Fahrig, L. 2003. Effects of habitat fragmentation on biodiversity. Annual Review of Ecology, Evolution, and Systematics 34:487–515.

Fahrig, L. 2007. Estimating minimum habitat for population persistence. Pages 64–80 in D. B. Lindenmayer and R. J. Hobbs, editors. Managing and designing landscapes for conservation: moving from perspectives to principles. Blackwell, Oxford, UK.

Fahrig, L. 2017. Ecological responses to habitat fragmentation per se. Annual Review of Ecology, Evolution, and Systematics 48:1–23.

Fahrig, L, V. Arroyo-Rodríguez, J. R. Bennett, V. Boucher-Lalonde, E. Cazetta, D. J. Currie, F. Eigenbrod, A. T. Ford, S. P. Harrison, J. A. G. Jaeger, et al. 2019. Is habitat fragmentation bad for biodiversity? Biological Conservation 230:179–186.

Fahrig, L., J. H. Pedlar, S. E. Pope, P. D. Taylor, and J. F. Wegner. 1995. Effect of road traffic on amphibian density. Biological Conservation 73:177–182.

Fazey, I., J. Fischer, and D. B. Lindenmayer. 2005. What do conservation biologists publish? Biological Conservation 124:63–73.

Federal Highway Administration. 2016. Highway statistics 2013. https://www.fhwa.dot.gov/policyinformation/statistics/2013/. Accessed 3 Mar 2016.

Fischer, J., and D. B. Lindenmayer. 2002. Small patches can be valuable for biodiversity conservation: two case studies on birds in southeastern Australia. Biological Conservation 106:129–136.

Fletcher, R. J., Jr., R. K. Didham, C. Banks-Leite, J. Barlow, R. M. Ewers, J. Rosindell, R. D. Holt, A. Gonzalez, R. Pardini, E. I. Damschen, et al. 2018. Is habitat fragmentation good for biodiversity? Biological Conservation 226:9–15.

Food and Agriculture Organization. 2003. World agriculture: towards 2015/2030: an FAO perspective. Earthscan, London, UK.

Fuller, R. J., L. R. Norton, R. E. Feber, P. J. Johnson, D. E. Chamberlain, A. C. Joys, F. Mathews, R. C. Stuart, M. C. Townsend, W. J. Manley, et al. 2005. Benefits of organic farming to biodiversity vary among taxa. Biology Letters 1:431–434.

Government of Canada. 2016. Species at risk public registry. https://www.canada.ca/en/environment-climate-change/services/species-risk-public-registry.html. Accessed 4 Apr 2016.

Haddad, N. M., L. A. Brudvig, J. Clobert, K. F. Davies, A. Gonzalez, R. D. Holt, T. E. Lovejoy, J. O. Sexton, M. P. Austin, C. D. Collins, et al. 2015. Habitat fragmentation and its lasting impact on Earth's ecosystems. Science Advances 1:e1500052.

Hadley, A. S., and M. G. Betts. 2016. Refocusing habitat fragmentation research using lessons from the last decade. Current Landscape Ecology Reports 1:55–66.

Haila, Y., and I. K. Hanski. 1984. Methodology for studying the effect of habitat fragmentation on land birds. Annales Zoologici Fennici 21:393–397.

Harding, E. K., and S. Gomez. 2006. Positive edge effects for arboreal marsupials: an assessment of potential mechanisms. Wildlife Research 33:121–129.

Hinam, H. L., and C. C. St. Clair. 2008. High levels of habitat loss and fragmentation limit reproductive success by reducing home range size and provisioning rates of northern saw-whet owls. Biological Conservation 141:524–535.

Holland, J. D., L. Fahrig, and N. Cappuccino. 2005. Fecundity determines the extinction threshold in a Canadian assemblage of longhorned beetles (Coleoptera: Cerambycidae). Journal of Insect Conservation 9:109–119.

Homan, R. N., B. S. Windmiller, and J. M. Reed. 2004. Critical thresholds associated with habitat loss for two vernal pool-breeding amphibians. Ecological Applications 14:1547–1553.

Huffaker, C. B. 1958. Experimental studies on predation: dispersion factors and predator-prey oscillations. Hilgardia 27:795–835.

Ikin, K., R. M. Beaty, D. B. Lindenmayer, E. Knight, J. Fischer, and A. D. Manning. 2013. Pocket parks in a compact city: how do birds respond to increasing residential density? Landscape Ecology 28:45–56.

Intergovernmental Panel on Climate Change. 2014. Climate change 2014: synthesis report. Contribution of working groups I, II and III to the fifth assessment report of the Intergovernmental Panel on Climate Change. Intergovernmental Panel on Climate Change, Geneva, Switzerland.

Jack, J., T. Rytwinski, L. Fahrig, and C. M. Francis. 2015. Influence of traffic mortality on forest bird abundance. Biodiversity and Conservation 24:1507–1529.

Jackson, N. D., and L. Fahrig. 2014. Landscape context affects genetic diversity at a much larger spatial extent than population abundance. Ecology 95:871–881.

Jackson, N. D., and L. Fahrig. 2016. Habitat amount, not habitat configuration, best predicts population genetic structure in fragmented landscapes. Landscape Ecology 31:951–968.

Johnson, C. A., J. M. Fryxell, I. D. Thompson, and J. A. Baker. 2009. Mortality risk increases with natal dispersal distance in American martens. Proceedings of the Royal Society B: Biological Sciences 276:3361–3367.

Kajtoch, L., M. Żmihorski, and P. Wieczorek. 2015. Habitat displacement effect between two competing owl species in fragmented forests. Population Ecology 57:517–527.

Karsai, I., and G. Kampis. 2011. Connected fragmented habitats facilitate stable coexistence dynamics. Ecological Modelling 222:447–455.

Keyser, A. J. 2002. Nest predation in fragmented forests: landscape matrix by distance from edge interactions. Wilson Bulletin 114:186–191.

Kurki, S., A. Nikula, P. Helle, and H. Lindén. 2000. Landscape fragmentation and forest composition effects on grouse breeding success in boreal forests. Ecology 81:1985–1997.

Lahti, D. C. 2001. The "edge effect on nest predation" hypothesis after twenty years. Biological Conservation 99:365–374.

Lindenmayer, D. B., and J. Fischer. 2007. Tackling the habitat fragmentation panchreston. Trends in Ecology and Evolution 22:127–132.

Macreadie, P. I., J. S. Hindell, M. J. Keough, G. P. Jenkins, and R. M. Connolly. 2010. Resource distribution influences positive edge effects in a seagrass fish. Ecology 91:2013–2021.

Martin, A. E., and L. Fahrig. 2016. Reconciling contradictory relationships between mobility and extinction risk in human-altered landscapes. Functional Ecology 30:1558–1567.

Martin, A. E., S. L. Graham, M. Henry, E. Pervin, and L. Fahrig. 2018. Flying insect abundance declines with increasing road traffic. Insect Conservation and Diversity 11:608–613.

Matthies, D., I. Bräuer, W. Maibom, and T. Tscharntke. 2004. Population size and the risk of local extinction: empirical evidence from rare plants. Oikos 105:481–488.

Matthysen, E., and D. Currie. 1996. Habitat fragmentation reduces disperser success in juvenile nuthatches Sitta europaea: evidence from patterns of territory establishment. Ecography 19:67–72.

McGarigal, K., and B. J. Marks. 1995. FRAGSTATS: spatial pattern analysis program for quantifying landscape structure. US Department of Agriculture, Pacific Northwest Research Station, Dolores, Colorado, USA.

Novacek, M. J., and E. E. Cleland. 2001. The current biodiversity extinction event: scenarios for mitigation and recovery. Proceedings of the National Academy of Sciences of the United States of America 98:5466–5470.

Öckinger, E., K.-O. Bergman, M. Franzén, T. Kadlec, J. Krauss, M. Kuussaari, J. Pöyry, H. G. Smith, I. Steffan-Dewenter, and R. Bommarco. 2012. The landscape matrix modifies the effect of habitat fragmentation in grassland butterflies. Landscape Ecology 27:121–131.

Parker, M., and R. Mac Nally. 2002. Habitat loss and the habitat fragmentation threshold: an experimental evaluation of impacts on richness and total abundances using grassland invertebrates. Biological Conservation 105:217–229.

Pimm, S. L., C. N. Jenkins, R. Abell, T. M. Brooks, J. L. Gittleman, L. N. Joppa, P. H. Raven, C. M. Roberts, and J. O. Sexton. 2014. The biodiversity of species and their rates of extinction, distribution, and protection. Science 344(6187):1246752.

Prevedello, J. A., and M. V. Vieira. 2010. Does the type of matrix matter? A quantitative review of the evidence. Biodiversity and Conservation 19:1205–1223.

Quesnelle, P. E., K. E. Lindsay, and L. Fahrig. 2015. Relative effects of landscape-scale wetland amount and landscape matrix quality on wetland vertebrates: a meta-analysis. Ecological Applications 25:812–825.

Rahmstorf, S. 2007. A semi-empirical approach to projecting future sea-level rise. Science 315:368–370.

Ribeiro, M. C., J. P. Metzger, A. C. Martensen, F. J. Ponzoni, and M. M. Hirota. 2009. The Brazilian Atlantic Forest: how much is left, and how is the remaining forest distributed? Implications for conservation. Biological Conservation 142:1141–1153.

Riitters, K. H., and J. D. Wickham. 2003. How far to the nearest road? Frontiers in Ecology and the Environment 1:125–129.

Samson, F., and F. Knopf. 1994. Prairie conservation in North America. BioScience 44:418–421.

Scheffer, M., G. J. van Geest, K. Zimmer, E. Jeppesen, M. Søndergaard, M. G. Butler, M. A. Hanson, S. Declerck, and L. De Meester. 2006. Small habitat size and isolation can promote species richness: second-order effects on biodiversity in shallow lakes and ponds. Oikos 112:227–231.

Schloss, C. A., T. A. Nuñez, and J. J. Lawler. 2012. Dispersal will limit ability of mammals to track climate change in the Western Hemisphere. Proceedings of the National Academy of Sciences of the United States of America 109:8606–8611.

Seto, K. C., M. Fragkias, B. Güneralp, and M. K. Reilly. 2011. A meta-analysis of global urban land expansion. PLOS One 6:e23777.

Soulard, C. E., W. Acevedo, R. F. Auch, T. L. Sohl, M. A. Drummond, B. M. Sleeter, D. G. Sorenson, S. Kambly, T. S. Wilson, J. L. Taylor, et al. 2014. Land cover trends dataset, 1973–2000 [data set]. US Geological Survey Data Series 844, Reston, Virginia, USA.

Swift, T. L., and S. J. Hannon. 2010. Critical thresholds associated with habitat loss: a review of the concepts, evidence, and applications. Biological Reviews 85:35–53.

Tilman, D., J. Fargione, B. Wolff, C. D'Antonio, A. Dobson, R. Howarth, D. Schindler, W. H. Schlesinger, D. Simberloff, and D. Swackhamer. 2001. Forecasting agriculturally driven environmental change. Science 292:281–284.

Truett, J. C., M. Phillips, K. Kunkel, and R. Miller. 2001. Managing bison to restore biodiversity. Great Plains Research 11:123–144.

Trzcinski, M. K., L. Fahrig, and G. Merriam. 1999. Independent effects of forest cover and fragmentation on the distribution of forest breeding birds. Ecological Applications 9:586–593.

Tuck, S. L., C. Winqvist, F. Mota, J. Ahnström, L. A. Turnbull, and J. Bengtsson. 2014. Land-use intensity and the effects of organic farming on biodiversity: a hierarchical meta-analysis. Journal of Applied Ecology 51:746–755.

Vetter, D., M. M. Hansbauer, Z. Végvári, and I. Storch. 2011. Predictors of forest fragmentation sensitivity in Neotropical vertebrates: a quantitative review. Ecography 34:1–8.

Virkkala, R., R. K. Heikkinen, N. Leikola, and M. Luoto. 2008. Projected large-scale range reductions of northern-boreal land bird species due to climate change. Biological Conservation 141:1343–1353.

Watson, J. E. M., R. J. Whittaker, and T. P. Dawson. 2004. Habitat structure and proximity to forest edge affect the abundance and distribution of forest-dependent birds in tropical coastal forests of southeastern Madagascar. Biological Conservation 120:311–327.

Wilcove, D. S., C. H. McLellan, and A. P. Dobson. 1986. Habitat fragmentation in the temperate zone. Pages 237–256 in M. E. Soulé, editor. Conservation biology: the science of scarcity and diversity. Sinauer Associates, Sunderland, Massachusetts, USA.

Wirsenius, S., C. Azar, and G. Berndes. 2010. How much land is needed for global food production under scenarios of dietary changes and livestock productivity increases in 2030? Agricultural Systems 103:621–638.

With, K. A., and D. M. Pavuk. 2012. Direct versus indirect effects of habitat fragmentation on community patterns in experimental landscapes. Oecologia 170:517–528.

Wretenberg, J., Å. Lindström, S. Svensson, and T. Pärt. 2007. Linking agricultural policies to population trends of Swedish farmland birds in different agricultural regions. Journal of Applied Ecology 44:933–941.

Zuckerberg, B., and W. F. Porter. 2010. Thresholds in the long-term responses of breeding birds to forest cover and fragmentation. Biological Conservation 143:952–962.

8

JACQUELINE L. FRAIR AND
GUILLAUME BASTILLE-
ROUSSEAU

Data Collection and Quantitative Considerations for Studying Pattern–Process Relationships on Landscapes

Introduction

Over the past few decades, there have been advances in the technologies available to track wildlife and their habitats, providing ever-finer observations over ever-larger regions of space. Worldwide, geo-referenced data representing various aspects of the Earth's surface and climate (i.e., landscape data) are produced and made publicly available by government agencies, academic, and private sources. Wildlife biologists readily exploit such landscape data to explore animal–environment relationships, often with the goal of inferring population responses to altered landscape conditions such as through habitat management or changing land use patterns (Gallant 2009). Landscape data enable spatially explicit models that can be used to locate favorable habitat (McComb et al. 2002, Niemuth 2004, Nielsen et al. 2006, Myatt and Krementz 2007); evaluate the proximity and potential connectivity of habitat (Kramer-Schadt et al. 2004, Courbin et al. 2014); assess the vulnerability of wildlife populations to disturbance, pollution, or climate change (Davidson 2004, Russell et al. 2009, Virkkala et al. 2013, Parent et al. 2016); relate landscape characteristics to wildlife diversity (Joly and Myers 2001, Gottschalk et al. 2005);

and evaluate land use alternatives that maintain desired habitat conditions (Marzluff et al. 2002, Frair et al. 2008, Beguin et al. 2015).

Increasing availability of data does not necessarily translate into greater knowledge of animal–environment relationships (Hebblewhite and Haydon 2010). Arguably, growing volumes of data, the widening range of spatiotemporal scales being investigated, and recent analytical advances have pushed the wildlife ecology field to greater statistical rigor, with contemporary models increasingly able to provide unbiased and precise estimates from typically correlated, structured, and nested forms of landscape and wildlife data. Nevertheless, gaining reliable and meaningful knowledge of how wildlife will respond to future land use change and associated stressors, and how managers might mitigate the negative effects of those stressors on wildlife populations, remains a fundamental challenge (Turner and Gardner 2015). This challenge is influenced in large part by real-world difficulties in achieving experiments with effective controls and replication in field studies, difficulties that likewise have plagued the field of landscape ecology.

Landscape ecologists have increasingly used models (Fig. 8.1) to gain reliable insights into the effects

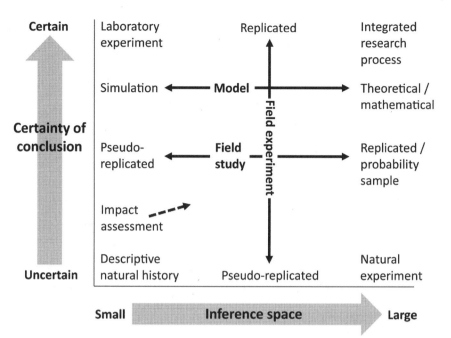

Figure 8.1. The potential for alternative study designs to produce conclusions with high certainty (i.e., few alternative hypotheses) and widespread applicability (i.e., inferences applicable to a diversity of populations). Reproduced from Garton et al. (2005:fig. 2).

of landscape patterns on ecological processes and to identify reliable approaches for quantifying and comparing landscape patterns across space and time. Herein, we aim to make key insights from the field of landscape ecology more accessible to the practicing wildlife manager through a review of available sources of landscape data, methods, and tools used to quantify landscape pattern, and common ways in which landscape patterns are related to wildlife population processes. Throughout we provide examples from the wildlife literature and guidance for avoiding pitfalls that might lead to uncertain study conclusions and, by extension, ineffective conservation action. Although it may be "difficult to make predictions, especially about the future" (Steincke 1948:227), we believe that revisiting fundamental data and study design lessons from landscape ecology would go a long way toward helping managers better identify and maintain effective wildlife habitat in our ever-changing world.

Available Landscape Data

Previous chapters established that spatial patterns on the landscape are associated with habitat features that influence population processes (Perotto-Baldivieso, Chap. 4, this volume). Relationships are commonly sought between a population process (the response variable, e.g., wildlife density) and landscape covariates (the pattern variables, e.g., percentage forest cover). Landscape covariates typically include local (point- or cell-based) values for a landscape characteristic and measures integrated over some surrounding area. Landscape covariates measured at a specific location might thus include point estimates (e.g., elevation), characteristics of the patch of vegetation the site falls within (e.g., patch area), or the relationship between the site and other features of interest (e.g., proximity of forest edge). Comparable metrics summarized over the surrounding area might include terrain complexity, mean patch size, and edge density (McGarigal and Marks 1995, Turner and Gardner 2015). Metrics that quantify landscape patterns might be calculated from empirical field data (e.g., plot or transect data), but more commonly they are derived from publicly accessible data visualized in a geographic information system (GIS; Table 8.1). Geographic information system data are stored in vector (point, line, or polygon) or raster (cell-based) formats, and landscape pattern metrics may be derived from each. Before quantifying landscape pattern, GIS data might be visualized, clipped (i.e., limited) to a specific study area, edited (e.g., reclassified to a simpler set of cover classes), and analyzed using a variety of commercial or open source software (GISGeography 2019).

Importantly, each data source has errors stemming from the coverage, resolution, and positional accuracy of the data; age of the data; data representation (i.e., crenulated raster edges versus smooth linear edges); artificial boundaries (e.g., political, study area); misclassification of content; and data smoothing, scaling, and aggregation processes (Turner and Gardner 2015). Metadata (i.e., data about data) accompany landscape data to indicate the source and age of the original data and the processes used to manipulate those data into the final product. The use of a specific data set should be carefully considered and include field- or image-based assessment as appropriate. Large-area land use or landcover maps may take years to complete, increasing the likelihood that something will have changed in the landscape between the dates of image acquisition and publication of the final product (Gallant 2009). Moreover, classification systems like land use and landcover data schemes (Anderson et al. 1976) are powerful for organizing and summarizing expectations about the landscape but are also imperfect (Gallant 2009). When using a classified product, completing a confusion matrix (or other similar validation analyses; Diamond and Elliott, Chap. 6, this volume) can increase the certainty of study conclusions because it lays an important foundation for decisions to either smooth the data set or aggregate classes prior to calculating pattern metrics (Thogmartin et al. 2004).

Alternatively, when using a vector product, such as forest inventory polygons with stand-level variables (e.g., age, height, density, canopy cover), assessing the accuracy of each variable will help guide effective landscape characterizations derived from these data (Dussault et al. 2001). Importantly, subjective decisions made in the preprocessing of a land use and landcover data set exerts a nontrivial influence on resulting pattern metrics (Fig. 8.2) and, by extension, on conclusions regarding the relationship between landscape patterns and wildlife population processes.

Quantifying Landscape Pattern
Patch-Based Metrics for Categorical Landscape Data

Two general classes of pattern metrics exist: patch-based metrics derived from categorical data and spatial statistics derived from continuously distributed data. Patch-based metrics can further be divided into metrics of landscape composition (e.g., the percentage of area covered by specific land use or landcover classes) and configuration (e.g., patch size, perimeter-area relationships, edge density, contagion, connectivity; Fig. 8.3). Landscape composition and configuration metrics are most commonly derived from land use and landcover data that are stored in a raster format where each cell takes an integer value representing a single cover type from an explicit set of types. Landcover type is not synonymous with habitat (Lindenmayer et al. 2008). Habitat is generally considered to be the combination of resources and conditions that support occupancy of a site by the species, including their survival and reproduction (Hall et al. 1997). In contrast, landcover classes are determined by energy reflectance values for a given location as recorded by satellites and parsed into classes that can be meaningfully discriminated by image interpreters (Homer et al. 2015). For the purposes of classification, cells of a given cover type are presumed to be homogeneous

Table 8.1. Publicly Available Sources of Landscape Data Used to Characterize Spatial Patterns Affecting the Distribution and Dynamics of Wildlife Populations

Landscape Characteristic	Description	Data Source	Spatial Resolution and Extent	Temporal Resolution	Availability
Land use, landcover	Classes defined using the United Nations Land Cover Classification System	GlobCover[a]	300 m, global	Static	2005–2006, 2009–2010
	Classes primarily defined by the International Geosphere Biosphere Programme	Moderate Resolution Imaging Spectroradiometer (MODIS)[b]	500 m, global	Annually	2001–2018
	National Land Cover Database (NLCD)	NLCD[c]	30 m, USA	Intermittent	2001–2016 (intermittent)
Forest cover	Percentage tree cover	MODIS Vegetation Continuous Fields (MOD44B)[b]	250 m, global	Annually	2000–2017
		Landsat Global Land Cover Facility (GLCF)[d]	30 m, global		2000, 2005, 2010, 2015, 2019
Greenness	Two different measures available: Normalized Difference Vegetation Index (NDVI; values correlate with vegetation biomass) and Enhanced Vegetation Index (EVI; values are more sensitive than NDVI to canopy type and architecture)	National Oceanic and Atmospheric Administration-Advanced Very High Resolution Radiometer (NOAA-AVHRR)[e]	1 km, global	Daily	1989 to present
		Landsat Thematic Mapper (TM)[f]	1–30 m, global	16 days	1984 to present
		MODIS Terra or Aqua[b]	250 m to 1 km, global	Daily	2000 to present
		Satellite pour l'Observation de la Terre-Vegetation (SPOT-VGT)[g]	1 km, global	10 days	1998 to present
		Advanced Spaceborne Thermal Emission and Reflection Radiometer (ASTER)[h]	15–90 m, global	1–2 days	2000 to present
		Airborne Visible / Infrared Imaging Spectrometer (AVIRIS)[i]	1–12 m, global	Patchy	1992 to present
		Hyperion Earth-Observing One (EO-1)[j]	30 m, global	16 days	2003–2017
Wetness	Normalized Difference Water Index (NDWI): water content in vegetation canopies	Landsat TM, MODIS[k]	30–500 m, global	Variable	2000 to present
	Soil moisture: an integrated product based on satellite and ground observation networks	Global Land Data Assimilation System (GLDAS)[l]	0.25°, global	3 hours	2000 to 2010
Climate	Precipitation	Tropical Rainfall Measuring Mission (TRMM)[m]	0.25°, 50° N-S	3 hours	1998–2015
		Integrated Multi-Satellite Retrievals for Global Precipitation Measurement (GPM [IMERG])[n]	0.1°, global	3 hours	2014 to present

Table 8.1. continued

Landscape Characteristic	Description	Data Source	Spatial Resolution and Extent	Temporal Resolution	Availability
	Temperature, rainfall, snowfall, and related variables	GLDAS[i]	0.25°, global	3 hours	2000 to present
	Land surface temperature (LST)	MODIS-MYD11A2[l]	1 km, global	8 days	2002 to present
		ASTER[h]	90 m, global	1–2 days	2000 to present
Topography	Digital Elevation Model (DEM), from which terrain indices (e.g., slope, aspect, ruggedness) can be derived	ASTER[h]	30 m, global		
		Shuttle Radar Topography Mission (SRTM)[o]	30 m, global		2000
		Digital Elevation Model-3D Elevation Program (DEM-3DEP)[p]	1 m, USA		2017
Human disturbance	Human population density	Griddled Population of the World (GPW)[q]	1 km, global		2000, 2005, 2010, 2015, 2020
	Road network	Global Roads Open Access Data Set (gROADS)[r]	1:250,000, global		1980–2010

[a] European Space Agency (2009).

[b] US Geological Survey (2018).

[c] Multi-Resolution Land Characteristics Consortium (2016).

[d] National Aeronautics and Space Administration (2019a).

[e] US Geological Survey (2019a).

[f] US Geological Survey (2019b).

[g] Airbus (2019).

[h] National Aeronautics and Space Administration (2019b).

[i] National Aeronautics and Space Administration (2019c).

[j] US Geological Survey (2019c).

[k] Google Earth Engine (2019).

[l] National Aeronautics and Space Administration (2016).

[m] National Aeronautics and Space Administration (2015a).

[n] National Aeronautics and Space Administration (2019d).

[o] National Aeronautics and Space Administration (2000).

[p] US Geological Survey (2017).

[q] National Aeronautics and Space Administration (2015b).

[r] National Aeronautics and Space Administration (2010).

across the landscape. Yet most certainly those cells are not perceived as homogeneous to the animals that encounter them across the landscape because other biotic (e.g., grazing, predation risk) and abiotic (e.g., terrain, climate) influences also create spatial patterns to which animals likely respond (Bowyer et al. 1998, Brotons et al. 2005). For example, predation-sensitive patterns of foraging and aggregation have been observed in fish (Werner and Hall 1988), songbirds (Cresswell 2008), rodents (Newman et al. 1988, Bouskila 1995, Kotler 1997), lagomorphs (Arias-Del Razo et al. 2012), marsupials (Banks 2001), large herbivores (Creel et al. 2005, Iribarren and Kotler 2012, Kuijper et al. 2013), and carnivores (Mukherjee et al. 2009). As a result, multiple landscape characteristics are generally considered when relating landscape patterns to population processes.

Ecologists have described hundreds of patch-based statistics (Fig. 8.3), which are readily calculated using freely available software (Table 8.2). No single metric will be sufficient to capture all relevant aspects of landscape pattern, but one should also not calculate dozens of landscape metrics simply because it is easy to do so. Importantly, landscape composition, in particular the proportion of the landscape covered by a single cover type, influences the resulting range of values calculated for landscape configuration metrics (Turner and Gardner 2015). In fact, many configuration metrics include the proportion of the landscape covered by a given cover type as part of their equation (Turner and Gardner

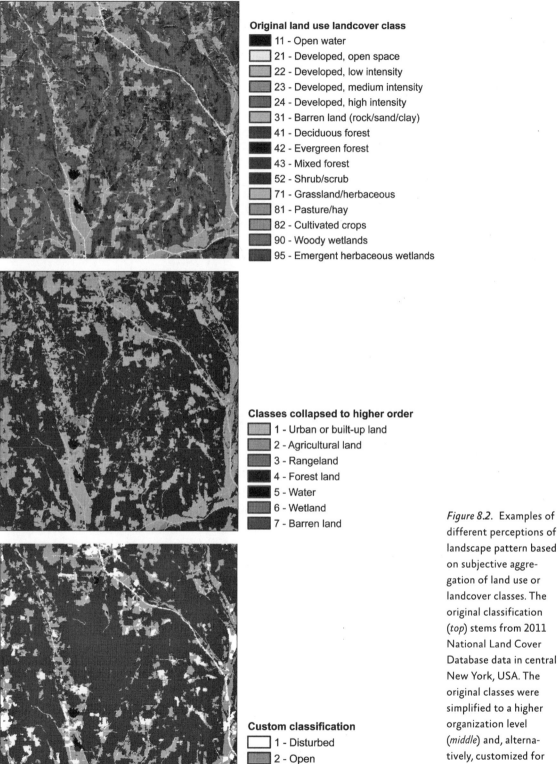

Original land use landcover class

- 11 - Open water
- 21 - Developed, open space
- 22 - Developed, low intensity
- 23 - Developed, medium intensity
- 24 - Developed, high intensity
- 31 - Barren land (rock/sand/clay)
- 41 - Deciduous forest
- 42 - Evergreen forest
- 43 - Mixed forest
- 52 - Shrub/scrub
- 71 - Grassland/herbaceous
- 81 - Pasture/hay
- 82 - Cultivated crops
- 90 - Woody wetlands
- 95 - Emergent herbaceous wetlands

Classes collapsed to higher order

- 1 - Urban or built-up land
- 2 - Agricultural land
- 3 - Rangeland
- 4 - Forest land
- 5 - Water
- 6 - Wetland
- 7 - Barren land

Custom classification

- 1 - Disturbed
- 2 - Open
- 3 - Forested
- 4 - Water

Figure 8.2. Examples of different perceptions of landscape pattern based on subjective aggregation of land use or landcover classes. The original classification (*top*) stems from 2011 National Land Cover Database data in central New York, USA. The original classes were simplified to a higher organization level (*middle*) and, alternatively, customized for a hypothetical wildlife application (*bottom*).

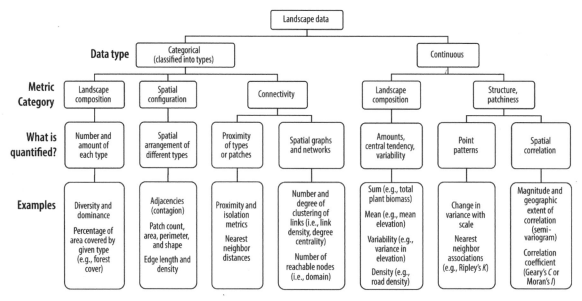

Figure 8.3. General classification of landscape pattern metrics useful in the study and management of wildlife habitat.

2015). For this reason, most patch-based metrics are correlated, rendering their information redundant and statistical analyses of them complicated. Multivariate, state-space, and other modeling approaches have been undertaken to identify uncorrelated sets of metrics that capture unique elements of the landscape (Li and Reynolds 1995, McGarigal and Marks 1995, Riitters et al. 1995). Cushman et al. (2008) narrowed the range of configuration metrics one might consider to 14 that displayed highly universal and consistent behavior across >150 different landscapes. Resulting metrics were divided between class-level (type) metrics, such as nearest neighbor distance and patch shape complexity, and metrics integrated across cover types at the landscape level, such as diversity, interspersion, juxtaposition, and edge contrast (Cushman et al. 2008). In practice, 14 metrics is still a lot to consider, especially when more than one scale may be considered for each metric and nonlinear responses or interactions among variables are expected, which quickly produces an unwieldy number of candidate models under consideration. Nevertheless, knowledge of uncorrelated sets of metrics will help winnow the list of suitable metrics to be considered in a given study.

Fundamental to all patch-based metrics is the definition of a *patch*, which implies an area that appears different from its surroundings (Forman and Godron 1986) and might be formally derived from a categorical map as "a contiguous group of cells of the same mapped category" (Turner et al. 2001:106), where contiguity is typically defined using an 8-cell neighborhood (Turner and Gardner 2015). While this structural classification of patches dominates the literature, a specific application may instead require a classification based on how the patch is expected to function for different species or processes (Li and Reynolds 1995, Fahrig et al. 2011, Kupfer 2012). For example, the degree to which the landscape facilitates or impedes the movement of organisms (i.e., connectivity) is often studied structurally and functionally. Structural connectivity is commonly investigated on the basis of the proximity of cells or patches and the matrix (inter-patch) permeability (Metzger and Décamps 1997; Fig. 8.3). Structural connectivity can be quantified at the cell, patch, or landscape level (Tischendorf and Fahrig 2000, Calabrese and Fagan 2004, Fall et al. 2007, Kindlmann and Burel 2008), typically assuming a specific movement capacity for the species of inter-

Table 8.2. Sample of Software Tools and Capabilities for Quantifying Landscape Pattern

Software	Description
FRAGSTATS,[a] Patch Analyst (A),[b] GuidosToolbox,[c] landscapemetrics (R)[d]	Calculates attributes of patches (e.g., patch size, perimeter) and landscapes (e.g., patch, edge density) from categorical data.
ArcGIS,[e] QGIS,[f] gstat (R),[g] spatialEco (R)[h]	Geostatistical tools enable consideration of autocorrelation in continuous data using variograms, spatial interpolation, and kriging. Neighborhood statistics summarize the landscape context around a cell within a defined search radius.
spatsat (R)[i]	Spatial point pattern analysis.
Circuitscape,[j] Linkage Mapper,[k] and Gnarly Landscape Utilities[l]; Pathmatrix,[m] gdistance (R)[n]	Tools for quantifying landscape connectivity (e.g., least cost paths).
Conefor,[o] Graphab,[p] igraph (R),[q] moveNT[r]	Evaluates importance of habitat "nodes" and "links" using spatial graphs.

Note: Codes R and A distinguish packages within the R statistical program and extensions within ArcGIS, respectively, from standalone programs.

[a]McGarigal et al. (2012).
[b]Rempel (2016).
[c]Vogt and Riitters (2017).
[d]Hesselbarth et al. (2019).
[e]Environmental Systems Research Institute (2020).
[f]QGIS Development Team (2020).
[g]Pebesma and Graeler (2020).
[h]Evans (2020).
[i]Baddeley and Turner (2005).

[j]McRae et al. (2013a).
[k]McRae et al. (2013b).
[l]Shirk and McRae (2013).
[m]Ray (2005).
[n]Van Etten (2017).
[o]Saura and Torné (2009).
[p]Foltéte et al. (2012).
[q]Csárdi (2019).
[r]Bastille-Rousseau et al. (2018).

est. Alternatively, functional connectivity explores actual wildlife movements or gene flow (Uezu et al. 2005). While reconciling potential (structural) versus actual (functional) landscape connectivity remains an active frontier of research in landscape ecology, in practice there are numerous software tools available to facilitate structural (Table 8.2) and functional (Table 8.3) assessments.

Network analyses, or spatial graphs, provide other means of studying habitat connectivity. Stemming from graph theory, a network is represented by a series of nodes (i.e., points, typically representing the location of a habitat patch) connected by a series of links among the nodes (Urban and Keitt 2001, Fall et al. 2007, Minor and Urban 2008). Various metrics quantify the relative influence of each node and link on the entire network (Rayfield et al. 2011). Applications of network theory have helped identify critical areas for protection of endangered species (O'Brien et al. 2006), unanticipated consequences of land management activities on predator-prey in-

teractions (Courbin et al. 2014), and core use areas important for connectivity (Bastille-Rousseau et al. 2018). Being relatively new, the statistical and ecological limitations of network-based assessments of connectivity are not yet as well understood as more traditional assessments, forming another active area of research in landscape ecology (Turner and Gardner 2015, Jacoby and Freeman 2016).

It is worth restating that the number and type of categories in the data set used to derive patch-based pattern metrics exert a strong influence on the numerical results of any patch-based pattern analysis (Fig. 8.2) and, by extension, on the conclusions drawn regarding the relationship between landscape pattern and wildlife population processes. Below, we discuss specific effects that scale decisions have on landscape pattern metrics, but it remains incumbent upon the analyst to explore, understand, and report the implications of subjective decisions in the preprocessing of land use, landcover, and other landscape data on their study conclusions.

Table 8.3. **Common Analyses and Software Tools for Quantifying Wildlife Space Use Patterns and Relating Landscape Covariates to Wildlife Population Processes**

Population Process	Software	Description
Explore species-environment associations from presence-only data	BIOMOD,[a] Maxent,[b] maxLike (R)[c]	Modeling and visualization tools relating landscape variables to species occurrence data.
Evaluate resource selection by animals by comparing sites used by animals to unused or available sites	adehabitatHS (R),[d] ResourceSelection (R),[e] TwoStepCLogit (R),[f] amt[g]	Tools for resource selection functions, resource selection probability functions, and step selection functions, typically involving animal tracking data.
Estimate proportion of area occupied (site occupancy probability) from repeat surveys of a site	Presence,[h] unmarked (R)[i]	Accounts for probability of species detection; landscape covariates inform both detection and site occupancy probabilities. With follow-up surveys, it can provide colonization and extinction probabilities informed by landscape pattern.
Estimate movement and space use patterns from a time series of relocation data on individual animals	adehabitatLT (R),[j] BBMM (R),[k] gdistance (R),[l] lsmnsd (R),[m] moveHMM,[n] moveNT[o]	Tools to extract paths from discrete location data, calculate movement parameters, conduct movement analyses (e.g., First Passage Time, Network analysis), and simulate movement through heterogeneous landscapes using least cost paths, Brownian bridges, and constrained random walks.
Quantify how landscape pattern influences animal density	Mark,[p] secr (R),[q] DENSITY,[r] SPACECAP[s]	Capture-recapture models estimate detection probability and animal abundance. Robust models yield estimates of apparent survival and population growth. Spatial capture-recapture models yield a spatially explicit estimate of animal density.
	unmarked (R)[i]	Fits n-mixture models to animal counts to estimate detection probability and animal abundance.
	Distance[t]	Estimates detection probability and animal abundance from counts of animals in point or transect surveys.
Assess influence of landscape on gene flow	Geneland (R),[u] Genetic Landscapes GIS Toolbox (A),[v] SPAGeDI,[w] MEMGENE[x]	A suite of spatial analysis tools for genetic data.

Note: Codes R and A distinguish packages within the R statistical program and extensions within ArcGIS, respectively, from standalone programs.

[a]Thuiller et al. (2009).

[b]Phillips et al. (2004).

[c]Royle et al. (2012).

[d]Calenge and Basille (2020).

[e]Lele et al. (2019).

[f]Radu et al. (2016).

[g]Signer et al. (2020).

[h]Hines (2006).

[i]Fiske and Chandler (2011).

[j]Calenge et al. (2020).

[k]Nielson et al. (2013).

[l]Van Etten (2017).

[m]Bastille-Rousseau (2016).

[n]Michelot et al. (2019).

[o]Bastille-Rousseau et al. (2018).

[p]White and Burnham (1999).

[q]Efford (2020).

[r]Efford et al. (2004).

[s]Gopalaswamy et al. (2012).

[t]Thomas et al. (2010).

[u]Guillot et al. (2005).

[v]Vandergast et al. (2011).

[w]Hardy and Vekemans (2002).

[x]Galpern et al. (2014).

Spatial Statistics for Continuously Distributed Landscape Data

In contrast to landcover classes, many landscape characteristics that influence wildlife are recorded as continuously distributed values (e.g., percentage slope, forage biomass) rather than discrete categories. Continuous variables are commonly related to wildlife population processes by measures of sums, central tendency, and variability (Fig. 8.3) calculated within defined areas (i.e., buffers) around locations where animals occur. But landscape metrics that specifically assess spatial pattern within continuously distributed data also exist and are useful to the wildlife manager. McGarigal et al. (2009) suggested a suite of metrics based on the texture of the landscape surface (e.g., surface roughness and shape of the surface height distribution) that can be calculated in FRAGSTATS but have not yet been widely applied. More commonly, point patterns and spatial statistics are used to quantify the change in variance with scale and the spatial structure of correlation, respectively, in continuously distributed landscape data (Fig. 8.3). These approaches depend on point-based records for some property of interest that is spatially distributed across the landscape (Turner and Gardner 2015). Spatial statistics quantify spatial dependence among sample points in the measured landscape characteristic, or more specifically the degree to which the value of the landscape characteristic at one location is correlated with the value at another location. If there is spatial dependence, knowing the value of that characteristic at one location enables one to infer its value at nearby locations (via interpolation or kriging). Spatial statistics quantify the overall amount of variance in the data, the proportion of that variance that is spatially dependent, and the distances (or scales) over which that spatial dependence extends (Turner and Gardner 2015). For example, Boyce et al. (2003) documented in Yellowstone National Park, USA, that abiotic covariates (e.g., elevation, terrain ruggedness) were correlated over much larger scales than biotic co-variates (e.g., vegetation type, richness), with <50% of the total park-wide variability in elevation being sampled at the scale of home ranges for elk (*Cervus canadensis*). A practical implication of this observation is that spatial autocorrelation in abiotic features may restrict our ability to accurately quantify the effect of those features on wildlife population processes when values are sampled within spatial scales finer than the full range of correlation. This observation underscores the importance of considering more than one scale when assessing animal-environment relationships.

An important assumption with spatial statistics is stationarity, which assumes the mean and variance for the measured property do not change over space. This assumption is likely violated when faced with an underlying gradient, when limited sampling extents fail to capture changes in variables over relevant scales, or given marked directionality (i.e., anisotropy; Turner and Gardner 2015). Topographic relief can impose strong gradients and directionality in spatially structured data, violating the assumption of stationarity. When a lack of stationarity is detected, the trend may be removed from the data prior to analysis or accounted for directly within the analysis itself (Turner and Gardner 2015). Tools to calculate spatial statistics are commonly available in GIS software packages (Table 8.2).

Faced with so many possible pattern metrics, how should one choose among them? Landscape ecologists say to be selective and strategic, and they provide the following advice: start with a well-conceived question or rationale for the analysis; carefully consider what qualities of spatial pattern are relevant to your question; ensure that available landscape data are appropriate for the analysis in terms of resolution, coverage, classification scheme, and accuracy; and choose an uncorrelated set of metrics that capture different aspects of relevant spatial patterns (Riitters et al. 1995, Cushman et al. 2008, Turner and Gardner 2015). Turner and Gardner (2015) additionally advise to be wary of complicated metrics, and that metrics quantified for individual cover types will be of greater use for management than metrics integrated across the landscape as a whole. For wildlife practitioners, manageable aspects of landscape pattern will likely be of greatest use, such as metrics focused on the amount of specific cover types, patch size and shape, cover type complementation, edge contrast, and connectivity (Lindenmayer et al. 2008). Importantly, for whatever metrics are chosen, sampled values should be distributed over the full range of potential values to guard against biased inference (Turner and Gardner 2015).

Scale Matters

Critically, landscape pattern metrics are scale dependent and may change unpredictably with changes in either grain (i.e., resolution, minimum mapping area) or extent (i.e., total area considered; Turner et al. 1989, Moody and Woodcock 1995, Wickham and Riitters 1995, Wu 2004, Ostapowicz et al. 2008). Generally, with patch-based metrics, apparent landscape heterogeneity decreases as grain size increases because rare types diminish and patch boundaries become less detailed (Turner and Gardner 2015; Fig. 8.4). Grain size is equally problematic for spatial statistics, where grain refers to the minimum distance between sample points (Fig. 8.4). Using spatial statistics to detect the characteristic scale of pattern in continuous data requires that the grain size be equal to or smaller than the range over which values are spatially dependent (Turner and Gardner 2015), as was apparent in the assessment of terrain conditions across Yellowstone National Park (Boyce et al. 2003). One technical consideration when calculating spatial statistics is that software may automatically re-bin distance classes for statistical convenience, modifying the effective grain and, in so doing, potentially obscure the detection of pattern (Turner and Gardner 2015). For this reason, and in point of fact wherever subjective decisions enter an analysis, one should test the effect of subjective decisions on the resulting inferences.

Comparing pattern metrics across different study areas, among different animals within the same study area, or even among relocations of the same animal over time requires that landscape data be sampled using a comparable grain and extent. Should grain vary among areas, finer-grained data are typically resampled to match coarser-grained data prior to statistical analysis. Should extent vary among areas, landscape metrics may be normalized by area, provided the extent has not induced serious bias (Turner and Gardner 2015). Bias can be introduced through the artificial truncation of patches and thus would tend to be greater for smaller extents compared to larger ones (Fig. 8.4). Whenever possible, strive to standardize the extent across your sample locations. For example, to understand landscape influences of home range size in elk, Anderson et al. (2005) quantified landscape patterns within concentric circles centered on elk home ranges, which maintained a constant extent across animals whose home ranges varied in size by several orders of magnitude. Spatial statistics are similarly affected by truncation of the range of distances compared (Fig. 8.4), with points lying near boundaries being inordinately influenced by truncation effects. To guard against truncation effects in spatial statistics, sampling extent should be more than twice the maximum distance of interest.

How should one choose the relevant scale(s) over which to measure pattern metrics for a specific ecological process? Addicott et al. (1987) provide a framework for identifying appropriate scales based on the ecological process of interest, the timescale appropriate to the process, and activity of the organism during each time period (Ball 2002). This framework has been applied to identify characteristic spatiotemporal scales of foraging behavior by large ungulates that can help guide study designs (Senft et al. 1987, Bailey et al. 1996). In defining the spatiotemporal nature of animal activities occurring at each level, this framework guides choices of relevant grain and extent for sampling the ecological process of interest. In terms of landscape data to be related to that ecological process, the choice of scale typically equates to choice of extent(s) because grain

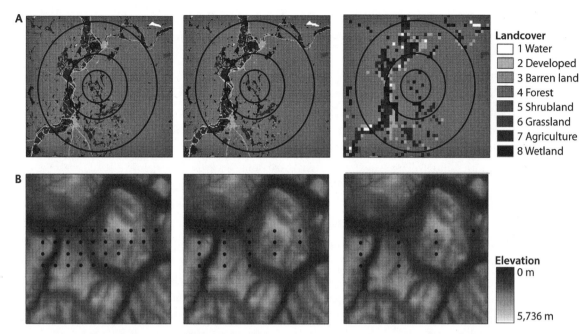

Landcover
☐ 1 Water
☐ 2 Developed
☐ 3 Barren land
☐ 4 Forest
■ 5 Shrubland
■ 6 Grassland
■ 7 Agriculture
■ 8 Wetland

Elevation
0 m

5,736 m

Figure 8.4. Visualization of how changes in grain and sampling extent might influence the quantification of landscape pattern. For categorical data (*A*), concentric circles delineate alternative extents (2.5, 5, and 10 km) within which pattern metrics might be calculated. Grain sizes increase from left to right by 4 and 16 orders of magnitude from the original 30-m resolution. For continuous data (*B*), sampling distances (lags) increase by 2 and 4 orders of magnitude from left to right.

is determined fundamentally by the available landscape data and rarely resampled to larger resolutions. Although coarser-grained sensors tend to have larger fields of view (larger extents) than finer-grained sensors, today even relatively fine-grained sensors (e.g., 30-m resolution land use/landcover data) provide seamless coverage over entire continents (Table 8.1). As a result, managers tend to choose the finest-resolution landscape data, keep the resolution fixed, and arbitrarily modify the extent over which pattern metrics are quantified. With this in mind, focal extents may be guided by natural history observations such as perceptual distances, home range size, or movement capabilities for either the species of interest or one with similar habitat needs (Lima and Zollner 1996, Thornton and Fletcher 2013). For example, when the process under study involves foraging movements by large herbivores, extents might be chosen on the basis of average movement distances over the course of hours, weeks, or months

(Johnson et al. 2002, Frair et al. 2007, Decesare et al. 2012) to correspond with the different spatiotemporal domains over which foraging decisions are made (Senft et al. 1987, Bailey et al. 1996). Additionally, spatial statistics such as first passage time (Fauchald and Tveraa 2003), fractal dimension (Nams 2005), and coarse-graining methods (Mayor et al. 2007), among others, may be used to identify the dominant or characteristic scale(s) at which wildlife respond to their environment.

Once a focal scale (or extent) is chosen for the ecological process of interest, at least two more scales should be considered: a broader scale to understand the landscape context that constrains outcomes at the focal level and a finer scale to understand mechanisms leading to the patterns observed at the focal level (Turner and Gardner 2015). In practice, wildlife studies routinely involve variables quantified within many different spatial extents, in part because of uncertainty regarding what the right scale is, but also to account for the multiscale nature by which wildlife perceive and respond to landscape pattern (Fuhlendorf et al. 2002, Holland et al. 2004, Johnson et al. 2004a, Frair et al. 2005, Boscolo and Metzger 2009). A review of studies of the distribution and abundance of 954 avian, mammalian, and herpetofaunal species showed the majority to be influenced by at least one level of landscape structure, with multilevel investigations often necessary to understand species responses in heterogeneous landscapes (Thornton et al. 2011). For illustration, a foraging animal might choose its foraging location based simultaneously on the abundance of vegetation biomass within their immediate vicinity (≤ 100 m^2) and the risk of encountering predators in the larger landscape ($\geq 1,000$ m^2). Moreover, context-sensitive responses are common in which wildlife responses to one landscape characteristic (e.g., forage abundance) might differ depending on the broader-scale distribution and abundance of that or another landscape characteristic such as human disturbance (Mysterud and Ims 1998, Matthiopoulos et al. 2011, Moreau et al. 2012).

Models that include environmental variables measured over more than one spatial scale have been reported to have greater predictive power than models that include variables measured across a common scale only (Cushman and McGarigal 2004, Graf et al. 2005, Meyer and Thuiller 2006, Boscolo and Metzger 2009, Doherty et al. 2010). Identifying the optimal scale (or extent) for each landscape metric might begin by quantifying metrics within alternative extents, such as within concentric buffers around animal locations (Fig. 8.4). Statistical models are then fit to relate landscape metrics to the population process variable (Table 8.3). Information-theoretic and other model selection approaches may be used to first identify the best scale for a given landscape metric (e.g., the model having the lowest value of Akaike's Information Criterion; Burnham and Anderson 2002) and second to identify the most parsimonious combination of landscape metrics, each measured at their best scale for the process of

interest. Similar approaches have been applied to understand the multiscale nature of resource selection (Leblond et al. 2011, Decesare et al. 2012), site occupancy (Elith and Leathwick 2009), animal density (Borchers and Efford 2008, Fuller et al. 2016), and survival (Johnson et al. 2004b, Frair et al. 2007, Smith et al. 2014; Table 8.3).

Following a review of multiscale model comparisons, Jackson and Fahrig (2012) warned that identifying as optimal either the smallest or largest extents tested may indicate that too few or too narrow a range of plausible extents were considered. Recently, Chandler and Hepinstall-Cymerman (2016) suggested a means of circumventing the need to predefine scales of analysis by integrating spatial statistics with maximum likelihood methods to simultaneously estimate the characteristic scale of influence of different landscape characteristics on animal density. Yet most wildlife applications still involve definition of sampling scales a priori, and selecting the appropriate scale for a given analysis remains a nontrivial endeavor. Bissonette (2017) argued that the problem of selecting an optimal sampling extent, especially in the case of resource selection analyses where the extent defines what is available for the animal to use, is largely intractable owing to the modifiable area unit problem (Openshaw 1983). Whether or not identification of an optimal scale lies beyond reach, understanding the sensitivity of inferences to the choice of scale remains a critically important aspect of wildlife studies. Ultimately, the value of any model for guiding management actions should be gauged by validation using an independent set of observations for the ecological process of interest.

In addition to consideration of spatial scale, comparing multiple temporal scales is also important for understanding wildlife responses to landscape pattern (Cushman et al. 2005, Boyce 2006, Bastille-Rousseau et al. 2017). Mismatches between fine-grained animal responses to dynamic environmental conditions and the comparatively coarse-grained, static data used to represent landscape patterns (Hebblewhite and Haydon 2010; Table 8.1), in addition to the study design issues described below, can mask the effects of landscape pattern on wildlife population processes. A priori delineation of the spatiotemporal scale at which the ecological process of interest is operating will help guide the choice of landscape data most relevant to that process (Boyce et al. 2003, Johnson et al. 2004a). Moreover, augmentation of publicly available landscape data with field-based data will provide greater insight into the landscape, patch, and within-patch factors influencing wildlife populations on daily, seasonal, and interannual time scales (Hebblewhite and Haydon 2010, Thornton et al. 2011).

Additional Design Considerations

Proper experimentation, with randomized, replicated, and controlled treatments, plays as important a role in wildlife ecology and management as it does in landscape ecology (Fig. 8.1). Nevertheless, the sampling of large spatial extents, application of landscape-level treatments, and tracking of long-term responses of wildlife populations to landscape changes impose serious constraints on study designs. As a result, it is perhaps more appropriate to describe wildlife-landscape studies as correlations of wildlife distributions or behavior (i.e., population processes) to observed landscape patterns, while attempting to statistically control for design imbalances and uncontrolled heterogeneity that may mask or confound responses to landscape patterns (Cunningham and Lindenmayer 2017). In this regard, common pitfalls of landscape studies to be aware of include pseudo-replication and autocorrelation, multicollinearity, inappropriate sampling extent leading to truncation of the range of variation in sampled metrics, and broadscale spatial gradients that can confound or bias severely inferences.

Pseudo-replication and autocorrelation stem from lack of consideration of spatial structure in either the landscape or wildlife data, especially when samples reflect repeated observations on a given animal (e.g., telemetry data) or clustered sampling of

sites (Boyce 2006). Failing to account for autocorrelation will artificially inflate statistical significance and, by extension, how certain we are regarding the effects of landscape pattern on wildlife populations (Legendre and Fortin 1989, Legendre 1993, Gillies et al. 2006). When measuring landscape pattern within buffers around animal locations, as is common in wildlife studies, sampling specifically to eliminate spatial overlap among buffers is not sufficient to ensure statistical independence among sample locations (Zuckerberg et al. 2012). Spatial statistics might be used a priori to define a sampling interval over which animal locations or plots become independent. But remember that spatial dependence is not a problem per se, as it can prove to be informative to ecological processes such as identifying and managing for nested scales of variability (Nielsen et al. 2002, Ficetola and De Bernardi 2004, Frair et al. 2005, Moore and Swihart 2005). Moreover, when buffering animal locations, autocorrelation does not necessarily increase given an increasing amount of overlap among neighboring buffers (Zuckerberg et al. 2012). Detection of autocorrelation in the residuals of a model requires adjusting the model to account for spatial dependence (Legendre 1993, Fortin and Payette 2002, Dormann et al. 2007, Ishihama et al. 2010), such as through the inclusion of hierarchical, nested, or random effects (Guisan and Thuiller 2005, Gillies et al. 2006, Fieberg et al. 2010, Schielzeth and Nakagawa 2013) or fitting an alternative model such as geographically weighted regression (Fotheringham et al. 2002). As these statistical models become more commonly used, tools that facilitate their application in wildlife studies become more readily available (Table 8.3).

Multicollinearity in a statistical model occurs when two or more covariates are linearly related (Dormann et al. 2013). Multicollinearity causes unstable parameter estimates, inflated standard errors, and potentially biased inference regarding individual covariate effects (Dormann et al. 2013). As an extreme example, the inextricable nature by which changing land use simultaneously affects the amount and arrangement of wildlife habitat confounded understanding of the relative importance of habitat loss (reduced habitat amount) versus fragmentation (arrangement of remnant habitat patches) on wildlife for decades (Fahrig 1997, 2003, Ewers and Didham 2005). Although collinearity and correlation are independent statistical concepts, a high coefficient of correlation ($r > 0.7$) can reliably indicate collinear variables prior to fitting models (Dormann et al. 2013). In multiple regression models, however, collinearity also can arise from interaction terms and the inclusion of polynomial terms used to capture nonlinear responses to a landscape pattern metric. Centering and standardizing landscape variables prior to model fitting are simple and efficient means of reducing collinearity issues (Schielzeth 2010). An added benefit of centering and standardizing landscape variables is that doing so facilitates more efficient evaluation of the relative effect of different landscape pattern metrics on the response variable when metrics are measured at different scales or in different units. But there are ecologically reasonable situations where it is necessary to include covariates that are collinear in a model. Regression methods allowing collinear input variables include sequential regression (Graham 2003) and hierarchical, nested, or random effect models (Dormann et al. 2013, Schielzeth and Nakagawa 2013).

To avoid another common design pitfall when relating landscape patterns to population processes, it is important to sample a broad range of values for each landscape metric (Boyce 2006, Eigenbrod et al. 2011). Statistically detecting, and properly quantifying, ecological responses to a landscape pattern metric requires sampling broadly across the range of values that metric can take (Brennan et al. 2002). In practice, however, wildlife-landscape studies are rarely designed to capture the full variation in landscape patterns of interest and are thus often limited in the range of values that landscape pattern data might take. This could be particularly problematic in the case of non-monotonic responses, such as browsing animals preferentially using patches of

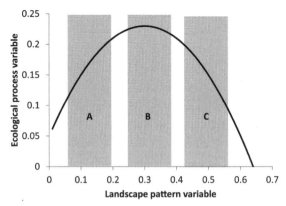

Figure 8.5. Graphic depicting three different samples (gray boxes) taken across the potential range of values for a hypothetical landscape pattern metric, leading to conflicting conclusions of a positive (A), null (B), or negative (C) relationship between the pattern metric and the ecological process of interest.

intermediate stem density in contrast to lower and higher values, where truncated sampling of the stem density metric might lead to contradictory conclusions regarding the effects of stem density on animal browsing patterns (Fig. 8.5). Sampling across the full range of values for a landscape covariate may not be plausible for every study, and certainly not for every relevant landscape covariate in every study. In this case, Eigenbrod et al. (2011) recommended that investigators sample the largest possible range of the variable within their study area, clearly indicate the conditions over which their inferences may be sound (i.e., caution against extrapolation to values of the variable beyond those sampled), and discuss the effects of their limited sampling of that variable on the inferences drawn. McGarigal and Cushman (2002) additionally recommend complimentary use of empirical and simulation modeling studies to improve our understanding of habitat fragmentation effects.

Summary

A profusion of remotely sensed data has become publicly available in recent decades, providing a rich foundation for quantifying landscape patterns and relating those patterns to wildlife population processes. Different data types, and how these data are manipulated, have a large influence on observed landscape patterns. Although the issues described herein may seem overwhelming, thoughtful study design, careful data management, and strategic choices of landscape pattern metrics are key to ensuring that reliable knowledge is achieved. At the outset of a study, landscape data should be critically evaluated, and validated as appropriate, to identify and correct for potential sources of error and bias prior to quantifying landscape patterns.

A suite of patch-based metrics (involving categorical data) and spatial statistics (involving continuously distributed data) are available to quantify the composition and configuration of landscape characteristics related to wildlife habitat, and a number of analytical tools are available to aid in that effort. The specific question, species, and system will influence the choice of landscape pattern metrics, and a non-correlated set of metrics that can be field manipulated will be of most utility for guiding habitat management. The scale (extent and resolution) within which pattern metrics are calculated will influence the results. Techniques exist to identify the characteristic scale of environmental covariates or the ecological process of interest, and these can be used to guide otherwise arbitrary scale decisions. Typically, more than one scale (or more specifically extent) should be considered to account for the multiscale nature in which wildlife respond to landscape patterns, with wildlife expected to respond to different landscape characteristics on different scales, and with the optimal scale for each landscape covariate identifiable using model selection techniques. Finally, study designs that take care to manage potential issues of pseudo-replication, autocorrelation, and multicollinearity (Fig. 8.1) are important when planning habitat management actions based on comparisons between landscape pattern metrics and wildlife population processes.

We leave the practitioner with the recollection that "essentially, all models are wrong, but some are useful" (Box and Draper 1987:424) and emphasize that proper study design, thoughtful inclusion and assessment of covariates, and careful model validation will help ensure effective conservation actions.

LITERATURE CITED

Addicott, J. F., J. M. Aho, M. F. Antolin, D. K. Padilla, J. S. Richardson, and D. A. Soluk. 1987. Ecological neighborhoods: scaling environmental patterns. Oikos 49:340.

Airbus. 2019. Satellite pour l'observation de le Terre (SPOT), vegetation data. http://www.intelligence-airbusds.com/optical-and-radar-data/#spot. Accessed 15 Aug 2019.

Anderson, D. P., J. D. Forester, M. G. Turner, J. L. Frair, E. H. Merrill, D. Fortin, J. S. Mao, and M. S. Boyce. 2005. Factors influencing female home range sizes in elk (*Cervus elaphus*) in North American landscapes. Landscape Ecology 20:257–271.

Anderson, J. R., E. E. Hardy, J. T. Roach, and R. E. Witmer. 1976. A land use and land cover classification system for use with remote sensor data. US Geological Survey Professional Paper 964, Washington, DC, USA.

Arias-Del Razo, I., L. Hernández, J. W. Laundré, and L. Velasco-Vázquez. 2012. The landscape of fear: habitat use by a predator (*Canis latrans*) and its main prey (*Lepus californicus* and *Sylvilagus audubonii*). Canadian Journal of Zoology 90:683–693.

Baddeley, A., and R. Turner. 2005. Spatstat: an R package for analyzing spatial point pattern analysis. Journal of Statistical Software 12:1–42.

Bailey, D. W., J. E. Gross, E. A. Laca, L. R. Rittenhouse, M. B. Coughenour, D. M. Swift, and P. L. Sims. 1996. Mechanisms that result in large herbivore grazing distribution patterns. Journal of Range Management 49:386–400.

Ball, L. C. 2002. A strategy for describing and monitoring bat habitat. Journal of Wildlife Management 66:1148–1153.

Banks, P. B. 2001. Predation-sensitive grouping and habitat use by eastern grey kangaroos: a field experiment. Animal Behaviour 61:1013–1021.

Bastille-Rousseau, G. 2016. lsmnsd, R package to analyze animal movement using latent-state model and net squared displacement. https://github.com/Bastille Rousseau/lsmnsd. Accessed 7 May 2020.

Bastille-Rousseau, G., I. Douglas-Hamilton, S. Blake, J. M. Northrup, and G. Wittemyer. 2018. Applying network theory to animal movements to identify properties of landscape use. Ecological Applications 28:854–864.

Bastille-Rousseau, G., J. P. Gibbs, C. B. Yackulic, J. L. Frair, F. Cabrera, L.-P. Rousseau, M. Wikelski, F. Kümmeth, and S. Blake. 2017. Animal movement in the absence of predation: environmental drivers of movement strategies in a partial migration system. Oikos 126:1004–1019.

Beguin, J., E. J. B. McIntire, and F. Raulier. 2015. Salvage logging following fires can minimize boreal caribou habitat loss while maintaining forest quotas: an example of compensatory cumulative effects. Journal of Environmental Management 163:234–245.

Bissonette, J. A. 2017. Avoiding the scale sampling problem: a consilient solution. Journal of Wildlife Management 81:192–205.

Borchers, D. L., and M. G. Efford. 2008. Spatially explicit maximum likelihood methods for capture-recapture studies. Biometrics 64:377–385.

Boscolo, D., and J. P. Metzger. 2009. Is bird incidence in Atlantic forest fragments influenced by landscape patterns at multiple scales? Landscape Ecology 24:907–918.

Bouskila, A. 1995. Interactions between predation risk and competition: a field study of kangaroo rats and snakes. Ecology 76:165–178.

Bowyer, R. T., J. G. Kie, and V. Van Ballenberghe. 1998. Habitat selection by neonatal black-tailed deer: climate, forage, or risk of predation? Journal of Mammalogy 79:415–425.

Box, G. E. P., and N. R. Draper. 1987. Empirical model-building and response surfaces. John Wiley and Sons, New York, New York, USA.

Boyce, M. S. 2006. Scale for resource selection functions. Diversity and Distributions 12:269–276.

Boyce, M. S., M. G. Turner, J. Frxell, and P. Turchin. 2003. Scale and heterogeneity in habitat selection by elk in Yellowstone National Park. Ecoscience 10:421–431.

Brennan, J., D. J. Bender, T. A. Contreras, and L. Fahrig. 2002. Focal patch landscape studies for wildlife management: optimizing sampling effort across scales. Pages 68–91 in J. Liu and W. W. Taylor, editors. Integrating landscape ecology into national resource management. Cambridge University Press, Cambridge, UK.

Brotons, L., A. Wolff, G. Paulus, and J. L. Martin. 2005. Effect of adjacent agricultural habitat on the distribution of passerines in natural grasslands. Biological Conservation 124:407–414.

Burnham, K. P., and D. R. Anderson. 2002. Model selection and multimodel inference: a practial information-theoretic approach. Second edition. Springer-Verlag, New York, New York, USA.

Calabrese, J. M., and W. F. Fagan. 2004. A comparison-shopper's guide to connectivity metrics. Frontiers in Ecology and the Environment 2:529–536.

Calenge, C., and M. Basille. 2020. adehabitatHS: Analysis of Habitat Selection by Animals. http://cran.r-project.org/web/packages/adehabitatHS/index.html. Accessed 7 May 2020.

Calenge, C., S. Dray, and M. Royer. 2020. adehabitatLT: analysis of animal movements. http://cran.r-project.org/web/packages/adehabitatLT/index.html. Accessed 7 May 2020.

Chandler, R., and J. Hepinstall-Cymerman. 2016. Estimating the spatial scales of landscape effects on abundance. Landscape Ecology 31:1383–1394.

Courbin, N., D. Fortin, C. Dussault, and R. Courtois. 2014. Logging-induced changes in habitat network connectivity shape behavioral interactions in the wolf-caribou-moose system. Ecological Monographs 84:265–285.

Creel, S., J. Winnie, B. Maxwell, K. Hamlin, and M. Creel. 2005. Elk alter habitat selection as an antipredator response to wolves. Ecology 86:3387–3397.

Cresswell, W. 2008. Non-lethal effects of predation in birds. Ibis 150:3–17.

Csárdi, G. 2019. igraph: R package for network analysis and simulation. http://www.igraph.org/r/. Accessed 7 May 2020.

Cunningham, R. B., and D. B. Lindenmayer. 2017. Approaches to landscape scale inference and study design. Current Landscape Ecology Reports 2:42–50.

Cushman, S. A., M. Chase, and C. Griffin. 2005. Elephants in space and time. Oikos 109:331–341.

Cushman, S. A., and K. McGarigal. 2004. Patterns in the species-environment relationship depend on both scale and choice of response variables. Oikos 105:117–124.

Cushman, S. A., K. McGarigal, and M. C. Neel. 2008. Parsimony in landscape metrics: strength, universality and consistency. Ecological Indicators 8:691–703.

Davidson, C. 2004. Declining downwind: amphibian population declines in California and historical pesticide use. Ecological Applications 14:1892–1902.

Decesare, N. J., M. Hebblewhite, F. Schmiegelow, D. Hervieux, G. J. Mcdermid, L. Neufeld, M. Bradley, J. Whittington, K. G. Smith, L. E. Morgantini, et al. 2012. Transcending scale dependence in identifying habitat with resource selection functions. Ecological Applications 22:1068–1083.

Doherty, K. E., D. E. Naugle, and B. L. Walker. 2010. Greater sage-grouse nesting habitat: the importance of managing at multiple scales. Journal of Wildlife Management 74:1544–1553.

Dormann, C. F., J. Elith, S. Bacher, C. Buchmann, G. Carl, G. Carré, J. R. G. Marquéz, B. Gruber, B. Lafourcade, P. J. Leitão, et al. 2013. Collinearity: a review of methods to deal with it and a simulation study evaluating their performance. Ecography 36:27–46.

Dormann, C. F., O. Schweiger, I. Augenstein, D. Bailey, R. Billeter, G. De Blust, R. DeFilippi, M. Frenzel, F. Hendrickx, F. Herzog, et al. 2007. Effects of landscape structure and land use intensity on similarity of plant and animal communities. Global Ecology and Biogeography 16:774–787.

Dussault, C., R. Courtois, J. Huot, and J-P. Ouellet. 2001. The use of forest maps for the description of wildlife habitats: limits and recommendations. Canadian Journal of Forest Research 31:1227–1234.

Efford, M. G. 2020. secr: Spatially Explicit Capture-Recapture. http://cran.r-project.org/web/packages/secr /index.html. Accessed 7 May 2020.

Efford, M. G., D. K. Dawson, and C. S. Robbins. 2004. DENSITY: software for analysing capture–recapture data from passive detector arrays. Animal Biodiversity and Conservation 271:217–228.

Eigenbrod, F., S. J. Hecnar, and L. Fahrig. 2011. Sub-optimal study design has major impacts on landscape-scale inference. Biological Conservation 144:298–305.

Elith, J., and J. R. Leathwick. 2009. Species distribution models: ecological explanation and prediction across space and time. Annual Review of Ecology, Evolution, and Systematics 40:677–697.

Environmental Systems Research Institute. 2020. ArcGIS Desktop Release 10.8. Environmenal Systems Research Institute, Redlands, CA.

European Space Agency. 2009. GlobCover 2009 (Global Land Cover Map). http://dup.esrin.esa.int/page_glob cover.php. Accessed 7 May 2020.

Evans, J. 2020. spatialEco: R package for spatial analysis and modelling of ecological systems. http://github.com /jeffreyevans/spatialEco. Accessed 7 May 2020.

Ewers, R. M., and R. K. Didham. 2005. Confounding factors in the detection of species responses to habitat fragmentation. Biological Reviews 81:117–142.

Fahrig, L. 1997. Relative effects of habitat loss and frag-mentation on population extinction. Journal of Wildlife Management 61:603–610.

Fahrig, L. 2003. Effects of habitat fragmentation on biodiversity. Annual Review of Ecology, Evolution, and Systematics 34:487–515.

Fahrig, L., J. Baudry, L. Brotons, F. G. Burel, T. O. Crist, R. J. Fuller, C. Sirami, G. M. Siriwardena, and J. L. Martin. 2011. Functional landscape heterogeneity and animal biodiversity in agricultural landscapes. Ecology Letters 14:101–112.

Fall, A., M.-J. Fortin, M. Manseau, and D. O'Brien. 2007. Spatial graphs: principles and applications for habitat connectivity. Ecosystems 10:448–461.

Fauchald, P., and T. Tveraa. 2003. Using first-passage time in the analysis of area-restricted search and habitat selection. Ecology 84:282–288.

Ficetola, G. F., and F. De Bernardi. 2004. Amphibians in a human-dominated landscape: the community structure is related to habitat features and isolation. Biological Conservation 119:219–230.

Fieberg, J., J. Matthiopoulos, M. Hebblewhite, M. S. S. Boyce, and J. L. L. Frair. 2010. Correlation and studies of habitat selection: problem, red herring or opportu-nity? Philosophical Transactions of the Royal Society B: Biological Sciences 365:2233–2244.

Fiske, I., and R. Chandler. 2011. Unmarked: an R package for fitting hierarchical models. Journal of Statistical Software 43:1–23.

Foltéte, J. C., C. Clauzel, and G. Vuidel. 2012. A software tool dedicated to the modelling of landscape networks. Environmental Modelling and Software 38:316–327.

Forman, R. T. T., and M. Godron. 1986. Landscape ecology. Wiley, New York, New York, USA.

Fortin, M. J., and S. Payette. 2002. How to test the signif-icance of the relation between spatially autocorrelated data at the landscape scale: a case study using fire and forest maps. Ecoscience 9:213–218.

Fotheringham, A. S., C. Brunsdon, and M. Charlton. 2002. Geographically weighted regression: the analysis of spatially varying relationships. Wiley, New York, New York, USA.

Frair, J. L., E. H. Merrill, J. R. Allen, and M. S. Boyce. 2007. Know thy enemy: experience affects elk translocation success in risky landscapes. Journal of Wildlife Manage-ment 71:541–554.

Frair, J. L., E. H. Merrill, H. L. Beyer, and J. M. Morales. 2008. Thresholds in landscape connectivity and mortal-ity risks in response to growing road networks. Journal of Applied Ecology 45:1504–1513.

Frair, J. L., E. H. Merrill, D. R. Visscher, D. Fortin, H. L. Beyer, and J. M. Morales. 2005. Scales of movement by elk (Cervus elaphus) in response to heterogeneity in forage resources and predation risk. Landscape Ecology 20:273–287.

Fuhlendorf, S. D., A. J. W. Woodward, D. M. Leslie, and J. S. Shackford. 2002. Multi-scale effects of habitat loss and fragmentation on lesser prairie-chicken populations of the US southern Great Plains. Landscape Ecology 17:617–628.

Fuller, A. K., C. S. Sutherland, J. A. Royle, and M. P. Hare. 2016. Estimating population density and connectivity of American mink using spatial capture-recapture. Ecologi-cal Applications 26:1125–1135.

Gallant, A. L. 2009. What you should know about land-cover data. Journal of Wildlife Management 73:796–805.

Galpern, P., P. R. Peres-Neo, J. Polfus, and M. Manseau. 2014. MEMGENE: spatial pattern detection in genetic distance data. Methods in Ecology and Evolution 5:1116–1120.

Garton, E. O., J. T. Ratti, and J. H. Guidice. 2005. Research and experimental design. Pages 43–71 in C. Braun, editor. Techniques for wildlife investigations and management. Sixth edition. The Wildlife Society, Bethesda, Maryland, USA.

Gillies, C. S., M. Hebblewhite, S. E. Nielsen, M. A. Krawchuk, C. L. Aldridge, J. L. Frair, D. J. Saher, C. E. Stevens, and C. L. Jerde. 2006. Application of random effects to the study of resource selection by animals. Journal of Animal Ecology 75:887–898.

GISGeography. 2019. Mapping out the GIS software landscape. gisgeography.com/mapping-out-gis-software-landscape/. Accessed 15 Aug 2019.

Google Earth Engine. 2019. Multiple day Landsat-TM or MODIS composite data. https://explorer.earthengine.google.com/#search/NDWI. Accessed 15 Aug 2019.

Gopalaswamy, A. M., J. A. Royle, J. E. Hines, P. Singh, D. Jathanna, N. S. Kumar, and K. U. Karanth. 2012. Program SPACECAP: software for estimating animal density using spatially explicit capture-recapture models. Methods in Ecology and Evolution 3:1067–1072.

Gottschalk, T. K., F. Huettmann, and M. Ehlers. 2005. Review article: thirty years of analysing and modelling avian habitat relationships using satellite imagery data: a review. International Journal of Remote Sensing 26:2631–2656.

Graf, R. F., K. Bollmann, W. Suter, and H. Bugmann. 2005. The importance of spatial scale in habitat models: Capercaillie in the Swiss Alps. Landscape Ecology 20:703–717.

Graham, M. H. 2003. Confronting multicollinearity in ecological multiple regression. Ecology 84:2809–2815.

Guillot, G., F. Mortier, and A. Estoup. 2005. GENELAND: a computer package for landscape genetics. Molecular Ecology Notes 5:712–715.

Guisan, A., and W. Thuiller. 2005. Predicting species distribution: offering more than simple habitat models. Ecology Letters 8:993–1009.

Hall, L. S., P. R. Krausman, and M. L. Morrison. 1997. The habitat concept and a plea for standard terminology. Wildlife Society Bulletin 25:173–182.

Hardy, O. J., and X. Vekemans. 2002. SPAGeDi: a versatile computer program to analyse spatial genetic structure at the individual or population levels. Molecular Ecology Notes 2:618–620.

Hebblewhite, M., and D. T. Haydon. 2010. Distinguishing technology from biology: a critical review of the use of GPS telemetry data in ecology. Philosophical Transactions of the Royal Society B: Biological Sciences 365:2303–2312.

Hesselbarth, M. H. K., M. Sciaini, K. A. With, K. Wiegand, and J. Nowosad. 2019. Landscapemetrics: an open-source R tool to calculate landscape metrics. Ecography 42:1–10.

Hines, J. E. 2006. PRESENCE—software to estimate patch occupancy and related parameters. US Geological Survey. http://www.mbr-pwrc.usgs.gov/software/presence.html. Accessed 7 May 2020.

Holland, J. D., D. G. Bert, and L. Fahrig. 2004. Determining the spatial scale of species' response to habitat. BioScience 54:227–233.

Homer, C. G., J. A. Dewitz, L. Yang, S. Jin, P. Danielson, G. Xian, J. Coulston, N. D. Herold, J. D. Wickham, and K. Megown. 2015. Completion of the 2011 National Land Cover Database for the conterminous United States—representing a decade of land cover change. Photogrammetric Engineering and Remote Sensing 81:345–354.

Iribarren, C., and B. P. Kotler. 2012. Foraging patterns of habitat use reveal landscape of fear of Nubian ibex Capra nubiana. Wildlife Biology 18:194–201.

Ishihama, F., T. Takeda, H. Oguma, and A. Takenaka. 2010. Comparison of effects of spatial autocorrelation on distribution predictions of four rare plant species in the Watarase wetland. Ecological Research 25:1057–1069.

Jackson, H. B., and L. Fahrig. 2012. What size is a biologically relevant landscape? Landscape Ecology 27:929–941.

Jacoby, D. M. P., and R. Freeman. 2016. Emerging network-based tools in movement ecology. Trends in Ecology and Evolution 31:301–314.

Johnson, C. J., M. S. Boyce, R. Mulders, A. Gunn, R. J. Gau, H. D. Cluff, and R. L. Case. 2004a. Quantifying patch distribution at multiple spatial scales: applications to wildlife–habitat models. Landscape Ecology 2004:869–882.

Johnson, C. J., M. S. Boyce, C. C. Schwartz, and M. A. Haroldson. 2004b. Modeling survival: application of the Andersen-Gill model to Yellowstone grizzly bears. Journal of Wildlife Management 68:966–978.

Johnson, C. J., K. L. Parker, D. C. Heard, and M. P. Gillingham. 2002. Movement parameters of ungulates and scale-specific responses to the environment. Journal of Animal Ecology 71:225–235.

Joly, K., and W. L. Myers. 2001. Patterns of mammalian species richness and habitat associations in Pennsylvania. Biological Conservation 99:253–260.

Kindlmann, P., and F. Burel. 2008. Connectivity measures: a review. Landscape Ecology 23:879–890.

Kotler, B. P. 1997. Patch use by gerbils in a risky environment: manipulating food and safety to test four models. Oikos 78:274–282.

Kramer-Schadt, S., E. Revilla, T. Wiegand, and U. Breitenmoser. 2004. Fragmented landscapes, road mortality and patch connectivity: modelling influences on the dispersal of Eurasian lynx. Journal of Applied Ecology 41:711–723.

Kuijper, D. P. J., C. de Kleine, M. Churski, P. van Hooft, J. Bubnicki, and B. Jędrzejewska. 2013. Landscape of fear in Europe: wolves affect spatial patterns of ungulate browsing in Białowieża Primeval Forest, Poland. Ecography 36:1263–1275.

Kupfer, J. A. 2012. Landscape ecology and biogeography: rethinking landscape metrics in a post-FRAGSTATS landscape. Progress in Physical Geography 36:400–420.

Leblond, M., J. Frair, D. Fortin, C. Dussault, J. P. Ouellet, and R. R. Courtois. 2011. Assessing the influence of resource covariates at multiple spatial scales: an application to forest-dwelling caribou faced with intensive human activity. Landscape Ecology 26:1433–1446.

Legendre, P. 1993. Spatial autocorrelation: trouble or new paradigm? Ecology 74:1659–1673.

Legendre, P., and M. J. Fortin. 1989. Spatial pattern and ecological analysis. Vegetation 80:107–138.

Lele, S. R., J. L. Keim, and P. Solymos. 2019. Resource-Selection: resource selection (probability) functions for use-availability data. http://cran.r-project.org/web/packages/ResourceSelection/index.html. Accessed 7 May 2020.

Li, H., and J. F. Reynolds. 1995. On definition and quantification of heterogeneity. Oikos 73:280–284.

Lima, S. L., and P. A. Zollner. 1996. Towards a behavioral ecology of ecological landscapes. Trends in Ecology and Evolution 11:131–135.

Lindenmayer, D., R. J. Hobbs, R. Montague-Drake, J. Alexandra, A. Bennett, M. Burgman, P. Cale, A. Calhoun, V. Cramer, P. Cullen, et al. 2008. A checklist for ecological management of landscapes for conservation. Ecology Letters 11:78–91.

Marzluff, J. M., J. J. Millspaugh, K. R. Ceder, C. D. Oliver, J. Withey, J. B. McCarter, C. L. Mason, and J. Comnick. 2002. Modeling changes in wildlife habitat and timber revenues in response to forest management. Forest Science 48:191–202.

Matthiopoulos, J., M. Hebblewhite, G. Aarts, and J. Fieberg. 2011. Generalized functional responses for species distributions. Ecology 92:583–589.

Mayor, S. J., J. A. Schaefer, D. C. Schneider, and S. P. Mahoney. 2007. Spectrum of selection: new approaches to detecting the scale-dependent response to habitat. Ecology 88:1634–1640.

McComb, W. C., M. T. McGrath, T. A. Spies, and D. Vesely. 2002. Models for mapping potential habitat at landscape scales: an example using northern spotted owls. Forest Science 48:203–216.

McGarigal, K., and S. A. Cushman. 2002. Comparative evaluation of experimental approaches to the study of habitat fragmentation effects. Ecological Applications 12:335–345.

McGarigal, K., and B. J. Marks. 1995. FRAGSTATS: spatial analysis program for quantifying landscape structure. US Department of Agriculture General Technical Report PNW-GTR-351, Portland, Oregon, USA.

McGarigal, K., S. Tagil, and S. A. Cushman. 2009. Surface metrics: an alternative to patch metrics for the quantification of landscape structure. Landscape Ecology 24:433–450.

McGarigal, K., S. A. Cushman, and E. Ene. 2012. FRAGSTATS v4: spatial pattern analysis program for categorical and continuous maps. http://www.umass.edu/landeco/research/fragstats/fragstats.html. Accessed 7 May 2020.

McRae, B., V. Shah, and T. Mohapatra. 2013a. Circuitscape 4 user guide. The Nature Conservancy, Fort Collins, CO. http://www.circuitscape.org.

McRae, B. et al. 2013b. Linkage Mapper. Open source software. http://www.circuitscape.org/linkagemapper. Accessed 7 May 2020.

Metzger, J.-P., and H. Décamps. 1997. The structural connectivity threshold: an hypothesis in conservation biology at the landscape scale. Acta Oecologica 18:1–12.

Meyer, C. B., and W. Thuiller. 2006. Accuracy of resource selection functions across spatial scales. Diversity and Distributions 12:288–297.

Michelot, T., R. Langrock, T. Patterson, B. McClintock, and E. Rexstad. 2019. moveHMM: animal movement modelling using hidden Markov models. http://cran.r-project.org/web/packages/moveHMM/index.html. Accessed 7 May 2020.

Minor, E. S., and D. L. Urban. 2008. A graph-theory framework for evaluating landscape connectivity and conservation planning. Conservation Biology 22:297–307.

Moody, A., and C. E. Woodcock. 1995. The influence of scale and the spatial characteristics of landscapes on land-cover mapping using remote sensing. Landscape Ecology 10:363–379.

Moore, J. E., and R. K. Swihart. 2005. Modeling patch occupancy by forest rodents: incorporating detectability and spatial autocorrelation with hierarchically structured data. Journal of Wildlife Management 69:933–949.

Moreau, G., D. Fortin, S. Couturier, and T. Duchesne. 2012. Multi-level functional responses for wildlife conservation: the case of threatened caribou in managed boreal forests. Journal of Applied Ecology 49:611–620.

Mukherjee, S., M. Zelcer, and B. P. Kotler. 2009. Patch use in time and space for a meso-predator in a risky world. Oecologia 159:661–668.

Multi-Resolution Land Characteristics Consortium. 2016. MRLC homepage. https://www.mrlc.gov/index.php. Accessed 15 Aug 2019.

Myatt, N. A., and D. G. Krementz. 2007. Fall migration and habitat use of American woodcock in the central United States. Journal of Wildlife Management 71:1197–1205.

Mysterud, A., and R. A. Ims. 1998. Functional responses in habitat use: availability influences relative use in trade-off situations. Ecology 79:1435–1441.

Nams, V. O. 2005. Using animal movement paths to measure response to spatial scale. Oecologia 143(2):179–188.

National Aeronautics and Space Administration. 2000. Shuttle Radar Topographic Mission data. http://srtm.csi .cgiar.org/. Accessed 7 May 2020.

National Aeronautics and Space Administration. 2010. NASA Global Roads (gROADS) homepage. http://sedac .ciesin.columbia.edu/data/collection/groads. Accessed 7 May 2020.

National Aeronautics and Space Administration. 2015a. Tropical Rainfall Measuring Mission (TRMM) data. https://gpm.nasa.gov/. Accessed 7 May 2020.

National Aeronautics and Space Administration. 2015b. NASA Gridded Population of the World (GPW) data. http://sedac.ciesin.columbia.edu/data/collection/gpw-v4. Accessed 7 May 2020.

National Aeronautics and Space Administration. 2016. Global Land Data Assimilation System (GLDAS) data. http://ldas.gsfc.nasa.gov/gldas/. Accessed 7 May 2020.

National Aeronautics and Space Administration. 2019a. Global 30m Landsat Tree Canopy Version 4. https:// landsat.gsfc.nasa.gov/global-30m-landsat-tree-canopy -version-4-released/. Accessed 7 May 2020.

National Aeronautics and Space Administration. 2019b. Advanced Spaceborne Thermal Emission and Reflection Radiometer (ASTER) data. https://asterweb.jpl.nasa .gov/. Accessed 7 May 2020.

National Aeronautics and Space Administration. 2019c. Airborne Visible / Infrared Imaging Spectroradiometer (AVIRIS) data. http://aviris.jpl.nasa.gov/. Accessed 7 May 2020.

National Aeronautics and Space Administration. 2019d. Global Precipitation Measurement Mission (GPM) data. https://pmm.nasa.gov/data-access/downloads /gpm. Accessed 7 May 2020.

Newman, J., G. Recer, S. Zwicker, and T. Caraco. 1988. Effects of predation hazard on foraging "constraints": patch-use strategies in grey squirrels. Oikos 53:93–97.

Nielsen, S. E., M. S. Boyce, G. B. Stenhouse, and R. H. M. Munro. 2002. Modeling grizzly bear habitats in the Yellowhead Ecosystem of Alberta: taking autocorrelation seriously. Ursus 13:45–56.

Nielsen, S. E., G. B. Stenhouse, and M. S. Boyce. 2006. A habitat-based framework for grizzly bear conservation in Alberta. Biological Conservation 130(2):217–229.

Nielson, R. M., H. Sawyer, and T. L. McDonald. 2013. BBMM: Brownial Bridge Movement Model. http://www .cran.r-project.org/web/packages/BBMM/index.html. Accessed 7 May 2020.

Niemuth, N. D. 2004. Identifying landscapes for greater prairie chicken translocation using habitat models and GIS: a case study. Wildlife Society Bulletin 31:145–155.

O'Brien, D., M. Manseau, A. Fall, and M. J. Fortin. 2006. Testing the importance of spatial configuration of winter habitat for woodland caribou: an application of graph theory. Biological Conservation 130:70–83.

Openshaw, S. 1983. The modifiable area unit problem. Concepts and Techniques in Modern Geography 38:1–41.

Ostapowicz, K., P. Vogt, K. H. Riitters, J. Kozak, and C. Estreguil. 2008. Impact of scale on morphological spatial pattern of forest. Landscape Ecology 23:1107–1117.

Parent, C. J., F. Hernández, L. A. Brennan, D. B. Wester, F. C. Bryant, and M. J. Schnupp. 2016. Northern bobwhite abundance in relation to precipitation and landscape structure. Journal of Wildlife Management 80:7–18.

Pebesma, E., and B. Graeler. 2020. gstat: spatial and spatio-temporal geostatistical modelling, prediction and simulation. http://cran.r-project.org/web/packages/gsta t/index.html. Accessed 7 May 2020.

Phillips, S. J., M. Dudík, and R. E. Schapire. 2004. A maximum entropy approach to species distribution modeling. Page 83 in Twenty-first international conference on machine learning. ACM Press, New York, New York, USA.

QGIS Development Team. 2020. QGIS Geographic Information System. Open Source Geospatial Foundation Project. http://qgis.osgeo.org. Accessed 7 May 2020.

Radu, V. C., T. Duchesne, D. Fortin, and S. Baillargeon. 2016. TwoStepCLogit: conditional logistic regression. A two-step estimation method. http://cran.r-project/web /packages/TwoStepCLogit/index.html. Accessed 7 May 2020.

Ray, N. 2005. PATHMATRIX: a GIS tool to compute effective distances among samples. Molecular Ecology Notes 5:177–180.

Rayfield, B., M.-J. Fortin, and A. Fall. 2011. Connectivity for conservation: a framework to classify network measures. Ecology 92:847–858.

Rempel, R. 2016. Patch Analyst 5.2. http://patch-analyst .software.informer.com. Accessed 7 May 2020.

Riitters, K. H., R. V. O'Neill, C. T. Hunsaker, J. D. Wickham, D. H. Yankee, S. P. Timons, and K. B. Jones. 1995. A factor analysis of landscape pattern and structure metrics. Landscape Ecology 10:23–40.

Royle, J. A., R. B. Chandler, C. Yackulic, and J. D. Nichols. 2012. Likelihood analysis of species occurrence probability from presence-only data for modelling species distributions. Methods in Ecology and Evolution 3:545–554.

Russell, R. E., J. A. Royle, V. A. Saab, J. F. Lehmkuhl, W. M. Block, and J. R. Sauer. 2009. Modeling the effects of environmental disturbance on wildlife communities: avian responses to prescribed fire. Ecological Applications 19:1253–1263.

Saura, S., and J. Torné. 2009. Conefor Sensinode 2.2: a software package for quantifying the importance of habitat patches for landscape connectivity. Environmental Modelling and Software 24:135–139.

Schielzeth, H. 2010. Simple means to improve the interpretability of regression coefficients. Methods in Ecology and Evolution 1:103–113.

Schielzeth, H., and S. Nakagawa. 2013. Nested by design: model fitting and interpretation in a mixed model era. Methods in Ecology and Evolution 4:14–24.

Senft, R. L., M. B. Coughenour, D. W. Bailey, L. R. Rittenhouse, O. E. Sala, and D. M. Swift. 1987. Large herbivore foraging and ecological hierarchies. BioScience 37:789–795.

Shirk, A. J., and B. H. McRae. 2013. Gnarly landscape utilities: core mapper user guide. The Nature Conservancy, Fort Collins, CO, USA. http://circuitscape.org/gnarly-landcape-utilities/.

Signer, J., B. Reineking, B. Smith, U. Schlaegel, and S. LaPoint. 2020. amt: Animal Movement Tools. http://cran.r-project.org/web/packages/amt/index.html. Accessed 7 May 2020.

Smith, K. T., C. P. Kirol, J. L. Beck, and F. C. Blomquist. 2014. Prioritizing winter habitat quality for greater sage-grouse in a landscape influenced by energy development. Ecosphere 5:1–20.

Steincke, K. K. 1948. Farvel og tak: ogsaa en tilvaerelse IV (1935–1939). Publisher Fremad, Copenhagen, Denmark. [Translated to english and verified.] https://quoteinvestigator.com/2013/10/20/no-predict/#note-7474-1. Accessed 15 Aug 2019.

Thogmartin, W. E., A. L. Gallant, M. G. Knutson, T. J. Fox, and M. J. Suárez. 2004. A cautionary tale regarding use of the National Land Cover Dataset 1992. Wildlife Society Bulletin 32:970–978.

Thomas, L., S. T. Buckland, E. A. Rexstad, J. L. Laake, S. Strindberg, S. L. Hedley, J. R. B. Bishop, T. A. Marques, and K. P. Burnham. 2010. Distance software: design and analysis of distance sampling surveys for

estimating population size. Journal of Applied Ecology 47:5–14.

Thornton, D. H., L. C. Branch, and M. E. Sunquist. 2011. The influence of landscape, patch, and within-patch factors on species presence and abundance: a review of focal patch studies. Landscape Ecology 26:7–18.

Thornton, D. H., and R. J. Fletcher. 2013. Body size and spatial scales in avian response to landscapes: a meta-analysis. Ecography 37:454–463.

Thuiller, W., B. Lafourcade, R. Engler, and M. B. Araújo. 2009. BIOMOD—a platform for ensemble forecasting of species distributions. Ecography 32:369–373.

Tischendorf, L., and L. Fahrig. 2000. How should we measure landscape connectivity? Landscape Ecology 15:633–641.

Turner, M. G., and R. H. Gardner. 2015. Landscape ecology in theory and practice. Second edition. Springer-Verlag, New York, New York, USA.

Turner, M. G., R. H. Gardner, and R. V. O'Neill. 2001. Landscape ecology in theory and practice. Springer-Verlag, New York, New York, USA.

Turner, M. G., R. V. O'Neill, R. H. Gardner, and B. T. Milne. 1989. Effects of changing spatial scale on the analysis of landscape pattern. Landscape Ecology 3:153–162.

Uezu, A., J. P. Metzger, and J. M. E. Vielliard. 2005. Effects of structural and functional connectivity and patch size on the abundance of seven atlantic forest bird species. Biological Conservation 123:507–519.

Urban, D., and T. Keitt. 2001. Landscape connectivity: a graph-theoretic perspective. Ecology 82:1205–1218.

US Geological Survey. 2017. Digital Elevation Model (DEM). https://catalog.data.gov/dataset/usgs-national-elevation-dataset-ned-1-meter-downloadable-data-collection-from-the-national-map-. Accessed 7 May 2020.

US Geological Survey. 2018. Moderate Resolution Imaging Spectroradiometer (MODIS) data. https://lpdaac.usgs.gov/dataset_discovery/modis/modis_products_table. Accessed 7 May 2020.

US Geological Survey. 2019a. Advanced Very High Resolution Radiometer (AVHRR) Normalized Difference Vegetation Index (NDVI) data. https://www.usgs.gov/land-resources/eros/phenology/science/ndvi-avhrr?qt-science_center_objects=0#qt-science_center_objects. Accessed 7 May 2020.

US Geological Survey. 2019b. Landsat Thematic Mapper (TM). https://www.usgs.gov/land-resources/nli/landsat/. Accessed 7 May 2020.

US Geological Survey. 2019c. Hyperion data. https://archive.usgs.gov/archive/sites/eo1.usgs.gov/hyperion.html. Accessed 7 May 2020.

Vandergast, A. G., W. M. Perry, R. V. Lugo, and S. A. Hathaway. 2011. Genetic landscapes GIS Toolbox: tools to map patterns of genetic divergence and diversity. Molecular Ecology Resources 11:158–161.

Van Etten, J. 2017. R Package gdistance: distances and routes on geographical grids. Journal of Statistical Software 76:1–21.

Virkkala, R., R. K. Heikkinen, S. Fronzek, H. Kujala, and N. Leikola. 2013. Does the protected area network preserve bird species of conservation concern in a rapidly changing climate? Biodiversity and Conservation 22:459–482.

Vogt, P., and K. Riitters. 2017. GuidosToolbox: universal digital image object analysis. European Journal of Remote Sensing 50:352–361.

Werner, E. E., and D. J. Hall. 1988. Ontogenetic habitat shifts in bluegill: the foraging rate-predation risk trade-off. Ecology 69:1352–1366.

White, G. C., and K. P. Burnham. 1999. Program MARK: survival estimation from populations of marked animals. Bird Study 46:S120–S139. http://www.phidot.org/software/mark/.

Wickham, J. D., and K. H. Riitters. 1995. Sensitivity of landscape metrics to pixel size. International Journal of Remote Sensing 16:3585–3594.

Wu, J. 2004. Effects of changing scale on landscape pattern analysis: scaling relations. Landscape Ecology 19:125–138.

Zuckerberg, B., A. Desrochers, W. M. Hochachka, D. Fink, W. D. Koenig, and J. L. Dickinson. 2012. Overlapping landscapes: a persistent, but misdirected concern when collecting and analyzing ecological data. Journal of Wildlife Management 76:1072–1080.

9 — Part II Synthesis

Establishing a Landscape Foundation for Wildlife Managers

DAVID M. WILLIAMS

It would be naive to suggest that wildlife managers have only recently embraced a landscape perspective for management. Conservationists and managers have considered the effects of local and surrounding habitats for population growth and persistence. The theory behind those effects and the availability of data to inform those effects, however, were limited in the past. Today, nearly every government agency or nongovernmental organization that is tasked with managing land manages functional landscapes. While the specifics of how those landscapes are defined and what their management entails differ, there is no question that management of wildlife populations and ecological systems has continued to embrace a landscape perspective. Fortunately, the field of landscape ecology has benefitted management by providing the foundational theory and a suite of tools and approaches for assessing the rapidly advancing collection of landscape data. The chapters in Part II provide the wildlife manager with important terminology, context, and considerations for appropriate and successful integration of landscape ecology and wildlife management. I summarize the content of these chapters and highlight some important ideas raised therein.

Chapter 4 explores spatial heterogeneity, shows how scale affects the relationships between ecological processes and landscape patterns, describes multiple models for representing landscape features, and provides potential applications for management of wildlife habitat. Spatial heterogeneity, or the spatial distribution of relevant resources, can have positive and negative effects on conservation and management goals. Furthermore, those effects can differ at different scales. Scale describes the dimensions of a pattern or process and consists of two components: extent and grain. Perotto-Baldivieso's discussion of scale is of particular importance for the wildlife manager for a couple reasons. First, the term *scale* is frequently misused or casually used to primarily refer to spatial extent (grain, or the resolution of the data, is often not addressed) and is unfortunately used in terms of small and large or local and landscape to reference an extent relative to a human perspective. Thus management occurring across a region or county may be identified as large- or landscape-scale management, while activities conducted on a specific property or section might be referred to as small- or local-scale management. Unfortunately, these understandings of scale occur irrespective of the ecological process of interest. The wildlife manager would do well to consider scale as the combination of its parts (extent and grain) and identify scales relevant to the species and processes being managed.

Second, a clear understanding of scale is essential to the wildlife manager because processes of interest may operate at different scales. As the author points out, multiscale influences can have surprising implications for management and challenge preconceived understandings of ecology derived from convenient- or single-scale studies. Identifying the best scales for investigation and management can present unique challenges for the manager when ecologically relevant scales do not align with the scales of management practices, which are often dictated by property or political boundaries.

Those challenges are explored further in Chapter 5, which outlines the factors that influence the patterns of resources on the landscape. Veech reviews natural and anthropogenic factors that influence landscape patterns: soil, hydrology, fire, grazing, and forestry. Understanding those factors is an essential prerequisite for the development of wildlife–habitat- or wildlife-landscape-related management. Of particular interest is the discussion of additional challenges to linking wildlife management and landscape ecology. Designing experiments that quantify relationships between ecological processes of wildlife populations and the spatial heterogeneity of the landscape is difficult. There is seldom opportunity for appropriate controls, and limited resources can constrain the potential for suitable replication, particularly when questions involve large spatial extents and fine grain. Those limitations of research, and by extension theory and knowledge, add further uncertainty to our abilities to manage wildlife habitat at landscape scales. Nevertheless, Veech emphasizes that the manager should maintain a mindset that appreciates the ecological complexities of landscapes, has control over some of the processes that maintain and create spatial heterogeneity, uses that control to their advantage when possible, and recognizes that ecological processes connect managed lands to other areas of the landscape that might not be receiving direct management.

To begin to understand how to relate landscape features to ecological processes, Diamond and El-liot (Chap. 6) provide a primer to the relationships between landcover data and habitat features. This chapter is of particular importance for two reasons. First, landcover maps are necessary because they provide a means to quantify landscape patterns and integrate with other spatial data (e.g., soils, wildlife movements) that is impossible from ground-based observations or raw aerial imagery. Second, while the use of landcover data is widespread by wildlife managers, the data are often used with little consideration for their source, accuracy, or limitations. Diamond and Elliot provide a valuable summary of the types of satellite data and the workflow used to classify that imagery into landcover data sets (e.g., National Land Cover Database, LANDFIRE). They discuss the potential for further modifications of those data sets by coupling landcover with soil, hydrology, or on-the-ground knowledge and data. But the wildlife manager is reminded that there is no universally applicable landcover classification or map because goals, the ecological processes of interest, and planning regions will vary.

Having the capacity to quantify landscape characteristics with landcover data and accepting that many ecological processes of interest to the wildlife manager are influenced by landscape composition and configuration are only first steps toward establishing a landscape ecological foundation. Martin et al. (Chap. 7) address fragmentation, a fundamental concept of landscape ecology, and provide an important discussion for the wildlife manager: how habitat loss and fragmentation influence wildlife populations. From a landscape ecology perspective, habitat loss and fragmentation are disparate. Habitat loss is the reduction in amount of habitat, a compositional change to a landscape. Habitat loss has consistent negative effects on wildlife. Habitat fragmentation, however, can have positive or negative effects. Fragmentation is the breaking up of habitat, a change in configuration, and does not necessarily require habitat loss. Martin et al. argue that these differences require the wildlife or habitat manager to discriminate between habitat loss and fragmentation, lest it be assumed that all fragmentation is negative. Finally, several recommendations are provided for management consideration. First, because of consistent evidence for negative effects of habitat loss on wildlife, managers should focus on reducing loss and providing for restoration of habitats. Second, when habitat loss is inevitable, the effects of fragmentation should be considered to identify strategic recommendations for habitat removal. Third, species-based management will need to consider species-specific responses to fragmentation, but diversity-based management is more likely to meet objectives by managing for greater habitat fragmentation.

Fragmentation and habitat loss are two of many aspects of the pattern-process relationships from landscape ecology that should be important to the wildlife manager. Frair and Bastille-Rousseau (Chap. 8) extend the perspective of the wildlife manager for data collection and quantitative analyses that link landscape patterns to the ecological processes of wildlife more broadly. Landscapes are generally quantified and by implication later related to wildlife processes according to two primary categories: patch based or continuous data. Patches are typically defined structurally as groups of contiguous-like cells. The authors highlight two important considerations for working with patches. First, how patches are defined and categorized affects pattern-based analyses. Thus it is important for the analyst to be clear about the decisions made to define patches and to test the effects of those subjective decisions on potential inferences. Second, the wildlife manager need not be restricted to defining patches solely as structural. When assessing the effect of patch-based patterns on wildlife processes, patches identified by their functional role may be more useful. Continuous data are not discrete (e.g., percentage of a landcover type in a surrounding area) and cannot be quantified in the same ways as patch-based data. Continuous data are often related to ecological processes of wildlife by measures of sums, central tendency, and variability but may also be summarized using spatial statistics. While these spatial statistics can be used to identify the spatial structure of correlation and how the variance of these landscape metrics change with scale, relationships between these metrics and wildlife processes may be more difficult to interpret and implement within a land management framework. Nevertheless, the authors highlight an important but perhaps overlooked truth about spatial correlation among landscape metrics: if we sample ecological processes at scales finer than the full range of correlation of our landscape metrics, we may be unable to accurately quantify those relationships. Thus it is important to consider multiple scales when assessing wildlife–landscape relationships.

Wildlife practitioners who are seeking to quantify wildlife-landscape relationships should focus on manageable aspects of landscape pattern and remember that scale matters (preferably testing a focal, broader, and finer scale) for the patterns and the processes. Finally, Frair and Bastille-Rousseau provide a helpful outline for the wildlife manager who is seeking to evaluate wildlife–landscape relationships to inform management: begin with the question, choose qualities of the landscape relevant to the question, ensure landscape data are appropriate (e.g., scale, accuracy, classification scheme), and choose uncorrelated explanatory metrics.

The rapid development of the field of landscape ecology, potentially foreign jargon developed to describe specific aspects of landscape pattern and theory, and the seemingly innumerable options for landscape metrics and analyses relating those metrics to ecological processes may be overwhelming to the wildlife manager. Part II, however, provides the foundational vocabulary and landscape perspectives for the wildlife manager to critically engage the field of landscape ecology, to begin forming a landscape perspective for management, and to develop research that informs how landscape patterns are influencing wildlife processes of interest.

PART III • Establishing a Wildlife Management Foundation for Landscape Ecologists

— *10* — Managing Wildlife at Landscape Scales

John W. Connelly and
Courtney J. Conway

Introduction

Landscape or *landscape scale* has been defined many different ways. Two examples illustrate the range of variation in definitions and the ambiguity of existing definitions: "a zone or area as perceived by local people or visitors, whose visual features and character are the result of the action of natural and/or cultural factors" (European Landscape Convention 2000:6) and "action that covers a large spatial scale, usually addressing a range of ecosystem processes, conservation objectives and land uses" (Ahern and Cole 2012:9). A single definition, certainly one that mentions quantitative metrics, is not helpful because the size of an area (i.e., extent) that qualifies as landscape scale varies depending on the species in question and the seasonal or year-round resources that species requires (Liu and Taylor 2002, Ahern and Cole 2012). Yet any definition would undoubtedly encompass a spatial scale that supports more than just a few individuals of the focal species. Managing wildlife populations at large landscape extents is not new to the field of wildlife management. Leopold (1933) explored some of the early concepts of landscape-scale management more than 80 years ago, and Dalke (1937) proposed the use of cover maps and aerial photography to manage wildlife at

landscape scales. Natural resource managers have increasingly recognized the importance of managing landscapes for wildlife (Bissonette 1997, Sinclair et al. 2006). Nevertheless, managing wildlife populations at large spatial extents is challenging. Indeed, Sinclair et al. (2006) recognized the complexities of managing wildlife at landscape scales when they stated that population declines occur as a result of interacting factors involving the environment and biota and that anthropogenic change can further alter ecosystem dynamics and affect populations.

Despite the intervening years since Leopold's (1933) and Dalke's (1937) efforts to encourage wildlife management at landscape scales, a variety of barriers exist to effective management of wildlife species at such large scales. Moreover, some of these barriers have arisen (or became more impenetrable) in the 80 years since Leopold's writings on the topic. These barriers can be broadly classified as challenges (difficulties that can be overcome with improved communication, management, or research) and constraints (difficulties that cannot be overcome without administrative or policy changes), and further characterized by type of barrier (e.g., policy, administrative, biological, technological, academic).

State fish and wildlife agencies generally manage wildlife populations at politically determined extents, usually by game management unit (typically delineated on the basis of ecological criteria but with boundaries imposed by political lines or roads), county, or region despite the recognition that populations commonly range over multiple counties or regions. Thus competing management objectives for the same population can occur in parts of two adjoining counties, regions, or states. For example, a single breeding population of greater sage-grouse (*Centrocercus urophasianus*, sage-grouse) occupies parts of southeastern Idaho, northeastern Utah, and western Wyoming, USA (Cardinal 2015). Two of the three states allow sage-grouse harvest of this population (with different harvest limits), and Idaho and Utah consider the area occupied by this population to be core habitat in their state, affording substantial protection from development under each state's plan for conserving sage-grouse. In contrast, Wyoming has no special designation for the area occupied by this same population. Thus three different management approaches are applied to this single sage-grouse population. These administrative and policy barriers limit the effectiveness of management.

Wildlife population dynamics and the ability of a population to persist in an area play out at a landscape scale. But there is a spatial mismatch between population processes and management actions because wildlife management actions have traditionally occurred at small spatial extents. Moreover, research projects intended to inform management have typically focused on even smaller spatial extents. Research at large spatial extents has become more common (Barea-Azcón et al. 2007, Sawyer et al. 2009, Stevens and Conway 2020), but studies are typically correlative rather than mechanistic and examine pattern rather than process because experiments at landscape scales are often prohibitively difficult and expensive. Additionally, three to four agencies may be charged with managing wildlife habitat at varying spatial scales within a single state (e.g., state fish and wildlife agency, state department of lands, US Forest Service, US Bureau of Land Management [BLM], and US Fish and Wildlife

Service). For example, in the Curlew Valley of southern Idaho, Columbian sharp-tailed grouse (*Tympanuchus phasianellus columbianus*) use land managed by the BLM and US Forest Service and privately owned lands (Apa 1998). In this geography, federal agencies manage habitat, private landowners may or may not manage habitat, and the Idaho Department of Fish and Game is responsible for managing the grouse population. Thus implementing management actions across relatively large spatial extents can be difficult because one manager or agency may have little or no input on the management of habitats or populations managed by other agencies. We believe that this situation is typical for most wildlife species. Landscape ecologists, wildlife managers, and policymakers should be aware of such barriers and incorporate these unique challenges into their research questions, management plans, and policies.

The purpose of this chapter is to identify challenges, constraints, and opportunities to managing wildlife at landscape scales and discuss how resource agencies and landscape ecologists can collaborate to overcome challenges and constraints. We also explore potential management consequences of continuing to manage wildlife populations and habitats at small spatial extents. Finally, we provide a case study using the greater sage-grouse as an example of challenges to managing a species at a landscape scale.

Challenges to Managing Wildlife at Landscape Scales

Challenges to managing wildlife at landscape scales, although difficult, can be overcome by better planning, improved coordination, and emphasis on a landscape-scale approach to management and research (Table 10.1). Managers often face substantial challenges with managing wildlife populations and habitat at a landscape scale, including differing management goals and responsibilities of various state and federal agencies. State fish and wildlife agencies place a priority on managing wildlife populations,

Table 10.1. Barriers to Managing Wildlife on Landscapes

Barrier	Class	Type of Barrier	Description
Disparate agency mandates	Challenge or constraint	Policy	Wildlife habitat may be owned or managed by many landowners or agencies with different mandates and goals.
Scale mismatch	Challenge	Policy	Scale mismatch among research projects, management actions, and population dynamics.
Disjunct seasonal habitats	Challenge	Biological	Many populations depend on different areas for breeding, postbreeding or migration, and wintering, all of which can have different landowners.
Landscape scale varies among species	Challenge	Biological	Any area will have a suite of coexisting wildlife species; the appropriate landscape for management will likely differ among species.
Habitat loss	Challenge	Biological	Large expanses of some habitats can be rapidly lost to fire, energy development, and other factors.
Inability to quantify movements	Challenge	Technological	Until recently, we lacked the technical ability to follow individual animals as they disperse or migrate.
Lack of communication	Challenge	Academic	Lack of communication and a different vocabulary between landscape ecologists and management personnel.
Strongly held beliefs	Challenge	Academic	Lack of communication or education can lead to beliefs that are not supported by best available science.
State boundaries	Constraint	Administrative	A single wildlife population can straddle multiple state boundaries and may be managed differently in different states.
Agency boundaries	Constraint	Administrative	Within an agency, management is often assigned by area; management actions and goals may differ among suboffices, although one wildlife population can straddle multiple areas.
Anthropogenic development	Challenge or constraint	Policy	Various types of development can impede migration and reduce habitat.

federal land management agencies, such as the BLM and US Forest Service, have a mandate to promote multiple use, and state land agencies (e.g., Idaho Department of Lands, Arizona State Land Department) have a mandate to maximize economic return. Yet both multiple-use and maximizing economic return are management priorities that can cause declines in wildlife populations, thereby conflicting with management priorities of state fish and wildlife agencies. For example, timber harvest programs of federal or state land management agencies may improve elk (*Cervus canadensis*) habitat at one spatial extent by increasing forage, but road creation associated with these programs may shrink the useable landscape for elk and negatively affect the species at a larger spatial extent (Lyon and Christensen 2002). Although state wildlife agencies may support a timber harvest if roads are closed following the project, land

management agencies may leave these roads open to enhance public access. A lack of collaboration and adherence to divergent mandates among these agencies may adversely affect wildlife populations and may be challenging to resolve. Compromises and flexibility by all agencies and landowners are needed to help resolve these challenges.

Some migratory species provide templates for success, whereas others serve as troubling examples of the challenges associated with managing wildlife at a landscape scale. Development and successful implementation of the North American waterfowl management plan provide an example of how these challenges can be overcome for managing harvest at landscape or even continental scales (Williams et al. 1996), whereas the plight of the cerulean warbler (*Setophaga cerulea*) illustrates these difficulties where they have not yet been overcome. The cerulean warbler breeds in mature floodplain forests of eastern North America, a landscape that has experienced substantial habitat loss and fragmentation (Robbins et al. 1992). This species winters in the humid evergreen forest at the base of the Andes in South America, an area that has been intensively logged and cultivated (Robbins et al. 1992). From the mid-1960s to the mid-1980s, cerulean warbler populations declined more than any other North American warbler. Unless large landscapes providing breeding habitat in North America and wintering habitat in South America are conserved, the future of this warbler is in jeopardy (Robbins et al. 1992). Thus managers hoping to conserve this species must deal with landscape management and differing management goals of various agencies at an international scale.

The northern bobwhite (*Colinus virginianus*) is a popular and intensively managed game bird, and extensive research has been conducted on this species. Despite long-term management and research efforts, the northern bobwhite has suffered broad declines (Brennan 1991, Hernández et al. 2012). For many years, habitat management practices for northern bobwhite have been applied at relatively small extents (e.g., parts of plantations) and have used a variety of techniques (Rosene 1969). Williams et al. (2004), however, proposed that northern bobwhite research and habitat management have occurred at a spatial scale incompatible with the biological problem. They recommended expanding the extent of habitat management to incorporate preservation and creation of useable space at a regional scale, and these recommendations are being developed in numerous states (Harvey 2015, Palmer and Sisson 2017).

The spatial extent at which most research projects are conducted provides another challenge to managing wildlife on landscapes. Research is often conducted at local scales (e.g., within the winter home range of one population or small subplots within breeding territories), and inferences are made about the status of the population and its habitat. Depend-

ing on the species, that scale may be entirely insufficient for characterizing the conservation status of the species or identifying appropriate management recommendations (Williams et al. 2004). Moreover, the seasons when animals migrate and disperse have frequently been ignored compared to breeding and wintering seasons, often for logistical reasons; in the Web of Science, the ratio of published papers in ecology and management journals with dispersal in the title relative to those with breeding in their title was <5% from 1900 through 1940 but increased to 41% during 2011–2016 (Web of Science 2019). Since the mid-1990s, improved satellite-based technology has allowed researchers and managers to document individual animals moving long distances regardless of access, land use, or political and geographic barriers (Perras and Nebel 2012).

Many other examples of landscape-scale management challenges can be cited, ranging from American bison (*Bison bison*) in the Yellowstone ecosystem to trumpeter swans (*Cygnus buccinator*) in the Rocky Mountains, USA (Shea et al. 2002, Plumb et al. 2009). With a more complete understanding of these challenges, landscape ecologists can apply or consider them when developing future research questions. Some challenges reflect our relatively minimal knowledge (at least until recently) regarding the nature, extent, and variability of seasonal and annual movements of most wildlife populations. Landscape ecologists can help resource agencies address these challenges by developing research questions that economize resource agency efforts and stress the importance of ecological research at multiple spatial scales. Ultimately, resource agencies can implement an explicit focus on management of spatial heterogeneity (Turner 1989).

Constraints on Managing Wildlife at Landscape Scales

Constraints on managing wildlife at landscape scales can be thought of as barriers that cannot be overcome without a change in policy, legislation, or agency administrative structure (Table 10.1). Thus constraints are inherently more difficult for wildlife managers and landscape ecologists than challenges. Because managing wildlife at a landscape scale usually involves many administrative offices within multiple agencies and numerous private landowners, these constraints cannot be overcome without explicit efforts to do so. Land management agencies and private landowners may have different or competing objectives or policies for the land and differing philosophies on and approaches to decision making. Additionally, different interest groups may have varying influences on agency decisions and may desire different outcomes.

For example, multiple-use management might allow mining or energy development in critical wildlife habitat while wilderness management would exclude these uses. Differing land management policies can provide a substantial constraint on species that undergo seasonal migrations. A migration corridor will often traverse public and private ownership, complicating attempts for their maintenance. Lendrum et al. (2013) reported that ungulate migrations generally occur along migratory corridors that often are affected by anthropogenic disturbances. Thus reducing or eliminating threats to migratory wildlife may involve policy changes that alter or restrict allowable activities in these areas. Moreover, these policy changes may require agreements with private landowners to influence how portions of their lands are used or managed.

For example, in 1978 a private landowner in south-central Wyoming erected a fence on his land that cut off a pronghorn (*Antilocapra americana*) migration corridor. In 1983, approximately 700 pronghorns were prevented from migrating to their winter range and died of exposure or starvation along the landowner's fence (Ryder et al. 1984, Moody and Alldredge 1986). In 1985, a judge ordered that the fence be removed, but litigation lasted several years while the fence remained in place. This example illustrates several important points germane to the landscape-scale management paradigm: (1) the best possible stewardship of wildlife ranges are fruitless if the migration corridor between them is compromised, (2) effective wildlife management requires knowledge of migration and dispersal routes along with the location of seasonal ranges for all target populations, (3) protection of critical land areas will often require management coordination and agreements among many public agencies and private landowners, and (4) it is likely more efficient and effective to pursue the second and third issues listed above in advance of conflicts like the controversy in south-central Wyoming. Natural resource managers need detailed, year-round movement data on migration and dispersal for all wildlife populations of management importance. Without such data, it is difficult if not impossible to start discussing landscape-scale management because the landscape cannot be defined and partners adequately identified. Wildlife managers and landscape ecologists can collaborate to acquire this information and provide it to policymakers. Communication among scientists, managers, policymakers and the public, including educational outreach, is paramount to successfully managing at landscape scales and overcoming constraints.

Opportunities for Managing Wildlife at Landscape Scales

Landscape ecologists and wildlife managers can work together to create opportunities to improve management on the landscape. This requires collaboration within and among agencies, nongovernmental organizations, private landowners, and universities to identify shared goals and institutional obstacles. Landscape ecologists can help natural resource agencies prioritize management by making their models spatially explicit, recognizing many of the current barriers (e.g., economic, political, administrative) and helping agencies and private landowners understand the relevance of protected areas (e.g., wildlife management areas) to the species the area is designed to conserve.

A current example where opportunities exist to

better manage wildlife at a landscape scale through better collaboration among landscape ecologists and natural resources agencies is in the sagebrush (*Artemisia* spp.) steppe of western North America. The sagebrush steppe is considered one of the most threatened ecosystems in North America (Thompson 2007) in part because of extensive wildfires and invasion by exotic plants (Miller et al. 2011). Following a wildfire, the federal management agency (usually the BLM) develops a rehabilitation plan often aimed at first stabilizing soils and reducing invasion of exotics and then providing livestock grazing opportunities and wildlife habitat. A fundamental component of sage-grouse conservation is restoration of the sagebrush ecosystem damaged by fire (Pyke 2011). In addition to sage-grouse, degraded shrub steppe has caused declines in many other birds, reptiles, mammals, and plants (Rosentreter 1994, Ertter and Nosratinia 2016). Therefore restoration efforts of this ecosystem must be based on a strong scientific foundation and guided by individuals with substantial expertise in ecosystem management, plant ecology, and animal biology. Rehabilitation plans offer an opportunity for landscape, plant, and wildlife ecologists from various state and federal agencies, universities, environmental consulting firms, and private landowners to work together to achieve land management goals. Given the large size of many wildfires in sagebrush landscapes, limited agency resources, and limited seed sources, rehabilitation is currently possible only on small subsets of the affected area. Landscape ecologists can work with wildlife and range managers to focus rehabilitation efforts on key areas within the affected landscape. Managers can help landscape ecologists to improve understanding of landscape function by investing in post-rehabilitation monitoring whereby rehabilitated sites are compared to control and baseline sites over time. Without such comparisons, managers may be expending limited resources on approaches that may be ineffective.

Effects of livestock grazing and fire on sage-grouse populations have been sources of concern for management agencies in western North America. To address this issue, landscape ecologists, a state research biologist, and a range ecologist modeled landscape factors that could affect sage-grouse in an area that was grazed by domestic sheep and had experienced recent wildfire and prescribed burns (Pedersen et al. 2003). They concluded that large fires ($\geq$10% of breeding habitat) at high frequencies (17 years between fires) may lead to sage-grouse population extinction, whereas grazing will not by itself cause extinction but may contribute to population decline. This cooperative effort provided natural resource managers greater insight into this landscape-scale issue and support for revising management programs in ways to explicitly increase the likelihood of small, low-frequency fires and decrease the likelihood of large, high-frequency fires.

Copeland et al. (2014) suggested that conservation of migratory species offers unique challenges because conservation efforts hinge greatly on connectivity of the migration corridor. In some areas, protection can be afforded for greater sage-grouse and mule deer (*Odocoileus hemionus*) migration routes because the two species' ranges overlap, mule deer habitat is similar to sage-grouse habitat, and deer winter ranges and migration corridors are thus protected within the umbrella of sage-grouse conservation. This protection, however, is dependent on developments that might be allowed within these sage-grouse habitats not intersecting with high-priority mule deer stopovers or corridors, and land management policy, which discourages development on deer ranges outside of important sage-grouse areas (Copeland et al. 2014).

Landscape efforts to benefit wildlife populations are becoming more common. State and federal agencies have begun working together to rehabilitate ecosystems to benefit wildlife populations and to provide other ecosystem services. These large-scale, coordinated rehabilitation efforts are a step in the right direction, when done properly. But some vegetation rehabilitation efforts can do more harm than good in sagebrush-steppe landscapes. For example, researchers suggest that standard postfire rehabilitation is too intrusive in several ecosystems in western North America (Downs et al. 2010, Miller et al. 2012). Minimum till drills and seeding shrubs in sagebrush-steppe landscapes are likely more effective than extensive soil disturbance by rangeland drills that plow the soil and thus create an ideal seed bed for cheatgrass (*Bromus tectorum*) and other exotic species. These rangeland drills also disturb biological soil crusts (Miller et al. 2012) that damage mycorrhizal fungi. As a result, these sites must be re-inoculated with fungi; otherwise, reestablishment of shrubs is poor to nonexistent owing to the lack of fungi. Minimizing soil disturbance maximizes fungi survival (Wicklow-Howard 1998). Thus collaboration among landscape, wildlife, and range ecologists enables more effective planning and implementation of rehabilitation programs that target appropriate areas and minimizes likelihood of negative effects.

Greater Sage-Grouse: A Case Study

As a sagebrush-obligate species (Patterson 1952, Connelly et al. 2000b), sage-grouse depend largely on sagebrush for cover and forage throughout most of the year. They also are considered a landscape species (Knick and Connelly 2011) owing to their requirement for extensive and largely contiguous areas of sagebrush to meet their biological needs year round. Sage-grouse populations have declined in range and numbers over the last 60-plus years (Connelly et al. 2004, Schroeder et al. 2004, Garton et al. 2011) owing to loss, fragmentation, and alteration of the sagebrush-steppe landscape across the intermountain West (Ashe 2010, Connelly et al. 2011a).

Sage-grouse typically inhabit large, interconnected expanses of sagebrush (Connelly et al. 2004) and depend on migratory corridors to move among seasonal ranges (Connelly et al. 1988, 2000b). For example, sage-grouse in eastern Idaho during the late 1990s used an annual range of $\geq$2,764 km^2, continuing the same migratory movements first documented in the 1950s (Dalke et al. 1963, Leonard et al. 2000). Historically, sage-grouse distribution was closely tied to distribution of the sagebrush ecosystem (Wambolt et al. 2002, Schroeder et al. 2004). But sage-grouse populations have been extirpated from areas throughout their former range (Wambolt et al. 2002, Schroeder et al. 2004, Aldridge et al. 2008), and the species' current distribution is fragmented compared to that prior to European settlement.

Challenges

Some strongly held beliefs are embedded in natural resource management agencies that pose difficult challenges associated with managing landscape species like sage-grouse. These beliefs result in some management actions repeatedly being implemented without strong scientific support. For example, some land management agencies have issued directives to purportedly help guide conservation efforts for sage-grouse that called for prescribed burning of sagebrush despite many publications demonstrating negative effects of prescribed burning and fire on sage-grouse (Connelly et al. 2000a, Beck et al. 2009, Baker 2011, Hess and Beck 2014), and others have identified loss and fragmentation of sagebrush as a major impediment to conservation of the species (Connelly et al. 2011a, 2011b, Miller et al. 2011). As recently as 2013, land management agencies in the western United States issued directives for fire operations and fuels management related to sage-grouse conservation that did not provide guidance on size of fuels treatments with respect to the landscape being protected, or timing with respect to sage-grouse breeding activities. These directives included lists of best management practices and emphasized sagebrush treatments but did not mention that these treatments can potentially increase spread of invasive grasses, enhance human access into remote sagebrush steppe, and potentially result in more, not less, wildfire if new roads are created in the process (Miller et al. 2011). Future directives would benefit from more details regarding spatial scale and context of treatments so that they improve rather than damage sage-grouse habitat (Connelly 2013). Providing more detailed guidance on size and spatial scale of fire breaks, maintenance of fire breaks, spatial scale of the contiguous sagebrush cover, fuel treatment locations relative to sage-grouse habitat, invasive species control, maintenance of large contiguous blocks of sagebrush, a prohibition on prescribed burning in winter and breeding habitat, and explicit definitions of all terms used will help improve the likelihood that they achieve the intended objective (Connelly 2013). Policymakers and managers may not have time to read the scientific literature, and scientists may not have time, or believe it is their job, to ensure that the most current results are disseminated to managers and policymakers. How do we change that culture? Monthly or quarterly meetings or webinars that engage scientists with decision makers regularly and repeatedly on important wildlife conservation challenges could help research findings better inform policies (Vierling et al., Chap. 11, this volume).

Another example of a situation where strongly held beliefs are counterproductive occurs when government agencies use the word "success" when referring to habitat treatments but fail to provide a sufficient definition of that term. Connelly (2013) stated the ultimate definition of success of any project intended to benefit sage-grouse, or any species, must be a positive change in vital rates (e.g., increased populations or dampening population declines). Kilometers of fence marked, number of grazing systems employed, or hectares of junipers (*Juniperus* spp.) cut within a given landscape often occur at relatively small spatial extents and should not be considered successful unless they result in measurable increases or stabilization of sage-grouse numbers (Connelly 2013). Field experiments followed by peer-reviewed publications are necessary to verify that a given habitat treatment or management action improves the landscape for sage-grouse. Without such research, support for these assertions is unsubstantiated. Publishing peer-reviewed research

findings in literature that natural resource managers and private landowners read also is important for disseminating newly acquired knowledge and management options. Natural resource managers will benefit by seeking solid research with testable hypotheses. Landscape ecologists can work with resource managers to design and implement research that test these hypotheses; managers and policymakers should be included in research projects from start to finish, allowing general agreement on the success of a given project or management strategy.

Habitat loss from wildfire and conifer encroachment also poses a challenge for wildlife managers in sagebrush steppe ecosystems. Wildfires in sagebrush steppe commonly exceed 10,000 ha, and conifer encroachment affects thousands of hectares throughout the Great Basin (Kuchy 2008, Miller et al. 2011). Losses of sage-grouse habitat over large landscapes within the intermountain West and Great Basin have been attributed to increased fire frequency in lower-elevation sagebrush, often closely tied to invasion of annual grasses such as cheatgrass (Miller et al. 1994, Crawford et al. 2004, Davies et al. 2011, Balch et al. 2013, Bradley et al. 2018). Additionally, decreased fire frequency in higher-elevation habitats and effects from inappropriate livestock grazing practices and other factors have resulted in conifer encroachment and subsequent reduction of the herbaceous understory and sagebrush canopy cover over large areas (Miller and Rose 1995, Miller and Eddleman 2001, Crawford et al. 2004). Recent collaborative efforts by the state of Idaho to encourage the BLM to reduce response time to wildfire and involve ranchers in local fire protection districts across entire landscapes have helped overcome some threats posed by wildfire within sagebrush-steppe ecosystems; these actions were in part predicated on advice from landscape ecologists and wildlife biologists (Idaho State 2012).

The BLM recently released a Draft Programmatic Environmental Impact Statement for the Great Basin that proposes establishing >17,000 km of fuel (fire) breaks on BLM administered lands in California, Idaho, Nevada, Oregon, Utah, and Washington, USA, within the sagebrush-steppe ecosystem (Bureau of Land Management 2019). From a landscape perspective, the goal of a management strategy that seeks to benefit sage-grouse should not be to achieve some arbitrary numbers for length of fuel breaks created, but rather fuel breaks should be strategically placed in areas with degraded habitat prone to wildfire and along major highways (e.g., interstate highways). Fuel breaks will be most effective if they are placed where they not only make sense from a firefighting perspective, but also avoid further fragmenting sage-grouse habitat. Research has yet to show that fuel breaks have a net positive outcome on the landscape, and implementing fuel breaks to reduce negative effects of fire in sagebrush ecosystems is effectively an experiment that is not feasible to replicate at the appropriate scales (Shinneman et al. 2018). Thus involving landscape ecologists and wildlife biologists in the planning and ultimately evaluation of fuel breaks will most likely produce a scientifically defensible approach that has the best chance of reaching management goals.

Constraints

Causes of loss, fragmentation, and degradation of sage-grouse habitat in sagebrush-dominated landscapes vary but often include brush control by land management agencies (Klebenow 1970, Martin 1970, Wallestad 1975), inappropriate grazing practices (Beck and Mitchell 2000), energy development (Lyon and Anderson 2003), urbanization (Leu and Hanser 2011), and various types of infrastructure (Braun et al. 2002, Johnson et al. 2011). Unfortunately, few studies have rigorously examined any of these threats at the landscape level regarding their relative effect on sage-grouse populations and habitat. Where such research has been conducted, results suggest extensive changes in sagebrush cover can substantially affect sage-grouse populations. Specifically, on the Upper Snake River Plain in eastern Idaho, nearly 30,000 ha of sagebrush in one study

area were converted to cropland between 1975 and 1992. This change corresponded to a substantial decrease (>45%) in sage-grouse numbers over this 17-year period (Leonard et al. 2000). Comparable losses were documented in North Park, Colorado, USA, following plowing and spraying 28% of the study area with 2,4-D (Braun and Beck 1996). Similarly, Leu and Hanser (2011) reported increasing the human footprint on the sagebrush landscape decreased sagebrush landcover, and reduced mean size of sagebrush patches.

A recent study (Dahlgren et al. 2015) reported that sage-grouse lek counts initially increased (compared to surrounding populations) following implementation of many 100- to 200-ha sagebrush treatments (11,386 ha treated; ~1.5% per year, or 15% of all sagebrush treated) to reduce shrub cover and increase herbaceous understory. These higher lek counts continued for almost 15 years, but with continued sagebrush treatments and the onset of adverse winter conditions, lek counts declined to levels similar to those of surrounding areas (Dahlgren et al. 2015). A better understanding of the relationship between the treatments and overall sagebrush landscape may have prevented population decline. Through collaboration and more landscape-scale research, wildlife managers and landscape ecologists can better understand the effect of various types of treatments and other activities and provide this information to policymakers and the public.

Opportunities

Sage-grouse respond to environmental conditions at local and landscape scales. In central Wyoming, landscape characteristics that influenced nest-site selection included surface roughness, surface ruggedness, elevation, and slope (Jensen 2006). In Idaho, at varying spatial extents, the primary differences between successful and unsuccessful sage-grouse nests were amount of sagebrush (positive effect on nesting) and grass-forb-dominated habitat (negative effect on nesting) on the landscape (Shepherd et al. 2011a). The percentage of sagebrush cover was an important predictor of winter use by sage-grouse in Wyoming and Montana, and the strength of association with sagebrush cover was strongest at a large (4 km²) spatial scale (Doherty et al. 2008). In contrast, Shepherd et al. (2011b) concluded that during summer and early fall, sage-grouse response to landscape fragmentation and habitat loss was related to interspersion of habitats and not amount of sagebrush cover. Sage-grouse in North Park, Colorado, concentrated during winter in seven small areas that totaled 85 km²; these areas made up only 7% of the sagebrush in the entire study area (Beck 1977). These results illustrate that spatial extent and season are important in understanding sage-grouse use of landscapes and response to environmental conditions.

Although sage-grouse are considered a landscape species, no conclusive data are available on minimum patch sizes of relatively contiguous sagebrush necessary to support a viable population. In Wyoming, sage-grouse flocks ranged over several thousand square kilometers (Patterson 1952). Migratory populations of sage-grouse may use areas exceeding 2,700 km² (Connelly et al. 2000b, Leonard et al. 2000). In contrast, sagebrush patches used by broods averaged 86 ha in early summer in central Montana but diminished to 52 ha in late summer (Wallestad 1971). Brood use areas are relatively small compared to areas used on a year-round basis (Wallestad 1971). Typically, the home range of a sage-grouse population comprises lands managed by multiple agencies and privately owned lands. The BLM and US Forest Service manage 59% of sage-grouse habitat (Knick 2011). Moreover, the Natural Resources Conservation Service can influence management of thousands of hectares of private land (Connelly 2013). Landscape ecologists could collaborate with resource managers to design and implement research that would allow managers to better understand and manage appropriate patch size and landscape characteristics to meet sage-grouse needs while taking into account the challenges imposed by multiple agencies involved in managing each population.

Sage-grouse usually occupy areas with a diversity of species and subspecies of sagebrush but may also use other vegetative communities or lands that are often intermixed, including riparian meadows, farmland, and steppe dominated by native grasses and forbs, shrub willow (*Salix* spp.), and sagebrush steppe with some conifer or quaking aspen (*Populus tremuloides*) (Patterson 1952, Dalke et al. 1963, Savage 1969). Sage-grouse frequently use traditional migratory corridors that allow them to move among two or three seasonal ranges (Dalke et al. 1963, Connelly et al. 1988, 2000b, Beck et al. 2006, Jensen 2006). Because a large proportion of these important habitats are managed by federal agencies, natural resource managers and landscape ecologists have many opportunities to develop and implement interagency conservation programs that provide protection for these key areas.

Finally, Hanser and Knick (2011) assessed the efficacy of managing shrub-steppe landscapes for habitat characteristics suitable for sage-grouse, assuming that other sagebrush-dependent species also would benefit. They concluded that habitat management that focused on creating a narrow set of plot-scale conditions likely would not be as effective as efforts that recognized and managed for landscape-scale heterogeneity (Hanser and Knick 2011).

Management Implications

Available evidence demonstrates that conserving large landscapes with suitable habitat is important for conservation of sage-grouse (Eng and Schladweiler 1972, Connelly et al. 2004, Doherty et al. 2008). Fire kills sagebrush and thus is a major threat to sage-grouse conservation. For example, fire burned 8,374,864 ha of sage-grouse habitat from 1984 to 2013, or 21% of the species' western range (Brooks et al. 2015). Given the relatively large loss of sagebrush cover over the landscape, Woodward (2006) claimed that some portions of sage-grouse habitat may benefit from management actions that resulted in more herbaceous cover, but only if doing so did not cause any loss of sagebrush. A landscape-scale model that examined the factors that influence persistence of sage-grouse populations demonstrated that the most effective sage-grouse conservation efforts were to maintain large expanses of sagebrush and enhance the quality and connectivity of such patches (Aldridge et al. 2008)..

Connelly (2013) suggested that federal natural resource agencies can conserve sage-grouse populations by first stopping habitat loss. This can be accomplished by terminating all sagebrush control projects in sage-grouse breeding and wintering habitat. Area closures to motorized vehicles during peak fire season can likely aid in reducing ignition in all sagebrush areas within and adjacent to sage-grouse habitat (Connelly 2013). Additionally, federal agencies can benefit sage-grouse if they develop rural fire protection associations (Idaho State 2012), aggressively suppress fire in key sage-grouse areas, and strategically implement firebreaks to reduce the threat of large fires. Nevertheless, rural fire protection associations and firebreaks will likely be most effective if designed across entire landscapes in collaboration with knowledgeable university and state agency personnel, including landscape ecologists and wildlife biologists and then implemented in a manner that absolutely minimizes loss of sagebrush within critical sage-grouse habitat. In other words, all aspects of sage-grouse management—including fire management plans, fire suppression efforts, harvest, habitat management, research, and policy—will be most effective if they are implemented at a landscape scale and involve landscape ecologists and managers working together.

To improve wildlife management at landscape scales, we suggest that management actions be coordinated across multiple agencies and multiple field offices within agencies to ensure all management actions (to the extent possible) are replicates of others throughout the landscape and the effectiveness of those actions are documented thoroughly. We realize some management actions might need to be implemented without rigorous scientific studies that demonstrate their effectiveness, but committing to monitoring effectiveness of the management actions over many years (i.e., >10 years in many cases) will help assess efficacy of the actions. This commitment involves rethinking how research efforts are selected and designed as well as how management actions are implemented (Williams et al. 2004). It further underscores the need for collaboration among landscape ecologists, wildlife managers, and wildlife researchers in planning and implementing research so that effective management actions are more likely to be implemented at appropriate landscape scales.

Summary

Wildlife management at landscape scales can be more tractable if a concerted effort is made to identify species of management importance for which detailed year-round movement data are not complete and to prioritize such information as a management need. This effort is more likely to be successful if it includes all agencies or organizations that may be involved with managing habitat or populations of the species of interest and involves both wildlife and landscape scientists. Knowledge of year-round movements is imperative for discussing landscape-scale management because only then are all key habitats, migration corridors, and interest groups known.

Finally, staying current with research, monitoring efforts, and effectiveness of management actions across disciplines helps provide a more interdisciplinary approach to management options. Biannual "think tank" management sessions on individual species or small groups of related species (e.g., sagebrush-dependent species, waterbirds) that include wildlife and habitat managers, species experts, and landscape ecologists could improve information transfer, management needs, consensus management actions at the appropriate spatial extent, identification of research funding priorities, and research design. Meeting formats can be formulated such that an effective organization can be replicated for all such sessions (i.e., replicated across all species of management interest). The fields of landscape ecology and wildlife management need explicit mechanisms to work together regularly and more seamlessly. Landscape ecologists can benefit tremendously by showing how principles and concepts of landscape ecology can be used to recover declining species, and wildlife managers and policymakers can benefit by leaning on landscape ecologists to help them overcome the many challenges of managing populations at the spatial scale at which they interact with their environment: the landscape scale.

LITERATURE CITED

Ahern, K., and L. Cole. 2012. Landscape scale—towards an integrated approach. ECOS 33:6–12.

Aldridge, C. L., S. E. Nielsen, H. L. Beyer, M. S. Boyce, J. W. Connelly, S. T. Knick, and M. A. Schroeder. 2008. Range-wide patterns of greater sage-grouse persistence. Diversity and Distributions 14:983–994.

Apa, A. D. 1998. Habitat use and movements of sympatric sage and Columbian sharp-tailed grouse in southeastern Idaho. Dissertation, University of Idaho, Moscow, Idaho, USA.

Ashe, D. M. 2010. Endangered and threatened wildlife and plants: 12-month findings for petitions to list the greater sage-grouse (Centrocercus urophasianus) as threatened or endangered. 75 Fed. Reg. 13909 (23 Mar 2010).

Baker, W. L. 2011. Pre-Euro-American and recent fire in sagebrush ecosystems. Studies in Avian Biology 38:185–202.

Balch, J. K., B. A. Bradley, C. M. Dantonio, and J. Gómez-Dans. 2013. Introduced annual grass increases regional fire activity across the arid western USA (1980–2009). Global Change Biology 19:173–183.

Barea-Azcón, J. M., E. Virgós, E. Ballesteros-Duperón, M. Moleón, and M. Chirosa. 2007. Surveying carnivores at large spatial scales: a comparison of four broad-applied methods. Biodiversity and Conservation 16:1213–1230.

Beck, J. L., J. W. Connelly, and K. P. Reese. 2009. Recovery of greater sage-grouse habitat features in xeric sagebrush communities following prescribed fire. Restoration Ecology 17:393–403.

Beck, J. L., and D. L. Mitchell. 2000. Influences of livestock grazing on sage-grouse habitat. Wildlife Society Bulletin 28:993–1002.

Beck, J. L., K. P. Reese, J. W. Connelly, and M. B. Lucia. 2006. Movements and survival of juvenile greater sage-grouse in southeastern Idaho. Wildlife Society Bulletin 34:1070–1078.

Beck, T. D. I. 1977. Sage grouse flock characteristics and habitat selection during winter. Journal of Wildlife Management 41:18–26.

Bissonette, J. A. 1997. Wildlife and landscape ecology: effects of pattern and scale. Springer-Verlag, New York, New York, USA.

Bradley, B. A., C. A. Curtis, E. J. Fusco, J. T. Abatzoglou, J. K. Balch, S. Dadashi, and M. N. Tuanmu. 2018. Cheatgrass (Bromus tectorum) distribution in the intermountain western United States and its relationship to fire frequency, seasonality, and ignitions. Biological Invasions 20:1493–1506.

Braun, C. E., and T. D. I. Beck. 1996. Effects of research on sage grouse management. Transactions of the North American Wildlife and Natural Resources Conference 61:429–436.

Braun, C. E., O. O. Oedekoven, and C. L. Aldridge. 2002. Oil and gas development in western North America: effects on sagebrush steppe avifauna with particular emphasis on sage grouse. Transactions of the North American Wildlife and Natural Resources Conference 67:337–349.

Brennan, L. A. 1991. How can we reverse the northern bobwhite decline? Wildlife Society Bulletin 19:544–555.

Brooks, M. L., J. R. Matchett, D. J. Shinneman, and P. S. Coates. 2015. Fire patterns in the range of greater sage-grouse, 1984–2013—implications for conservation and management. US Geological Survey Open-File Report 2015-1167, Reston, Virginia, USA.

Bureau of Land Management. 2019. Draft programmatic environmental impact statement for the Great Basin. https://eplanning.blm.gov/epl-front-office/projects/nepa/71149/175038/212593/DraftFuelBreaksPEIS_VolumeI.pdf. Accessed 1 Aug 2019.

Cardinal, C. J. 2015. Factors influencing the ecology of greater sage-grouse inhabiting the Bear Lake Plateau and Valley, Idaho and Utah. Thesis, Utah State University, Logan, Utah, USA.

Connelly, J. W. 2013. Federal agency responses to greater sage-grouse and the ESA: getting nowhere fast. Northwest Science 88:61–64.

Connelly, J. W., H. W. Browers, and R. J. Gates. 1988. Seasonal movements of sage grouse in southeastern Idaho. Journal of Wildlife Management 52:116–122.

Connelly, J. W., S. T. Knick, C. E. Braun, W. L. Baker, E. A. Beever, T. J. Christiansen, K. E. Doherty, E. O. Garton, S. E. Hanser, D. H. Johnson, et al. 2011a. Conservation of greater sage-grouse: a synthesis of current trends and future management. Studies in Avian Biology 38:549–564.

Connelly, J. W., S. T. Knick, M. A. Schroeder, and S. J. Stiver. 2004. Conservation assessment of greater sage-grouse and sagebrush habitats. Western Association of Fish and Wildlife Agencies, Cheyenne, Wyoming, USA.

Connelly, J. W., K. P. Reese, R. A. Fischer, and W. L. Wakkinen. 2000a. Response of a sage grouse breeding population to fire in southeastern Idaho. Wildlife Society Bulletin 28:90–96.

Connelly, J. W., E. T. Rinkes, and C. E. Braun. 2011b. Characteristics of greater sage-grouse habitats: a landscape species at micro and macro scales. Studies in Avian Biology 38:69–84.

Connelly, J. W., M. A. Schroeder, A. R. Sands, and C. E. Braun. 2000b. Guidelines to manage sage grouse populations and their habitats. Wildlife Society Bulletin 28:967–985.

Copeland, H. E., H. Sawyer, K. L. Monteith, D. E. Naugle, A. Pocewicz, N. Graf, and M. J. Kauffman. 2014. Conserving migratory mule deer through the umbrella of sage-grouse. Ecosphere 5(9):117.

Crawford, J. A., R. A. Olson, N. E. West, J. C. Mosley, M. A. Schroeder, T. D. Whitson, R. F. Miller, M. A. Gregg, and C. S. Boyd. 2004. Ecology and management of sage-grouse and sage-grouse habitat. Journal of Range Management 57:2–19.

Dahlgren, D. K., R. T. Larsen, R. Danvir, G. Wilson, E. T. Thacker, T. A. Black, D. E. Naugle, J. W. Connelly, and T. A. Messmer. 2015. Greater sage-grouse and range management: insights from a 25-year case study in Utah and Wyoming. Rangeland Ecology and Management 68:375–382.

Dalke, P. D. 1937. The cover map in wildlife management. Journal of Wildlife Management 1:100–105.

Dalke, P. D., D. B. Pyrah, D. C. Stanton, J. E. Crawford, and E. F. Schlatterer. 1963. Ecology, productivity, and management of sage-grouse in Idaho. Journal of Wildlife Management 27:810–841.

Davies, K. W., C. S. Boyd, J. L. Beck, J. D. Bates, T. J. Svejcar, and M. A. Gregg. 2011. Saving the sagebrush sea: an ecosystem conservation plan for big sagebrush plant communities. Biological Conservation 144:2573–2584.

Doherty, K. E., D. E. Naugle, B. L. Walker, and J. M. Graham. 2008. Greater sage-grouse winter habitat selection and energy development. Journal of Wildlife Management 72:187–195.

Downs, J. L., M. A. Chamness, C. M. Perry, S. D. Powell, and A. M. Playter. 2010. Effects of wildfire on plant communities. Section 8.14.2.2 in T. M. Poston, J. P. Duncan, and R. L. Dirkes, editors. Hanford Site environmental report for calendar year 2009, Pacific Northwest National Laboratory, Richland, Washington, USA.

Eng, R. L., and P. Schladweiler. 1972. Sage grouse winter movements and habitat use in central Montana. Journal of Wildlife Management 36:141–146.

Ertter, B., and S. Nosratinia. 2016. A new variety of *Abronia mellifera* (Nyctaginaceae) of conservation concern in southwestern Idaho. Phytoneuron 20:1–4.

European Landscape Convention. 2000. Explanatory report to the European Landscape Convention, European Treaty Series No. 176. https://rm.coe.int/16800cce47. Accessed 20 Aug 2019.

Garton, E. O., J. W. Connelly, J. S. Horne, C. A. Hagen, A. Moser, and M. A. Schroeder. 2011. Greater sage-grouse population dynamics and probability of persistence. Studies in Avian Biology 38:293–382.

Hanser, S. E., and S. T. Knick. 2011. Greater sage-grouse as an umbrella species for shrubland passerine birds: a multiscale assessment. Studies in Avian Biology 38:473–487.

Harvey, T. 2015. Can we bring the quail back? Texas Parks and Wildlife Magazine. http://tpwmagazine.com/archive/2015/dec/ed_3_quail/index.phtml. Accessed 18 Apr 2016.

Hernández, F., L. A. Brennan, S. J. DeMaso, J. P. Sands, and D. B. Wester. 2012. On reversing the northern bobwhite population decline: 20 years later. Wildlife Society Bulletin 37:177–188.

Hess, J. E., and J. L. Beck 2014. Forb, insect, and soil response to burning and mowing Wyoming big sagebrush in greater sage-grouse breeding habitat. Journal of Environmental Management 53:813–822.

Idaho State. 2012. Federal alternative of Governor C. L. "Butch" Otter for greater sage-grouse management in Idaho. https://species.idaho.gov/wp-content/uploads/sites/82/2016/05/Idaho-Sage-Grouse-Alternative.pdf. Accessed 22 Apr 2015.

Jensen, B. M. 2006. Migration, transition range and landscape use by greater sage-grouse (*Centrocercus urophasianus*). Thesis, University of Wyoming, Laramie, Wyoming, USA.

Johnson, D. H., M. J. Holloran, J. W. Connelly, S. E. Hanser, C. L. Amundson, and S. T. Knick. 2011. Influences of environmental and anthropogenic features on greater sage-grouse populations, 1997–2007. Studies in Avian Biology 38:407–450.

Klebenow, D. A. 1970. Sage grouse versus sagebrush control in Idaho. Journal of Range Management 23:396–400.

Knick, S. T. 2011. Historical development, principal federal legislation, and current management of sagebrush habitats: implications for conservation. Studies in Avian Biology 38:13–32.

Knick, S. T., and J. W. Connelly. 2011. Greater sage-grouse and sagebrush: an introduction to the landscape. Studies in Avian Biology 38:1–9.

Kuchy, A. L. 2008. Impacts of climate variability on large fires across spatiotemporal scales in sagebrush steppe. Thesis, University of Idaho, Moscow, Idaho, USA.

Lendrum, P. E., C. R. Anderson Jr., K. L. Monteith, J. A. Jenks, and R. T. Bowyer. 2013. Migrating mule deer: effects of anthropogenically altered landscapes. PLOS One 8(5):e64548.

Leonard, K. M., K. P. Reese, and J. W. Connelly. 2000. Distribution, movements, and habitats of sage grouse Centrocercus urophasianus on the Upper Snake River Plain of Idaho: changes from the 1950's to the 1990's. Wildlife Biology 6:265–270.

Leopold, A. 1933. Game management. Charles Scribner's Sons, New York, New York, USA.

Leu, M., and S. E. Hanser. 2011. Influences of the human footprint on sagebrush landscape patterns. Studies in Avian Biology 38:253–272.

Liu, J., and W. W. Taylor. 2002. Integrating landscape ecology into natural resource management. Cambridge University Press, Cambridge, UK.

Lyon, A. G., and S. H. Anderson. 2003. Potential gas development impacts on sage-grouse nest initiation and movement. Wildlife Society Bulletin 31:486–491.

Lyon, L. J., and A. G. Christensen. 2002. Elk and land management. Pages 557–581 in D. E. Toweill and J. W. Thomas, editors. North American elk: ecology and management. Smithsonian Institution Press, Washington, DC, USA.

Martin, N. S. 1970. Sagebrush control related to habitat and sage grouse occurrence. Journal of Wildlife Management 34:313–320.

Miller, M. E., M. A. Bowker, R. L. Reynolds, and H. L. Goldstein. 2012. Post-fire land treatments and wind erosion—lessons from the Milford Flat Fire, UT, USA. Aeolian Research 7:29–44.

Miller, R. F., and L. L. Eddleman. 2001. Spatial and temporal changes of sage grouse habitat in the sagebrush biome. Oregon Agricultural Experiment Station Bulletin 151, Corvallis, Oregon, USA.

Miller, R. F., S. T. Knick, D. A. Pyke, C. W. Meinke, S. E. Hanser, M. J. Wisdom, and A. L. Hild. 2011. Characteristics of sagebrush habitats and limitations to long-term conservation. Studies in Avian Biology 38:145–184.

Miller, R. F., and J. R. Rose. 1995. Historic expansion of Juniperus occidentalis (western juniper) in southeastern Oregon. Great Basin Naturalist 55(1):37–45.

Miller, R. F., T. J. Svejcar, and N. E. West. 1994. Implications of livestock grazing in the intermountain sagebrush region: plant composition. Pages 101–146 in M. Vavra, W. A. Laycock, and R. D. Pieper, editors. Ecological implications of herbivory in the West. Society for Range Management, Denver, Colorado, USA.

Moody, D. S., and A.W. Alldredge. 1986. Red Rim—mining, fencing and some decisions. Proceedings of the 12th Pronghorn Antelope Workshop. Nevada Department of Wildlife, Reno, Nevada, USA.

Palmer, W. E., and D. C. Sisson. 2017. Tall Timbers' bobwhite quail management handbook. Tall Timbers Press, Tallahassee, Florida, USA.

Patterson, R. L. 1952. The sage grouse in Wyoming. Sage Books, Denver, Colorado, USA.

Pedersen, E. K., J. W. Connelly, J. Hendrickson, and W. E. Grant. 2003. Effect of sheep grazing and fire on sage grouse populations in southeastern Idaho. Ecological Modeling 165:23–47.

Perras, M., and S. Nebel. 2012. Satellite telemetry and its impact on the study of animal migration. Nature Education Knowledge 3(12):4.

Plumb, G. E., P. J. White, M. B. Coughenour, and R. L. Wallen. 2009. Carrying capacity, migration, and dispersal in Yellowstone bison. Biological Conservation 142:2377–2387.

Pyke, D. A. 2011. Restoring and rehabilitating sagebrush habitats. Studies in Avian Biology 38:531–548.

Robbins, C. S., J. W. Fitzpatrick, and P. B. Hamel. 1992. A warbler in trouble: Dendroica cerulea. Pages 549–562 in J. W. Hagen III and D. W. Johnston, editors. Ecology and conservation of neotropical migrant landbirds. Smithsonian Institution Press, Washington, DC, USA.

Rosene, W. 1969. The bobwhite quail: its life and management. Rutgers University Press, New Brunswick, New Jersey, USA.

Rosentreter, R. 1994. Displacement of rare plants by exotic grasses. Pages 170–175 in S. D. Monsen and S. G. Kitchen, compilers. Proceedings of the symposium: ecology and management of annual rangelands. US Forest Service General Technical Report INT-GTR-313, Ogden, Utah, USA.

Ryder, T. J., L. L. Irwin, and D. S. Moody. 1984. Wyoming's Red Rim pronghorn controversy: history and current status. Proceedings of the Pronghorn Antelope Workshop 11:195–206.

Savage, D. E. 1969. Relation of sage grouse to upland meadows in Nevada. Nevada Fish and Game Commission, Job Completion Report, Project W-39-R-9, Job 12, Reno, Nevada, USA.

Sawyer, H. S., M. J. Kauffman, R. M. Nielson, and J. Horne. 2009. Identifying and prioritizing ungulate migration routes for landscape-level conservation. Ecological Applications 19:2016–2025.

Schroeder, M. A., C. L. Aldridge, A. D. Apa, J. R. Bohne, C. E. Braun, S. D. Bunnell, J. W. Connelly, P. A. Deibert, S. C. Gardner, M. A. Hilliard, et al. 2004. Distribution of sage-grouse in North America. Condor 106:363–376.

Shea, R. E., H. K. Nelson, L. N. Gillette, J. G. King, and D. K. Weaver. 2002. Restoration of trumpeter swans in North America: a century of progress and challenges. Waterbirds 25 (Special Publication 1): 296–300.

Shepherd, J. F., J. W. Connelly, and K. P. Reese. 2011a. Modeling landscape-scale greater sage-grouse nesting and brood-rearing habitat. Studies in Avian Biology 39:137–150.

Shepherd, J. F., K. P. Reese, and J. W. Connelly. 2011b. Effects of landscape fragmentation on non-breeding greater sage-grouse. Studies in Avian Biology 39:77–88.

Shinneman, D. J., C. L. Aldridge, P. S. Coates, M. J. Germino, D. S. Pilliod, and V. M. Vaillant. 2018. A conservation paradox in the Great Basin—altering sagebrush landscapes with fuel breaks to reduce habitat loss from wildfire. US Geological Survey Open-File Report 2018-1034, Reston, Virginia, USA.

Sinclair, A. R. E., J. M. Fryxell, and G. Caughley. 2006. Wildlife ecology, conservation, and management. Second edition. Blackwell, Malden, Massachusetts, USA.

Stevens, B. S., and C. J. Conway. 2020. Predictive multi-scale occupancy models at range-wide extents: effects of habitat and human disturbance on distributions of wetland birds. Diversity and Distributions 26:34–48.

Thompson, J. 2007. Sagebrush in western North America: habitats and species in jeopardy. PNW Science Findings, Issue 91, Pacific Northwest Research Station, US Forest Service, Portland, Oregon, USA.

Turner, M. G. 1989. Landscape ecology: the effect of pattern on process. Annual Review of Ecology and Systematics 20:171–197.

Wallestad, R. O. 1971. Summer movements and habitat use by sage grouse broods in central Montana. Journal of Wildlife Management 35:129–136.

Wallestad, R. O. 1975. Male sage-grouse responses to sagebrush treatment. Journal of Wildlife Management 39:482–484.

Wambolt, C. L., A. J. Harp, B. L. Welch, N. Shaw, J. W. Connelly, K. P. Reese, C. E. Braun, D. A. Klebenow, E. D. McArthur, J. G. Thompson, et al. 2002. Conservation of greater sage-grouse on public lands in the western U.S.: implications of recovery and management policies. Policy Analysis Center for Western Public Lands, Policy Paper SG-02-02, Caldwell, Idaho, USA.

Web of Science. 2019. Web of Science homepage. https://www.webofknowledge.com/. Accessed 20 Aug 2019.

Wicklow-Howard, M. 1998. The role of mycorrhizae in rangelands. Ellen Trueblood Symposium: US Department of Interior, Idaho Bureau of Land Management Technical Bulletin 1998-01(January), Boise, Idaho, USA.

Williams, B. K., F. A. Johnson, and K. Wilkins. 1996. Uncertainty and the adaptive management of waterfowl harvests. Journal of Wildlife Management 60:223–232.

Williams, C. K., F. S. Guthery, R. D. Applegate, and M. J. Peterson. 2004. The northern bobwhite decline: scaling our management for the twenty first century. Wildlife Society Bulletin 32:861–869.

Woodward, J. K. 2006. Greater sage-grouse (Centrocercus urophasianus) habitat in central Montana. Thesis, Montana State University, Bozeman, Montana, USA.

—11— Improving Communication between Landscape Ecologists and Managers

Kerri T. Vierling,
Joseph D. Holbrook,
Jocelyn L. Aycrigg,
Teresa C. Cohn, and
Leona K. Svancara

Challenges and Opportunities

Introduction

Effective science communication is increasingly important in environmental decision making, whether decisions involve complex socio-scientific debates (Shome et al. 2009, Bowman et al. 2010) or individual choices (Fischhoff and Scheufele 2013, von Winterfeldt 2013). Science communication has the potential to inform public policy (Scheufele 2014), contribute to well-needed understanding of environmental processes (Lubchenco 1998), and garner public support for science (Treise and Weigold 2002). Perhaps most importantly, however, science communication offers the basic knowledge necessary to make informed choices amid increasingly interdisciplinary environmental concerns (Fischhoff 2013).

Effective science communication requires multiple approaches. First, knowing your audiences (Shome et al. 2009, Maibach et al. 2011) and framing information appropriately are critical to communicating well (Nisbet and Mooney 2007, Bubela et al. 2009). Participation and dialogue with an audience, rather than simply disseminating knowledge, enhance effective communication, particularly in collaborative processes (Bucchi 2008, Cooke et al. 2017). Different modes of learning are import-

ant to consider; the use of graphics and alternative written approaches increases the ability of scientists to communicate ideas to a variety of audiences (Kuehne and Olden 2015, Rodríguez Estrada and Davis 2015). Whether scientists are communicating through media, with stakeholder groups, or even among scientists of different disciplines, the above approaches support the interchange of knowledge essential in making informed choices (Fischhoff 2013).

The relationship between landscape ecologists and fish, wildlife, and land managers (i.e., managers) provides a unique setting from which to view effective communication. Landscape ecologists fundamentally work on issues of scale, and their products and discoveries are often shared with others through maps, summary statistics, and papers written for peer-reviewed journals. Landscape ecologists typically analyze spatial data and statistics to produce thematic maps representing the spatial arrangement of specific environmental features and changes in this arrangement through time. Thematic maps allow managers to better understand the spatial arrangement of critical landscape features that aid environmental decision making, such as the distribution of habitat for sensitive species. Additionally, thematic maps are effective ways for managers to communicate with the public about important issues requiring stakeholder input, and public participatory geographic information systems (PPGIS) practices are becoming more common (Sieber 2006, Brown and Kyttä 2014). In PPGIS approaches, stakeholders identify locations or draw polygon boundaries of place meanings or values (Lowery and Morse 2013). Such approaches provide scale-specific information on perceptions of the landscape that facilitate the incorporation of social perspectives into land management (e.g., environmental impact statements, or EISs; Brymer et al. 2016).

Thematic maps of appropriate scale and resolution are effective in assisting in management decisions, but clear communication between landscape ecologists and managers is critical to prevent misinterpretations of these maps. Misinterpretations occur when the spatial extent (study area size), thematic resolution (detail conveyed within maps), or grain (spatial resolution) is not clearly displayed or when these characteristics are not well matched with the management goal (Morrison et al. 2006). Similarly, misinterpretations arise when the limitations of the methods used to develop the maps are not clearly portrayed. Maps produced for one purpose may be inappropriate in a different context, and it is often difficult for practitioners, stakeholders, and the public to understand potential limitations of maps when viewed out of the original development context.

Ideally, landscape ecologists and managers work together from the beginning of a project through its completion on management relevant topics. Existing maps, however, are often used to address a management need regardless of its original purpose. Although existing maps may be appropriate for multiple uses, there is an increased risk of misinterpretation if the maps are used outside of their original intended purpose. For instance, the US Geological Survey Gap Analysis Program (GAP) develops landcover maps and species distribution models specifically for conducting analyses and identification of gaps in the conservation of biodiversity at state and regional levels (Scott et al. 1993). Products from GAP determinations are used for multiple management and conservation assessments at a variety of scales (Aycrigg et al. 2013, 2015a, 2015b, La Sorte et al. 2015) with varied success (Lorenz et al. 2015). Successfully applying GAP products depends on maintaining the extent of inference at the level (spatial, temporal, thematic scales) originally developed. Similarly, maps of landscape connectivity are often misunderstood and misapplied, increasing the risk of inappropriate management actions. Landscape connectivity maps and models are derived from spatial data representing the human footprint, originally defined as a quantitative evaluation of human influences on the land surface (Sanderson et al. 2002). Areas with the greatest landscape integrity (i.e., least amount of human footprint) are analyzed in various manners to determine linkages or corridors among sites of interest (e.g., protected areas). Mapped elements, developed by landscape ecologists, reflect the structural connectivity of an area but differ from the functional or realized connectivity (e.g., successful movement of a species from one area to another with subsequent survival and reproduction; Hess and Fischer 2001, Fremier et al. 2015). Potential misinterpretations of map elements can have consequences for broadscale conservation and restoration activities if mapping parameters are not effectively communicated or interpreted correctly.

Managers frequently use landscape ecology products; thus it is essential to examine how to effectively communicate landscape ecology concepts. We used communication research to highlight effective strategies among interdisciplinary teams of scientists, specifically between landscape ecologists and managers. We highlight three specific management case studies to illustrate applications of landscape ecology products in management and how communication between landscape ecologists and managers might improve. Each case study provides the management context, issues and challenges, and possible solutions. We selected a set of case studies representing the diversity of challenges that managers face. We summarize potential consequences of miscommunication between managers and landscape ecologists and provide a broad suite of suggestions to help improve communication.

Case Study 1: Closing the Communication Gap by Linking Landscape Research with On-the-Ground Needs of Managers

The context of this case study expects managers to make informed management decisions that include vertebrates and invertebrate species. Because different taxonomic groups operate at different spatial scales, they are differentially affected by landscape characteristics and management activities. The issues and challenges include access to and limited time to read peer-reviewed manuscripts related to the subject. Landscape ecologists also lack managers' training and experience in environmental assessment and decision making, which may be an obstacle to presenting research findings in a way that can be directly applied to on-the-ground activities. Possible solutions to these issues include encouraging landscape ecologists to publish in open access journals or in journals that include graphical abstracts and increasing opportunities for personal interactions between landscape ecologists and managers (e.g., professional conferences, participation in working groups), which could narrow the gap between research and management.

Wildlife managers often manage for multiple species that operate at different spatial scales (e.g., an invertebrate compared to an ungulate) and have different responses to management activities and landscape characteristics. For instance, a US Forest Service manager on a typical forest or ranger district is often expected to make informed management decisions on major taxa (e.g., fish, aquatic invertebrates, terrestrial invertebrates, birds, mammals, reptiles, amphibians) within a large jurisdictional area ($\geq 2,000$ km^2). It is not uncommon for forests and districts to contain enormous variation in landscape characteristics. The number of listed or sensitive indicator species that managers are tasked with managing can be large (e.g., >50 species). Furthermore, the projects that require input of managers are diverse and may span from fine-scale improvements of roads and trails to broadscale projects including timber sales and prescribed burns. Managers also are first in line to communicate with a diverse public, often with differing expectations on how landscapes and species should be managed. Public stakeholders include but are not limited to other natural resource agencies, hunters, cabin owners, recreationists (both motorized and nonmotorized), timber sales contractors, conservationists, and livestock operators with grazing rights.

Under these circumstances, it is not surprising that managers lack time to become experts on all species within their jurisdiction and how these species use landscape components. Graphical abstracts might be one time-saving mechanism through which landscape ecologists can share important findings with managers, and the use of graphical abstracts is a relatively recent development in science communication (Kim et al. 2018). Publishers like Elsevier suggest that researchers produce a visual summary of their results (a graphical abstract) that pictorially summarizes the take-home messages of the research article, which allows for readers to more quickly assess whether specific studies are of interest (Elsevier 2019). Some journals (e.g., *Forest Ecology and Management* and *Frontiers in Ecology and the Environment*) already have graphical abstracts (Fig. 11.1). Unlike the research articles themselves, these graphical abstracts are often freely available without a subscription, and they provide opportunities for landscape ecologists to describe the justification, application, and limitations of their research.

In addition to limited time, managers may also have limited access to research articles. Not all agencies pay membership and subscription costs to professional societies and their associated journals, and the prices for subscriptions can vary depending on the journal. Agencies working with multiple species may require access to multiple journals with differ-

Figure 11.1. This graphical abstract from Økland et al. (2016) is freely available from *Forest Ecology and Management*. It clearly and succinctly describes the main findings of the research paper and notes a potential limitation of accurately predicting the size of infestations, which minimizes the chance of misinterpretations. Reproduced from Forest Ecology and Management (2016).

ent foci, and annual costs can be high (e.g., $8,965 [USA] is the institutional price of an annual online subscription for *Forest Ecology and Management*). Open access articles are important sources of information for managers, but only a portion of articles published in peer-reviewed journals are open access. It is common for journals to charge a publication fee to the researcher (i.e., the authors), and if researchers would like articles to be open access, additional fees are often charged. An exception is the *Journal of Fish and Wildlife Management*, which has no publication fees and is open access. But often the fees to make an article fully open access can be approximately $3,000 (USA) for a single article (Wiley 2019), and while landscape ecologists would ideally like their research to be as accessible as possible, the charges associated with making research articles open access may be cost-prohibitive even when publications costs are included in the management or research budget.

Given the challenges listed above, solutions are needed to close the communication gap between landscape ecologists and managers. We offer two ways of improving the communication: encouraging landscape ecologists to use graphical abstracts and publish open access articles where possible, and increasing opportunities for personal interactions between landscape ecologists and managers. These solutions span individual, agency, and professional society decisions and cultures, and we thus acknowledge that these solutions may require initiative from multiple entities (agencies and professional societies) and individuals.

Where possible, we suggest that landscape ecologists consider publishing in journals that have open access options and use graphical abstracts. The costs associated with these types of articles can be substantial, but the effect on management—and thereby conservation—will be greater with improved access to research articles. The development of a graphical abstract represents a relatively small investment in time for the landscape ecologist but may help managers more quickly sift through numerous potentially relevant studies.

Second, encouraging personal interactions on shared interests can quickly and effectively open channels of communication. For instance, participation in interagency or professional organization working groups or in technical workshops can foster interdisciplinary networking between landscape ecologists and managers. The Wildlife Society (TWS) is considered the primary organization of wildlife professionals, while landscape ecologists might belong to the International Association for Landscape Ecology or the Ecological Society of America. Working groups address specific, specialized components

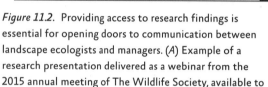

Figure 11.2. Providing access to research findings is essential for opening doors to communication between landscape ecologists and managers. (*A*) Example of a research presentation delivered as a webinar from the 2015 annual meeting of The Wildlife Society, available to society members. (*B*) Members of The Wildlife Society receive a weekly email with links to the *eWildlifer* newsletter, which contains two webinars to view. Reproduced from The Wildlife Society (2016).

within a larger organization. Currently, there are >20 working groups within TWS addressing issues as diverse as wildlife diseases, wetland conservation, invasive species, and molecular ecology (The Wildlife Society 2019). There are 11 working groups in the International Association for Landscape Ecology, including Biodiversity and Ecosystem Services Assessments, and Spatial Analysis of Organisms in the Environment (International Association for Landscape Ecology 2019). The Association of Fish and Wildlife Agencies (2019), Midwest Deer and Wild Turkey Study Group (2019), and Western Association of Fish and Wildlife Agencies (2019) committees and working groups represent additional opportunities where landscape ecologists and managers might work together on specific topics of interest, and these working groups may provide a list of experts who can be contacted about specific questions regarding landscape ecology, management, map development, and modeling. Participation in technical workshops can also help managers, administrators, and policymakers keep abreast of the latest tools and perspectives in a specific focal area.

These workshops provide additional opportunities for landscape ecologists and managers to meet and learn from each other.

Interactions between landscape ecologists and managers at professional meetings are beneficial, but organizing these interactions can be challenging. Providing online recorded symposia when attendance to meetings is not possible is one popular medium of information exchange. For example, TWS offers its members access to webinars on emerging new findings from their annual conference (Fig. 11.2). The use of interactive digital platforms and online tools in workshops, symposia, and working groups will likely expand the opportunities for landscape ecologists and managers to interact.

Implementing these solutions should help to improve communication between landscape ecologists and managers while helping to provide managers with the best and latest available science. These solutions vary in cost, and not all of them are within the control of individuals and agencies, but they do provide a starting point to improve communication between landscape ecologists and managers.

Case Study 2: Western Juniper Removal and Greater Sage-Grouse Habitat: Mapping Social Perceptions for Proposed Management Projects on Public Lands

The context of this case study focuses on management on public lands that may cause significant social or environmental impacts that require the development of an EIS. Maps are a critical component of the public land planning processes (e.g., EIS, resource management planning), and PPGIS approaches may be used to solicit the spatial distribution of social, ecological, and economic values from a diversity of stakeholders. The issues and challenges with this case study are that the spatial and temporal scales of project implementation are often unclear to many participants (e.g., landscape ecologists, managers, stakeholders), and diverse spatial scales (e.g., local watersheds and viewsheds) are important to different participants. Possible solutions include explicitly addressing issues of spatial and temporal scale early and often, matching the spatial and temporal scales of the maps with the spatial and temporal scales of the proposed project(s), and including additional maps and photos that capture the details of the proposed project(s) to help facilitate precise and clear discussion among the landscape ecologist, manager, and public stakeholders.

Managers of all public lands are faced with difficult challenges when public perception of the multiple-use paradigm conflicts with the conservation of endangered or threatened species. In general, multiple user groups are often in conflict with each other; some user groups might support increased use of the landscape while others support decreased use. This conflict among stakeholders can result in litigation when a management agency proposes a management project within areas occupied by the threatened or endangered species. For example, if the US Forest Service develops a vegetation alteration project within areas of known Canada lynx (*Lynx canandensis*) populations, it is likely that litiga-

tion will ensue during some part of the development of an EIS (Miner et al. 2010).

The National Environmental Policy Act (1969) requires an EIS to be developed when social or environmental effects are anticipated from federal management actions (activities of federal agencies or programs receiving federal funding or authorization). Several researchers noted that the effective inclusion of social perspectives to the planning process will help improve stakeholder understanding of a project while allowing for greater capacity for communication between stakeholders and managers (Vanclay 2002, Lowery and Morse 2013). Below we describe lessons learned from a PPGIS approach used to solicit input from stakeholders regarding proposed western juniper (*Juniperus occidentalis*) removal associated with efforts to improve greater sage-grouse (*Centorcerus urophasianus*; sage-grouse) habitat, the results of which incorporated spatial and social elements that were applied to EIS processes.

Sage-grouse are of conservation concern and were a candidate species for listing under the Endangered Species Act (US Fish and Wildlife Service 2010), largely as a result of >50% habitat loss since the mid-1800s (Knick et al. 2003). Conifer expansion reduces breeding habitat for sage-grouse (Baruch-Mordo et al. 2013). Consequently, the Bureau of Land Management proposed a project to improve or maintain sage-grouse habitat by removing western junipers encroaching on sage-grouse habitat within 708,200 ha in southwestern Idaho, USA.

Through frequent communication, landscape ecologists worked with managers to develop an approach to spatially map perceived values for the EIS that described the habitat restoration project for sage-grouse (Brymer et al. 2016). Multiple alternatives were considered within the EIS, including no action and various techniques to remove juniper trees (e.g., burning, mechanical treatments). Together, landscape ecologists and managers developed a qualitative approach to PPGIS, whereby the stakeholders were asked to draw polygons representing

Figure 11.3. Landscape ecologists collecting data at a public participatory geographic information systems (PPGIS) workshop with stakeholders. The PPGIS is increasingly applied to understand the spatial distribution of values and place meanings and can help inform the development of environmental impact statements on public lands. Photo by R. J. Niemeyer.

social (e.g., a place of deep social meaning), ecological (e.g., a biologically diverse area), and economic (e.g., an area important for personal or community revenue) values on a map of the proposed project area (Fig. 11.3). This qualitative and spatially explicit approach allowed managers to capture a general indication of the scale, shape, and extent of values perceived by the stakeholders. The stakeholders that participated in the mapping exercise covered a variety of occupations and user groups, which was important to capture the diversity of perspectives associated with the proposed management activities.

As expected, the mapping exercise provided a general depiction of the spatial extent and shape of social, ecological, and economic values across the proposed project area (Brymer et al. 2016). There was a clear increase in mapped values in project areas where wilderness and large rivers intersected. The nature of mapped values was diverse, including cultural traditions, ranching and farming industries, watershed and wildlife recreation, wildlife habitat, and tourism. The results from the PPGIS mapping exercise (Fig. 11.4) largely improved the social understanding within the EIS for sage-grouse habitat management. Additionally, this process provided insight concerning potential mitigation strategies to address the concerns of the stakeholders (e.g., implementation of juniper removal near wilderness;

Fig. 11.4). These data from PPGIS assisted managers in meeting the requirements of the National Environmental Policy Act (e.g., thoroughly assessing social and economic impacts) and helped refine the details of their project alternatives.

While the PPGIS process offers a good example of collaboration among landscape ecologists, managers, and stakeholder groups, it also illustrates several opportunities for improving the communication between landscape ecologists and managers, and between managers and stakeholders. For instance, some confusion occurred among all parties (landscape ecologists, managers, stakeholders) regarding the spatial and temporal scales of implementation for the juniper removal. This confusion may have influenced the precision of mapped values provided by the stakeholders, which could have consequences on the inferred patterns of spatial values. Different stakeholders also mapped values at extents that were personally relevant, resulting in some people mapping values at a local extent while others mapped values at a broader extent. The variation in extent may have influenced the precision of the mapping effort, which could affect the development of mitigation strategies by restricting the suite of alternatives considered within an EIS process.

A solution to dealing with these issues is being more explicit about spatial and temporal scales

A

B

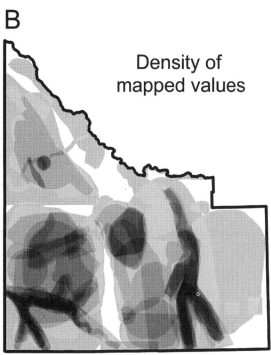

Density of
mapped values

Figure 11.4. Examples of maps used in public participatory geographic information systems (PPGIS) workshops. (*A*) Digitized polygons drawn by stakeholders within PPGIS workshops representing Owyhee County, Idaho, USA. (*B*) Density of polygons drawn by stakeholders (darker shades indicate increasing densities).

within the mapping process, which requires more precise communication between landscape ecologists and managers. The only extent presented in the stakeholder workshops for the mapping in Brymer et al. (2016) was a coarse and general extent of the proposed project area, which certainly overestimated the actual area affected. Specifically, the scale, shape, and extent of the polygons drawn by participants was directly tied to the scale of the map on which they were drawing (e.g., 1:24,000 topographic maps of specific areas will result in much different polygons than a statewide map at 1:100,000). A more detailed discussion between the landscape ecologist and manager on the extent and time line of the proposed project would have facilitated more effective mapping exercises from the stakeholders, such that the extent of treatments (e.g., <50 ha owing to logistical constraints) were more explicitly defined.

In addition, the use of multiple graphics (e.g., photos, diagrams, videos) may facilitate a better un-

derstanding of proposed actions and result in efficient and effective products that aid management decision making. For instance, the use of photographs that exhibit treatment characteristics would further clarify for stakeholders what areas will be affected on the ground and provide a mental image of the spatial and temporal scale of the project. Photos of similar areas might represent how treated areas appear immediately after treatment and years into the future. Graphics offer alternatives, or additions, to verbal communication and clarify information that would otherwise remain complex or vague (Frankel and DePace 2012, McIrney et al. 2014). Removing the vagaries associated with spatially explicit mapping would certainly increase the quality of the product generated.

Building the mapping exercises around these different aspects of spatial and temporal scale, coupled with the inclusion of many graphics, would likely provide a more complete and resolved characterization of stakeholder values across the landscape. To build a PPGIS that incorporates more explicit attention to scale issues, however, hinges on detailed communication early on between the manager and landscape ecologist. The number of meetings necessary to have everyone working together depends on the complexity of the proposed project and the degree to which complex concepts are clearly articulated. By increasing the lines of communication on applications, such as PPGIS during an EIS process, a trickle-down effect will ensue, clarifying for stakeholders the scope of the project (Brymer et al. 2016). When stakeholders clearly understand the scope of a project, the data generated from such exercises will likely become more detailed and informed, facilitating more intricate mitigation strategies that incorporate different social perceptions during the development of management projects on public lands.

Case Study 3: Incorporating Expert Review into Broadscale Predictive Species Distribution Models

The context of this case study addresses broadscale management issues that often use species distribution information, which requires input from multiple landscape ecologists and managers at coarse spatial scales (e.g., state, regional, continental scales). The species distribution models are developed for a wide variety of species that respond to different spatial scales (e.g., small riparian areas versus large forested areas) and have different seasonal responses to landscape characteristics. Hence species experts from multiple natural resource agencies across multiple states are needed to review the quality of species distribution models and to make recommendations on improving the data sets based on the extent of their experience and knowledge. The issues and challenges include communicating species distribution modeling methodologies, assumptions, and limitations, and thus appropriate applications to a variety of audiences with different backgrounds. There is often a mismatch in the spatial scale of species dis-

tribution models developed by landscape ecologists and expert knowledge of managers. Additionally, there is a lack of knowledge of the degree to which regional variability in landscape variables affects the outcome and uses of the species distribution models. Possible solutions include using consistent terminology and minimizing jargon to ensure the information is appropriate for the audience. We can encourage early interactions among managers with species expertise in specific geographical regions to help landscape ecologists understand the degree to which models need to be varied through space. Similarly, additional conversations early in the project enable managers to understand the limitations of modeling over large geographic ranges. Lastly, we can collaborate with the managers in organized face-to-face meetings as frequently as possible throughout the model development process.

As mentioned regarding the first case study, managers are typically expected to make informed management decisions on all major taxa and habitats under their jurisdiction. Traditional species-by-species or threat-by-threat piecemeal management approaches are often considered inadequate for conserving biodiversity and as being cost- and time-prohibitive. One approach to address this is gap analysis, which includes using landcover data, protected areas data, and species distribution models to identify areas or species that are underrepresented or unrepresented within lands managed for conservation across the United States (Scott et al. 1993, US Geological Survey 2019). The intent is to use these data to inform conservation planning and management, which inherently ties into landscape ecology. Multiple state and federal agency partners apply these spatial data to their planning processes, but ultimately the goal of developing these data is to conduct a gap analysis of the United States (Scott et al. 1993). Several hundred species distribution models are developed specifically for this task. Each of these models covers the entire range of the species with the same deductive (expert-driven) modeling approach applied for all species and uses a consis-

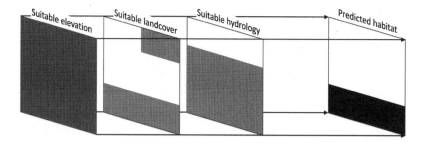

Figure 11.5. Schematic showing how spatial model variables are overlaid to predict a species distribution based on predicted habitat. This schematic was shown to managers to communicate the concept of how each species distribution model was created.

tent set of environmental data (e.g., landcover, elevation).

The models developed for gap analysis are often of interest to a variety of stakeholder groups who want to use these models for different management contexts. For instance, a Crucial Habitat Assessment Tool was developed for assessing areas for expansion of electrical grids in the western United States, and as part of this effort, it was important to have managers review the quality and accuracy of the regional species distribution models (Western Governors' Association 2013). Specifically, the aim of this partnership between landscape ecologists (i.e., species modelers) and this stakeholder group was to glean expert review from managers across 19 western states concerning the accuracy of predicted species distribution models.

Two communication challenges were identified in this multi-partner landscape-scale approach. First was finding and using a common language to clearly communicate the process used in model development, usually through the use of email, webinars, and conference calls. For example, we wrote an instructional guide to describe the process for developing species distribution models, including several definitions and a list of environmental variables used in the models along with a description of each environmental variable and its metadata (Fig. 11.5). Even with this instructional guide, managers were uncertain about how models were developed and therefore

did not always provide input that could be applied to improve the species model. It became apparent that either the instructional guide, which was written by landscape ecologists, was not clearly explaining the modeling methodology and environmental variables to the managers or the managers were not using the instruction guide.

Interestingly, in-person conversations between landscape ecologists and managers facilitated more progress compared to the other modes of communication. The input provided during these face-to-face meetings was more often used to improve a species distribution model, suggesting that when the review process was co-produced in one-on-one settings rather than disseminated from one group (i.e., landscape ecologists) to another group (i.e., managers), the input received was more useful and applicable to improving a species distribution model.

The second issue was related to the ability of managers to provide input on a species across its entire range. Most managers were experts within their state or region of interest but were not always able to comment on a species distribution outside of their state or sometimes even jurisdictional boundaries (Fig. 11.6). Unfortunately, because of the western extent of the modeling approach, input on a species distribution model in only a portion of its range could not be applied to the entire species model and therefore could not be used to improve the model; there was thus a mismatch in scale between the cur-

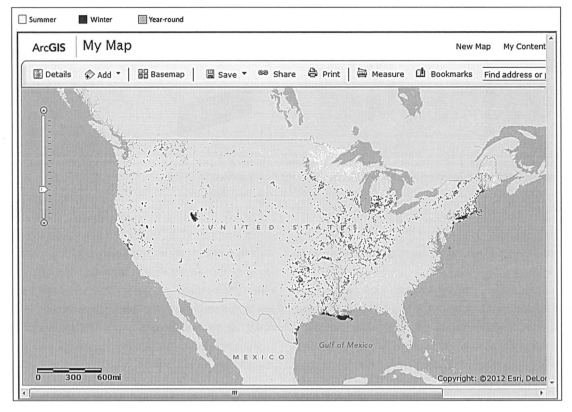

Figure 11.6. Interactive map developed by landscape ecologists to assist managers with providing expert input for a species distribution model. The managers could add spatial data layers or base maps, for example, to assist with their review of a species distribution model.

rent knowledge about a species and the extent over which a species distribution model was built.

Two important lessons about communicating science between landscape ecologists and managers are apparent in this case study. First, landscape ecologists should learn about their audience's background. Using face-to-face meetings to teach the modeling methodology would help to determine the background of the audience and refine content and whether different teaching approaches might be more useful. The challenge with this approach is that it is time consuming and can be costly (e.g., travel), but it is often worth the investment (Gilliam 2005). More knowledge and improved communication can be gleaned from the managers by meeting in person.

Second, an explicit discussion between managers and landscape ecologists about spatial scale may have assisted the process. Many managers are experts on a species within their state or region but do not feel confident in their knowledge of a species across the entire western United States, which was the extent of the project. Indeed, it might be challenging to find a single individual that has range-wide expertise on a widely dispersed species. A more explicit discussion could have resulted in the inclusion of managers providing input on a species model across the western United States; conversely, more explicit discussions could have focused on obtaining input from multiple managers at the same time on the same species. One way to address this challenge would be to incorporate the expertise of managers from regional technical working groups (e.g., US

Fish and Wildlife Service 2019). This would allow for co-producing knowledge among the multiple managers to improve a species distribution model.

Despite these challenges, the effort was successful in improving about 26 of the 100 species distribution models. Numerous errors in the models, unrelated to our communication challenges, were caught by the managers, and most managers agreed that the output of the species distribution models accurately reflected their distribution in the western United States. An additional benefit derived from this project was that many managers became more familiar with the species distribution models and modeling methods that were used in developing the Crucial Habitat Assessment Tool.

Summary

Landscape ecologists are typically conducting analyses and developing maps used in a variety of management activities. A lack of communication or understanding of the decisions made behind the mapped elements has implications for a variety of wildlife management activities. Based on our three case studies (which represent just a small fraction of the different activities that managers might engage in with landscape ecologists), we have identified multiple consequences of misunderstandings that might occur. First, managers may make decisions based on outdated information if they do not have access to current scientific literature. Second, miscommunication about spatial and temporal scales for maps used in stakeholder meetings may lead to misunderstandings about the potential effect of management activities on the social and ecological environment. Third, maps developed for one purpose may not be appropriate for different management contexts.

There are several solutions to these challenges. First, landscape ecologists are encouraged to publish in open access formats where possible and to use graphical abstracts. Face-to-face meetings, where possible, facilitate a better understanding and clarity about the strength, limitations, and development of the models and maps used by managers. Where geographical constraints exist, alternative forms of video communication may work well because many of the issues associated with communication center upon visual representation and interpretation of landscape elements.

Professional society involvement generally increases the accessibility of current information for managers. We acknowledge there might be financial limitations to professional society membership for managers, and many of the suggestions that could improve the scientific accessibility of content (e.g., the use of graphical abstracts) require changes at the level of the professional society and scientific journal. Nevertheless, increased access to professional society resources in general will improve the ability of managers to incorporate current findings into management activities. Activities that foster cross-fertilization between landscape ecologists and managers within this setting will likely be mutually beneficial and potentially include landscape ecologists joining professional societies such as TWS and managers joining professional societies such as the International Association for Landscape Ecologists. Additionally, participation in management-centric and technical working groups is also likely to foster strong relationships between landscape ecologists and managers.

It is important during these meetings to consistently clarify and define the spatial extents and terminology that relate to products used (and developed) by the landscape ecologists. This is particularly important when initiating a project, when there are multiple partners, or there is turnover of personnel within a project framework. For effective communication to occur, it is better to reiterate concepts and specialized terminology than to assume that all participants have the same background.

Finally, communicating early and often was noted to be important regardless of communication format. Ideally, research projects by landscape ecologists would be initiated by communication between landscape ecologists and managers. Such communication would ensure the transfer of the latest science, encourage appropriate uses of landscape products, and facilitate the co-development of research questions that have strong management relevance. Taking this time to exchange ideas, perspectives, and knowledge gaps will strengthen the collaborations between landscape ecologists and managers while simultaneously ensuring that research is truly serving on-the-ground needs for effective management. Frequent interactions after the initiation of a joint project will facilitate opportunities for the landscape ecologists and managers to continue to share perspectives regarding landscape ecology concepts, maps, models, and management needs. If projects are not initiated collaboratively, landscape ecologists must take the time to complete the (often dreaded) metadata such that future map users are aware of the data sets used to produce them.

By incorporating these ideas and approaches, we hope that communication will be enhanced between landscape ecologists and managers. Additionally, the same issues of communication concepts (e.g., communicating early and often, using graphics where possible, defining specialized terms) could be used in communication between managers and different stakeholder groups (e.g., members of the public involved in the public review process, legislators, state wildlife commission, others). With improved communication at multiple levels, the potential for disseminating sound science into society will increase and ultimately result in more efficient and effective on-the-ground management.

LITERATURE CITED

Association of Fish and Wildlife Agencies. 2019. AFWA homepage. https://www.fishwildlife.org. Accessed 22 July 2019.

Aycrigg, J., G. Beauvais, T. Gotthardt, F. Huettmann, S. Pyare, M. Andersen, D. Keinath, J. Lonneker, M. Spathelf, and K. Walton. 2015*a*. Novel approaches to modeling and mapping terrestrial vertebrate occurrence in the Northwest and Alaska: an evaluation. Northwest Science 89:355–381.

Aycrigg, J. L., R. T. Belote, M. S. Dietz, G. H. Aplet, and R. A. Fischer. 2015*b*. Bombing for biodiversity in the United States: response to Zentelis and Lindenmayer 2015. Conservation Letters 8:306–307.

Aycrigg, J. L., A. Davidson, L. Svancara, K. J. Gergely, A. McKerrow, and J. M. Scott. 2013. Representation of ecological systems within the protected areas network of the continental United States. PLOS One 8(1):e54689.

Baruch-Mordo, S., J. Evans, J. Severson, D. Naugle, J. Maestas, J. Kiesecker, M. Falkowski, C. Hagen, and K. Reese. 2013. Saving sage-grouse from the trees: a proactive solution to reducing a key threat to a candidate species. Biological Conservation 167:233–241.

Bowman, T. E., E. Maibach, M. E. Mann, R. C. Somerville, B. J. Seltser, B. Fischhoff, S. M. Gardiner, R. J. Gould, A. Leiserowitz, and G. Yohe. 2010. Time to take action on climate communication. Science 330:1044.

Brown, G., and M. Kyttä. 2014. Key issues and research priorities for public participation GIS (PPGIS): a synthesis based on empirical research. Applied Geography 46:122–136.

Brymer, A. L. B., J. D. Holbrook, R. J. Niemeyer, A. A. Suazo, J. D. Wulfhorst, K. T. Vierling, B. A. Newingham, T. E. Link, and J. L. Rachlow. 2016. A social-ecological impact assessment for public land management: application of a conceptual and methodological framework. Ecology and Society 21(3):9.

Bubela, T., M. C. Nisbet, R. Borchelt, F. Brunger, C. Critchley, E. Einsiedel, G. Geller, A. Gupta, J. Hampel, R. Hyde-Lay, and E. W. Jandciu. 2009. Science communication reconsidered. Nature Biotechnology 27:514–518.

Bucchi, M. 2008. Of deficits, deviations and dialogues: theories of public communication of science. Pages 57–76 *in* M. Bucchi and B. Trench, editors. Handbook of public communication of science and technology. Routledge, New York, New York, USA.

Cooke, S. J., A. J. Gallagher, N. M. Sopinka, V. M. Nguyen, R. A. Skubel, N. Hammerschlag, S. Boon, N. Young, and A. J. Danylchuk. 2017. Considerations for effective science communication. FACETS 2:233–248.

Elsevier. 2019. Graphical abstracts. https://www.elsevier.com/authors/journal-authors/graphical-abstract. Accessed 22 July 2019.

Fischhoff, B. 2013. The sciences of science communication. Proceedings of the National Academy of Sciences 110:14,033–14,039.

Fischhoff, B., and D. Scheufele. 2013. The science of science communication. Proceedings of the National Academy of Sciences 110:14,031–14,032.

Forest Ecology and Management. 2016. Graphical abstract.

https://www.sciencedirect.com/science/article/pii/S0378112715007343. Accessed 14 Aug 2019.

Frankel, F., and A. H. DePace. 2012. Visual strategies: a practical guide to graphics for scientists and engineers. Yale University Press, New Haven, Connecticut, USA.

Fremier, A. K., M. Kiparsky, S. Gmur, J. Aycrigg, R. K. Craig, L. K. Svancara, D. D. Goble, B. Cosens, F. W. Davis, and J. M. Scott. 2015. A riparian conservation network for ecological resilience. Biological Conservation 191:29–37.

Gilliam, B. 2005. Research interviewing: the range of techniques. Open University Press, New York, New York, USA.

Hess, G. R., and R. A. Fischer. 2001. Communicating clearly about conservation corridors. Landscape and Urban Planning 55:195–208.

International Association for Landscape Ecology. 2019. IALE working groups. https://www.landscape-ecology.org/working-groups.html. Accessed 22 July 2019.

Kim, S., E. Chung, and J. Y. Lee. 2018. Latest trends in innovative global scholarly journal publication and distribution platforms. Science Editing 5:92–104.

Knick, S. T., D. S. Dobkin, J. T. Rotenberry, M. A. Schroeder, W. M. Vander Haegen, and C. van Riper III. 2003. Teetering on the edge or too late? Conservation and research issues for avifauna of sagebrush habitats. The Condor 105:611–634.

Kuehne, L. M., and J. D. Olden. 2015. Opinion: lay summaries needed to enhance science communication. Proceedings of the National Academy of Sciences 112:3585–3586.

La Sorte, F. A., D. Fink, W. M. Hochachka, J. L. Aycrigg, K. V. Rosenberg, A. D. Rodewald, N. E. Bruns, A. Farnsworth, B. L. Sullivan, C. Wood, et al. 2015. Documenting stewardship responsibilities across the annual cycle for birds on U.S. public lands. Ecological Applications 25:39–51.

Lorenz, T. J., J. Aycrigg, J. Vogeler, J. Lonneker, and K. T. Vierling. 2015. Incorporation of shrub and snag specific LiDAR data into GAP wildlife models. Journal of Fish and Wildlife Management 6:437–447.

Lowery, D., and W. Morse. 2013. A qualitative method for collecting spatial data on important places for recreation, livelihoods, and ecological meanings: integrating focus groups with public participation geographic information systems. Society and Natural Resources 26:1422–1437.

Lubchenco, J. 1998. Entering the century of the environment: a new social contract for science. Science 279:491–497.

Maibach, E. W., A. Leiserowitz, C. Roser-Renouf, and C. K. Mertz. 2011. Identifying like-minded audiences for global warming public engagement campaigns: an audience segmentation analysis and tool development. PLOS One 6(3):e17571.

McInerny, G. J., M. Chen, R. Freeman, D. Gavaghan, M. Meyer, R. Rowland, D. J. Spiegelhalter, M. Stefaner, G. Tessarolo, and J. Hortal. 2014. Information visualisation for science and policy: engaging users and avoiding bias. Trends in Ecology and Evolution 29:148–157.

Midwest Deer and Wild Turkey Study Group. 2019. MDWTSG homepage. http://mdwtsg.org. Accessed 22 July 2019.

Miner, A. M. A, R. W. Malmsheimer, D. M. Keele, and M. J. Mortimer. 2010. Twenty years of forest service national environmental policy act litigation. Environmental Practice 12:116–126.

Morrison, M. L., B. G. Marcot, and R. W. Mannan. 2006. Wildlife–habitat relationships: concepts and applications. Third edition. Island Press, Washington, DC, USA.

National Environmental Policy Act. 1969. 42 U.S.C. § 4331 et seq.

Nisbet, M. C., and C. Mooney. 2007. Framing science. Science 316:56.

Økland, B., C. Nikolov, P. Krokene, and J. Vakula. 2016. Transition from windfall- to patch-driven outbreak dynamics of the spruce bark beetle Ips typographus. Forest Ecology and Management 363:63–73.

Rodríguez Estrada, F. C., and L. S. Davis. 2015. Improving visual communication of science through the incorporation of graphic design theories and practices into science communication. Science Communication 37(1):140–148.

Sanderson, E. W., J. Malanding, M. A. Levy, K. H. Redford, A. V. Wannebo, and G. Woolmer. 2002. The human footprint and the last of the wild. Bioscience 52:891–904.

Scheufele, D. A. 2014. Science communication as political communication. Proceedings of the National Academy of Sciences 111:13,585–13,592.

Scott, J. M., F. Davis, B. Csuti, R. Noss, B. Butterfield, C. Groves, H. Anderson, S. Caicco, F. D'Erchia, T. C. Edwards Jr., et al. 1993. Gap analysis: a geographic approach to protection of biological diversity. Wildlife Monographs 123:1–41.

Shome, D., S. Marx, K. Appelt, P. Arora, R. Balstad, K. Broad, A. Freedman, M. Handgraaf, D. Hardisty, D. Krantz, et al. 2009. The psychology of climate change communication: a guide for scientists, journalists, educators, political aides, and the interested public. Center for Research on Environmental Decisions, New York, New York, USA.

Sieber, R. 2006. Public Participation Geographic Information Systems: a literature review and framework. Annals of the Association of American Geographers 96:491–507.

The Wildlife Society. 2016. TWS homepage. http://wildlife.org. Accessed 5 Feb 2016.

The Wildlife Society. 2019. TWS working groups. http://wildlife.org/network/tws-local/working-groups/. Accessed 22 July 2019.

Treise, D., and M. F. Weigold. 2002. Advancing science communication: a survey of science communicators. Science Communication 23:310–322.

US Fish and Wildlife Service. 2010. Endangered and threatened wildlife and plants: 12-month findings for petitions to list the greater sage-grouse (Centrocercus urophasianus) as threatened or endangered. Federal Register 75:13,910–14,014.

US Fish and Wildlife Service. 2019. Flyways. https://www.fws.gov/birds/management/flyways.php. Accessed 22 July 2019.

US Geological Survey Gap Analysis Program. 2019. Gap Analysis Project. https://www.usgs.gov/core-science-systems/science-analytics-and-synthesis/gap. Accessed 22 July 2019.

Vanclay F. 2002. Conceptualising social impacts. Environmental Impact Assessment Review 22:183–211.

von Winterfeldt, D. 2013. Bridging the gap between science and decision making. Proceedings of the National Academy of Sciences 110:14,055–14,061.

Western Association of Fish and Wildlife Agencies. 2019. WAFWA committees and groups. https://www.wafwa.org/committees_groups/committees/. Accessed 22 July 2019.

Western Governors' Association. 2013. Crucial Habitat Assessment Tool. http://westgov.org/reports/crucial-habitat-assessment-tool. Accessed 23 July 2019.

Wiley. 2019. Article publication charges. https://authorservices.wiley.com/author-resources/Journal-Authors/open-access/article-publication-charges.html. Accessed 13 Aug 2019.

12

NEAL D. NIEMUTH,
MICHAEL E. ESTEY, AND
RONALD D. PRITCHERT

Developing Useful Spatially Explicit Habitat Models and Decision-Support Tools for Wildlife Management

Introduction

In this chapter we advocate an applied approach to landscape ecological research and conservation planning through the use of spatial models, where the end goal is informing and guiding on-the-ground conservation actions. We do so to provide guidance for powerful approaches that agencies and organizations often adopt but that are, in our experience, frequently not used to their full potential. We assume throughout this chapter that explicit conservation actions, or treatments, are to be implemented. Our target audience includes conservation planners and managers who are contemplating the use or development of spatially explicit models to guide such actions. A substantial information gap regarding the use of models often exists between scientists and managers (Stauffer 2002, Noon et al. 2009), and we present real-world considerations for planners and managers who may be unsure of how to proceed given the sometimes bewildering array of spatial methods that are available (Elith et al. 2006, Thompson and Millspaugh 2009, Franklin 2013, Phillips et al. 2017).

Landscape ecology is important to wildlife management because location is important to conser-

vation action. Landcover, vegetative communities, land use, landscape configuration, and land ownership vary across space (Forman 1995). Similarly, species distributions, densities, and demographics vary across space (Boyce and McDonald 1999), as do costs and opportunities for conservation action (Naidoo et al. 2006). In this chapter we present concepts and philosophies related to identifying the best place for each conservation treatment and the best conservation treatment for each place. Many of our examples come from migratory bird conservation efforts in the US Prairie Pothole Region (PPR), where spatial models guide annual expenditures of ~$70 million (USA) on acquisition of perpetual wetland and grassland easements held by the US Fish and Wildlife Service (USFWS) and also guide private lands management and programs such as the Farm Bill (Reynolds et al. 2006, Niemuth et al. 2008).

There are good reasons for adopting a landscape approach to conservation, and we assembled a list of the top 10 reasons we have seen advocated for use of spatial models (Table 12.1). Our top reason that agencies and organizations adopt a landscape approach to conservation (i.e., maps are pretty) is tongue-in-cheek, and we use it to underscore that although an attractive map is often the final product

Table 12.1. Top Ten Reasons for Adopting Use of Spatial Models in Wildlife Management

Number	Reason
10	Geographic information systems enable quick and efficient entry and processing of spatial data.
9	Theoretical advances in landscape ecology provide conceptual foundation for models.
8	Role of landscapes in sustaining wildlife populations is increasingly recognized.
7	Advances in statistical analysis enable development of models that accommodate complex data.
6	Increased computing power permits complex analyses.
5	Global positioning systems capacity permits easy acquisition of precise location data.
4	Remotely sensed data is more easily acquired and processed.
3	Many forms of response and predictor data are available for free over the Internet.
2	Programs promote use of spatial tools.
1	Maps are pretty.

of a landscape analysis, the myriad factors that went into producing the map may not be known or understood by some end users.

Maps are tools for conveying information and displaying patterns that would otherwise be difficult to comprehend (Monmonier 1996, Wiens 2002) and as such have strong appeal to managers, administrators, and conservation proponents. It is often difficult, however, if not impossible, to easily assess the quality of information in a map simply by looking at it, and the quality of data or inferences represented by the map may well be poor or misleading (Monmonier 1996, Wiens 2002, Anselin 2006). In addition, choices for displaying maps can be deceptive, as the selection of a value threshold or color palette for display can greatly influence what people perceive or focus on when viewing a map (Monmonier 1996, Krygier and Wood 2016, Morris et al. 2016). Consequently, a map runs the risk of being a shiny bauble that is used to draw attention to a program or effort but fails to provide useful direction for conservation. Worse, a poorly conceived and developed model can misdirect conservation actions to less efficient or effective locations. Our goal in this chapter is to pro-

vide guidance that will help users identify and avoid common pitfalls in developing spatial models and ensure that spatial analyses result in useful tools for conservation. The topics we present have repeatedly arisen in discussions with scientists, managers, and administrators as part of the targeting and delivery of conservation in the PPR, but the general concepts have broad applicability.

Definitions

Before proceeding, a few definitions are in order. We define a *model* as an articulation or representation of relationships. We use the terms *landscape model* and *spatial model* in a generic sense to describe models that are developed across broad spatial extents and then applied to geographic information system (GIS) data to create a map depicting the product of those relationships. These maps may depict a variety of biological responses, including species occurrence, density, survival, and reproductive success, as well as nonbiological responses.

Decision-support tools are derived from spatial models by integrating the spatial model with specific information about planned conservation actions and are used to determine the amount, type, or location of conservation treatments. Unlike spatial models, which are generally time consuming and expensive to develop, decision-support tools may be nothing more than a simple reclassification of a spatial model or models using treatment-specific thresholds. For example, the waterfowl thunderstorm map, so named because of its resemblance to a Doppler radar image of a thunderstorm crossing the Great Plains (Fig. 12.1A), is a simple decision-support tool used to guide upland conservation treatments for waterfowl. The thunderstorm map summarizes and establishes decision thresholds for output of regression and accessibility models for five species of upland-nesting ducks settling on >2 million wetlands of four water permanence classes over multiple years and states (Reynolds et al. 2006). The

A

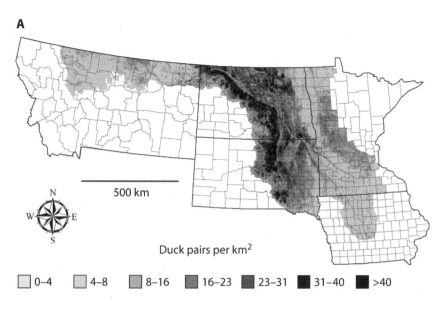

Duck pairs per km^2

0–4   4–8   8–16   16–23   23–31   31–40   >40

B

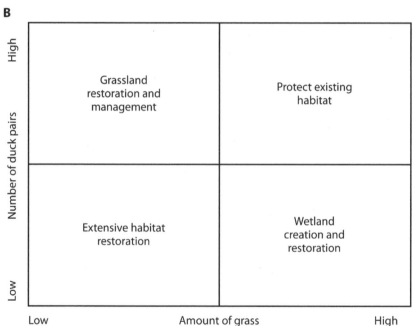

Figure 12.1. The water-fowl thunderstorm map (*A*) models the number of duck pairs with access to upland habitats (after Reynolds et al. 2006) in the Prairie Pothole Region of Iowa, Minnesota, Montana, North Dakota, and South Dakota, USA, 1987-2012. Used in conjunction with landcover data, a simple decision matrix (*B*) can be created to guide conservation actions. Thresholds for amounts of grass and number of duck pairs are determined by biology, funding, conservation goals, and availability of willing landowners.

thunderstorm map is the foundation of many upland conservation treatments in the region; when used in conjunction with landcover data, a simple decision matrix can be created to guide conservation actions (Fig. 12.1B). Options associated with this matrix are explored in "Identifying the Purpose," below.

Identifying the Purpose

Many spatial models have an unstated goal of identifying the best areas for conservation, but what these areas are best for is rarely specified. Land protection is often an implicit conservation treatment, but protection of existing habitat is not always an option and may be insufficient to conserve species whose habitat is limiting. Spatial models and decision-support tools will be more valuable to conservation if they are developed and applied in the context of specific, clearly articulated needs that explicitly consider intended conservation treatments, funding constraints, and species needs. For example, temperate grasslands are one of the most altered biomes on the planet, have the lowest rate of habitat protection of all major biomes (Hoekstra et al. 2005), and grassland loss in the PPR continues at a substantial rate (Stephens et al. 2008, Rashford et al. 2011, Lark et al. 2015). Grassland loss has major implications for waterfowl, as the majority of waterfowl in the PPR nest

in grasslands (Batt et al. 1989). Grassland loss in the region also has major implications for grassland birds, which have their highest diversity in northern plains and have a larger proportion of species that are decreasing than any other bird group in North America (Askins 1993, Peterjohn and Sauer 1999). Consequently, grassland conservation is a focus of many conservation programs in the region, with the goal of providing benefits for upland-nesting waterfowl and many other species of grassland-dependent wildlife.

Given the size of the US PPR, which extends 1,800 km from northwestern Montana to central Iowa, and uneven distribution of remaining grasslands (Fig. 12.2A), spatial decision-support tools are especially useful for guiding conservation actions.

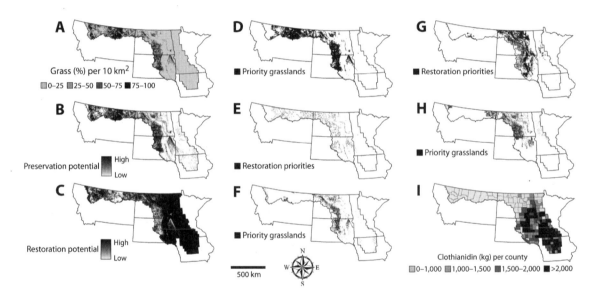

Figure 12.2. The amount of grassland in the Prairie Pothole Region (PPR) of Iowa, Minnesota, Montana, North Dakota, and South Dakota, USA, is generally lowest in the east and higher in the west (*A*), but Priority Areas for Conservation treatments differ greatly depending on intended treatments. Potential for grassland protection is highest where grass remains (*B*); potential for grassland restoration is highest where grass has been converted (*C*); areas with ≥35% grassland are located primarily in the western PPR (*D*); opportunities to restore grass to create areas with ≥35% grassland are more broadly distributed in the region (*E*); areas of existing grass with access to ≥16 pairs of upland-nesting ducks per square kilometer are located

primarily in the western portion of the PPR in North Dakota and South Dakota (*F*); areas that are not grass but have access to ≥16 pairs of upland-nesting ducks per square kilometer are located primarily in the central and eastern portion of the PPR in North Dakota and South Dakota (*G*); areas of existing grass within the range of LeConte's sparrow (*Ammospiza leconteii*) are located primarily in the north-central PPR (*H*); and county-level application rates of clothianidin in 2012 are highest in the southeastern portion of the region, which will influence treatment options related to grasslands and insects (*I*). Landcover data from the National Land Cover Database (Homer et al. 2015); clothianidin data from Baker and Stone (2015).

The location of priority grasslands, and therefore the location of conservation actions, will be highly dependent on the specific conservation goal and other factors. For example, areas with highest potential for grassland preservation are where grass presently exists (Fig. 12.2B). But many funding opportunities and conservation programs focus on habitat restoration, and areas with highest potential for grassland restoration are where the least grass exists (Fig. 12.2C). The selection of priority areas becomes far more complex, however, when waterfowl benefits, land use, cost, risk, benefits to other species, and other factors are considered.

For example, waterfowl nesting success increases with the amount of grassland in the landscape (Greenwood et al. 1995, Reynolds et al. 2001, Stephens et al. 2005), and approximately 35% grass in a 10-km² landscape should, in general, provide a nesting success of 15%, which is required for population maintenance (Cowardin et al. 1985, Klett et al. 1988, Reynolds et al. 2001). Therefore grassland protection in areas with >35% grass (Fig. 12.2D) may be a waterfowl conservation priority. If a 35% grassland threshold is also a priority for grassland restoration, conservationists may want to target areas with 29% grass in a 10-km² landscape (Fig. 12.2E), as addition of 65 ha of grass, which is a common size for land ownership and management in the area, will bring the surrounding landscape over the 35% threshold. The previous two examples demonstrate how different treatments (protection versus restoration) substantially change the location of priority areas when considering the same biological relationship between waterfowl nest success and amount of grass in the surrounding landscape. Similarly, grassland protection and restoration may be targeted to include areas with potential to provide nesting habitat for high numbers of dabbling ducks (Figs. 12.2F and 12.2G, respectively), changing the distribution yet again.

Priorities might change further if non-waterfowl species are considered; grassland protection targeted to benefit species of interest with a limited distribution, in this case Le Conte's sparrow (*Ammodramus leconteii*), shifts priority grasslands to the north-central portion of the PPR (Fig. 12.2H). Grassland-dependent pollinators might be another consideration for planning grassland protection or restoration, as populations of honeybees, native bees, moths, and prairie butterflies are declining (Potts et al. 2010, Farhat et al. 2014, Koh et al. 2016). Neonicotinoid insecticides have been implicated in the decline of these species (Godfray et al. 2014, Pecenka and Lundgren 2015), and a potential response would be to avoid butterfly conservation efforts in areas of high neonicotinoid use, which, in the PPR, would shift conservation efforts west (Fig. 12.2I). But this action would move conservation actions away from areas with the highest densities of some pollinator species, which are found in the southeastern portion of the PPR (Wassenaar and Hobson 1998).

All of the previous examples of factors influencing the purpose and application of a decision-support tool focused on biology, primarily species distributions and demographics, but socioeconomic considerations also can be key factors in determining where to do conservation and how to do it in a cost-effective manner (Haight and Gobster 2009). For example, the mean price per hectare of farmland in 2015 across the PPR ranged from a high of $19,770 in Iowa to a low of $2,200 in Montana, with intermediate values of $4,740 in North Dakota, $5,730 in South Dakota, and $11,610 in Minnesota (National Agricultural Statistics Services 2015). Additional factors, including soil productivity and risk of grassland conversion, are important when prioritizing locations for action (Olimb and Robinson 2019), and in a region dominated by private lands, acceptance by landowners is necessary (Fields 2017). Factors such as these will vary among regions, but the key point is that the results of spatial planning tools and the areas that are prioritized are dependent on the intended conservation treatments, and using the wrong criteria or no criteria may diminish the value of conservation efforts (Abrahms et al. 2017, Fields 2017). Even if the correct criteria are used, the manner in which they are applied must be carefully considered, as questionable math, hidden value judgments, and arbitrary scores can greatly influence results (Game et al. 2013).

Theoretical and Conceptual Foundations of Models

A thorough understanding of the pertinent mechanisms and issues—whether biological, social, or economic—that affect the system under consideration is necessary to ensure the relevance and effectiveness of spatial tools. For example, the PPR contains millions of wetlands that range in size and water permanence from ephemeral wetlands that cover tens of square meters to lakes that cover hundreds of square kilometers (Kantrud et al. 1989). The PPR has substantial conservation programs because of the importance of the region to wetland-dependent migratory birds (North American Waterfowl Management Plan Committee 1986, Beyersbergen et al. 2004, Niemuth et al. 2008) and extensive and ongoing loss of wetlands (Dahl 1990, Oslund et al. 2010). Island biogeographic theory (MacArthur and Wilson 1967) is frequently advocated as a framework for targeting conservation of prairie wetlands (Whited et al. 2000, Bertassello et al. 2018). Occurrence of many wetland-dependent species is indeed higher on large wetlands relative to small wetlands (Kantrud and Stewart 1984, Naugle et al. 2001, Niemuth et al. 2006), and connectivity is important to some ecological processes in pothole wetlands (Galatowitsch and van der Valk 1996). But island biogeography, though intellectually appealing and appropriate in some situations, may be a poor choice for guiding conservation of wetland-dependent migratory birds in the PPR for several reasons.

Risk of loss is greater on small wetlands because, in addition to being smaller and shallower than large wetlands, they also are generally higher in the local topographic neighborhood and therefore easier to drain or consolidate than large wetlands (Anteau 2012). Small wetlands have higher densi-

ties of breeding dabbling ducks than large wetlands because territorial behavior limits the number of individuals of a species on a wetland. Therefore, ten 1-ha wetlands will have more pairs of breeding dabbling ducks than one 10-ha wetland (Cowardin et al. 1995).

Many ecological processes are influenced by wetland size and depth, with the result that small wetlands provide better habitat for many species of breeding migratory birds, particularly waterfowl. Small wetlands are extremely productive because their shallow waters warm up early in spring, and their dynamic nature facilitates nutrient cycling and regeneration of vegetation (Harris and Marshall 1963, van der Valk and Davis 1978, Murkin et al. 1997). Small wetlands are more likely to support emergent vegetation communities that are absent from deeper wetlands (Kantrud et al. 1989). In addition, small wetlands often lack minnows, which forage on invertebrates and reduce invertebrate numbers, leading to reduced growth rates and survival of ducklings (Bouffard and Hanson 1997, Cox et al. 1998). Finally, proximity to other wetlands is not important for migratory birds to colonize a wetland, although wetland complexes may be important for providing foraging and brood-rearing opportunities throughout the breeding season. For all these reasons, a decision-support tool prioritizing large wetlands close to each other would be less efficient than tools based on pertinent issues and biology of the species under consideration, especially for conservation of upland-nesting dabbling ducks.

As always, conservation needs, purpose, and treatments must be considered. Water chemistry also varies with wetland size, as large, terminal wetlands generally have higher concentrations of salts and minerals than small wetlands (Kantrud et al. 1989). Resulting brackish water benefits some species, particularly shorebirds such as piping plover (*Charadrius melodus*) and American avocet (*Recurvirostra americana*), which breed on large or saline wetlands (Kantrud and Stewart 1984). For those species, large wetlands would be appropriate for conservation, assuming there was a threat to the wetlands, and the value of proximity to other wetlands would vary among species. Terrestrial applications of island biogeographic theory might be appropriate when considering resident species with limited mobility but would likely have unintended consequences and misdirect conservation actions for dabbling ducks given characteristics of PPR wetlands and many species of birds that use those wetlands.

Similarly, choice of response metric greatly affects how and where conservation will be implemented. Many spatial models focus on species richness, but unless the species and treatments are explicitly and well considered, richness metrics can have unintended consequences. First, the scale of species range or niche overlap and conservation treatment must be appropriate, as some species show congruent range at a broad scale but can actually be negatively correlated with each other at finer scales (Sherry and Holmes 1988), which is typically the scale at which management actions occur. Species richness models are often coarse grained, generally developed from occurrence data aggregated to large spatial units such as counties or summaries from broad surveys (Stohlgren et al. 2006, Distler et al. 2015). For these and other reasons, including spatially varying densities and annual rates of population change, overlays of species richness may be resistant to scaling up and scaling down (Conroy and Noon 1996). In addition, poorly considered richness measures often ignore rare species and promote generalists or weeds (Noss 1987, Rodda 1993) rather than the rarer and often declining specialists that are typically species of concern. Sites can have the same number of species but different species composition, which might require different conservation treatments or scales of treatment (Fischer et al. 2004). Species richness models can be sensitive to the species considered and their spatial distribution, which can cause a shift in the location of priorities (Van Horne 2002). Finally, areas of highest richness may not harbor every species under consideration; consequently, species not in the high-richness zones

might not receive and benefit from conservation actions.

Targeting species richness can be problematic from the standpoint of implementing treatments. At a conceptual level, priority areas include the zone of overlap for multiple species, which may exclude the best areas for the individual species under consideration (Fig. 12.3). This restricted zone of geographic overlap may also limit opportunities for implementing conservation treatments, which may be particularly important in landscapes dominated by private lands, where conservation action only occurs with willing landowners. If species richness is estimated without consideration of species' life histories, limiting factors, and habitat requirements, conservation treatments intended to benefit one species may reduce habitat quality for other species. For example, bird species richness in the northern Great Plains is higher in areas with trees than in nearby grasslands (Dieni and Jones 2002, Van Horne 2002), but preservation or expansion of high richness sites containing forest species would be deleterious to grassland birds, which are declining more broadly and rapidly

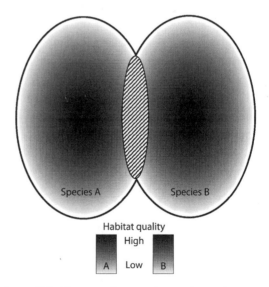

Figure 12.3. The zone of geographic overlap in the range of two hypothetical species excludes the best areas for the species under consideration and limits the area in which conservation treatments can be applied.

than any other bird group in North America (Knopf 1994, Peterjohn and Sauer 1999). One species richness decision-support tool that we reviewed included vertebrate, invertebrate, terrestrial, aquatic, rare, abundant, migratory, resident, generalist, specialist, breeding, nonbreeding, predator, and prey animals, as well as plants, in a variety of ecosystems under varying levels of human disturbance. Identifying a conservation treatment to improve the status of all species included in that model would be nearly impossible, with the possible exception, as a reviewer wryly noted, of broadscale human exclusion, as demonstrated by the response of wildlife in the Korean demilitarized zone (Kim 1997) and the Chernobyl exclusion zone (Deryabina et al. 2015).

Many of the shortcomings of species richness approaches can be avoided by targeting overlap of best areas rather than simple occurrence, or by addressing priority species individually or in small aggregates. If that is not possible, shortcomings can be circumvented by careful consideration of species biology and intended treatments and focusing on benefits for few similar species rather than richness of many disparate species (Fleishman et al. 2006). For example, many species of birds are area sensitive, where their occurrence, density, survival, or reproductive success is greater in large habitat patches than small habitat patches (Fig. 12.4A; Robbins 1979, Ribic et al. 2009). By meeting the requirements of the most area-sensitive species under consideration, requirements of less area-sensitive species also will be met, assuming that appropriate fine-grained features are also present (Fig. 12.4B). This approach uses the most area-sensitive species as a surrogate or umbrella for the less area-sensitive species; however, it has the consequence that smaller patches important to less area-sensitive species are dismissed, which limits management options and may prevent attainment of population objectives for those species.

The use of surrogate species for prioritizing areas for conservation poses similar problems in that biological outcomes in response to conservation practices are often unknown and may differ from what was intended (Carlisle et al. 2018). As mentioned above, protection of grassland and wetland complexes for waterfowl is a common conservation treatment in the PPR that has provided substantial benefits for other species (Niemuth et al. 2018). Waterfowl may be reasonable surrogates for those species of grassland birds in the PPR that respond positively to small wetlands with emergent vegetation; however, waterfowl will be poor surrogates for those species that prefer xeric sites and have different spatial distributions (Niemuth et al. 2017). The use of surrogates might be justified by a commitment to investigate assumptions of the approach, but by the time sufficient information is gathered to determine how well the surrogate tracks the target species, one could just as well work directly with the target species (i.e., if adequate information exists to justify the use of surrogates, then there is likely no need to use surrogates). If surrogate species are used, it is essential to explicitly state the surrogate approach being used and acknowledge that the use of surrogate species is a shortcut that will fail at some level, as species differ in distribution, ecological niches, and limiting factors (Che-Castaldo and Neel 2012, AMEC Environment and Infrastructure 2014, Carlisle et al. 2018). The bottom line is that decision-support tools based on richness or surrogate species will likely lack explicit, clearly articulated relationships that can be used to identify and guide specific conservation treatments to benefit recognized priority species.

Data Characteristics and Quality

Statisticians tend to think of data as information from which one can make inferences. With the advent of computers, the word "data" has taken on a more encompassing definition along the lines of information that is stored, regardless of its quality or ability to support inferences. Confusing the two meanings is easy in landscape-level investigations where large amounts of data (sometimes of the second meaning) are amassed, processed, and analyzed

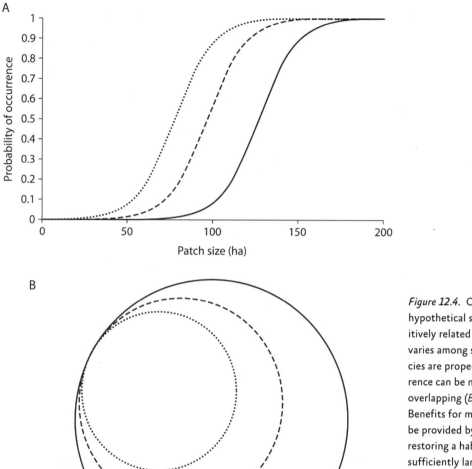

Figure 12.4. Occurrence of three hypothetical species (*A*) is positively related to patch size and varies among species. When species are properly selected, occurrence can be nested, rather than overlapping (*B*), as in Figure 12.3. Benefits for multiple species can be provided by protecting or restoring a habitat patch that is sufficiently large to harbor the most area-sensitive of the three species (solid line), thereby also benefitting less area-sensitive species. Dotted and dashed lines reflect patterns of area sensitivity shown in (*A*).

to create an end product, typically a map that is used to guide decisions. The inability of some of these data to support inferences, however, may be masked by poor or limited understanding of the study region, ecological processes, data, analyses, and the final map products.

Above all, model developers should be familiar with the landscape, the species of interest, and pertinent issues and programs. Spatial analyses rely heavily on remotely sensed data, and in many cases, people conducting spatial analyses are remote from the

study region and lack knowledge to catch problems with data. An example of this is a commonly used database that shows protected areas in the United States. The largest polygon identified as a protected area in North Dakota (Fig. 12.5) was classified as fee-title land held by the USFWS but is actually an administrative boundary for an area dominated by private land. This area is almost the size of Yellowstone National Park and should jump out at the most rudimentary level of data screening. Similarly, Native American tribal lands are depicted as encompassing

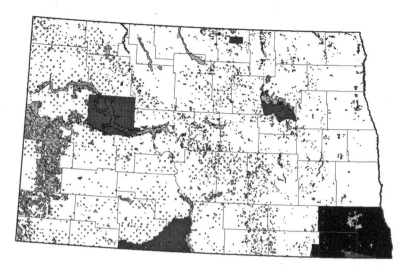

Figure 12.5. Protected lands (shaded areas) in North Dakota, USA, as identified by a commonly used database of protected areas in the United States. The black polygon in the southeastern portion of the state is listed as fee title land held by the US Fish and Wildlife Service (USFWS) but is actually an administrative boundary for a private lands easement acquisition program and represents no USFWS fee title lands. Medium-shaded polygons are identified as tribal lands but represent boundaries within which >50% of land is privately owned.

substantial portions of North Dakota (Fig. 12.5), but these polygons are again administrative boundaries, and >50% of land within these polygons is privately owned (Bureau of Indian Affairs 2019). Estimates of contributions of conservation efforts developed with these data will be inflated, and assessments of biodiversity of protected areas relative to unprotected areas could be biased. To the credit of the database compilers, erroneous ownership in the largest polygon (Fig. 12.5) has been corrected in recent iterations, but not before results of analyses using the erroneous data were published in at least one peer-reviewed journal (Wood et al. 2014).

Conservation practitioners should use data at resolutions and scales that are appropriate to the species, question, and treatment being considered (Wiens 1989, Boyce 2006, Mayor et al. 2009). For example, element occurrence analyses note records of a species' recorded occurrence (and, conversely, non-observation) in counties, watersheds, map quadrangles, major land resource areas, hexagons, or other coarse-grained units. The coarse resolution of many reporting units, which may be tens to thousands of times the size of intended treatments, precludes meaningful targeting of conservation actions (Fig. 12.6). In addition, element occurrence analyses do not identify or incorporate biological relationships that increase understanding and guide conservation treatments and therefore cannot incorporate appropriate scales for these treatments. Finally, counties, watersheds, major land resource areas, and many other units are variably sized, which contributes to biased estimates of species presence or richness owing to passive sampling (Connor and McCoy 1979), and data used in such analyses are generally opportunistic, often containing biases owing to species and observer location or interest (Anderson 2001, McKelvey et al. 2008, Niemuth et al. 2009).

Characteristics of environmental predictor data are frequently overlooked in spatial analyses, but predictors such as landcover and climate data are themselves the product of models. A primary assumption of regression analyses is that predictors are measured without error, which will almost never

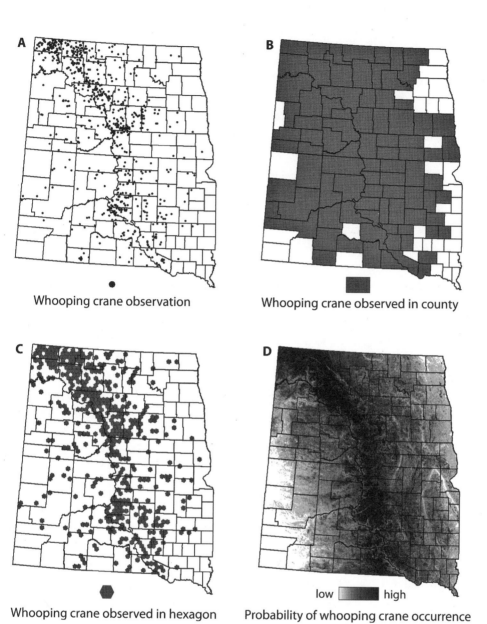

A

Whooping crane observation (•)

B

Whooping crane observed in county

C

Whooping crane observed in hexagon (⬡)

D

low ▦ high

Probability of whooping crane occurrence

Figure 12.6. Conservation guidance maps produced from a given data set can vary greatly depending on how data are processed and analyzed. Black dots (*A*) indicate observations from the Cooperative Whooping Crane Tracking Project database (Tacha et al. 2010) of whooping cranes (*Grus americana*) in North Dakota and South Dakota, USA, from 1955 to 2014. County-level element occurrence analysis (*B*), following species profile (US Fish and Wildlife Service 2019*b*), includes much non-habitat, identifies no biological relationships, and provides little ability to target specific actions. Element occurrence analysis (*C*) using nested hexagon grid (Western Association of Fish and Wildlife Agencies 2019) greatly reduces area of polygons identified as occupied, but again identifies no biological relationships and excludes much area that is likely habitat. Probability surface (*D*) developed using environmental predictors and generalized linear mixed models (Niemuth et al. 2018) captures biological relationships and enables quantitative ranking and evaluation of potential conservation sites but shares some limitations associated with use of opportunistic data.

be true when using many types of predictor data. Adding to the complexity of using these data is that classification accuracy varies throughout the data set because of differences in timing, availability of cloud-free images, and many other factors (Thogmartin et al. 2004, Gallant 2009). Accuracy of many landcover data sets can be improved through combining of similar classes where appropriate or the addition of ancillary data from other sources. For example, in the case of analyses where perennial cover is important, researchers might incorporate spatial data from the US Department of Agriculture National Agricultural Statistics Service that identify alfalfa fields (Boryan et al. 2011). Classification accuracy of landcover data should always be reported along with other pertinent quirks or characteristics that might affect interpretation and application of the models (Thogmartin et al. 2004, Gallant 2009). Characteristics of other forms of predictor data also should be assessed prior to analysis, and any limitations of the data to answer the question at hand should be acknowledged, paying particular attention to matters of accuracy and scale.

No data are perfect, whether they are used as predictors or response variables. Every analysis should begin with an assessment of the data and the need to account for biases, and data should be appropriate to the question. Data collection and analysis might differ even for the same species if different portions of its life cycle are being addressed. For example, monitoring and conservation of migrant or wintering waterfowl and waterbirds on large impoundments with moist-soil management along the Mississippi River, USA, will be different than monitoring and conservation of breeding waterfowl in small, dispersed prairie wetlands, and programs addressing the first question should not be forced into situations where they are not appropriate.

When acquiring or assessing data, whether it be for response or predictor variables, analysts should ask what questions could responsibly be addressed with the data rather than forcing inappropriate uses. Predictor and response data are readily available for landscape analyses, but many of them suffer from low quality caused by their opportunistic collection, sampling bias, classification error, or limited relevance to the landscape analysis under consideration (McKelvey et al. 2008, Lozier et al. 2009, Niemuth et al. 2009). Use of poor data is sometimes justified by saying the data are the best available information, but researchers should assess data quality prior to analysis, and not use data that cannot support inferences regarding the question of interest. In many cases, it might be preferable to develop a conceptual model that can capture important biological relationships (Clevenger et al. 2002, Johnson et al. 2010), or it might be necessary to collect appropriate data to address a specific question.

If data are collected when developing a spatial model, a sampling frame should be implemented that considers pertinent factors relative to the conservation need. Because landscape analyses cover broad spatial extents and a large range of landcover and environmental variation, it is imperative to consider the question being addressed prior to data collection to ensure that sample size is sufficiently large to cover the range of variation, support consideration of appropriate candidate predictor variables, and reduce extrapolation when models are applied to create maps. Again using the example of area-sensitive species, data should be collected along a continuum of patch size (or amount of habitat in the surrounding landscape) so that the analysis can identify relationships and inflection points. Too often, biologists only want to sample where they know a species occurs, which greatly reduces power to make inferences about habitat selection. Data collected as part of a properly designed study will simplify analysis and provide stronger inferences than data collected with little thought or in a manner that is convenient (Krebs 1989, Anderson 2001).

Analytical Considerations

After a treatment-specific purpose has been identified and quality of data assessed, an appropriate analytic technique must be chosen to identify and quantify the biological relationships that are the core of a useful model. A discussion of analytical techniques is beyond the scope of this chapter, but numerous resources provide in-depth guidance for analysis of spatial data (Shenk and Franklin 2001, Guisan and Thuiller 2005, Elith et al. 2006, Millspaugh and Thompson 2009, Zuur et al. 2009). In fact, the problem for analysis is generally not lack of an available method but the selection of an appropriate one (Jones-Farrand et al. 2011). Johnson (2001:113) wrote, "A model has value if it provides better insight, predictions, or control than would be available without the model." This philosophy leaves a lot of flexibility for model types and development, but Johnson noted that a model's value is determined in light of its intended application, once again reinforcing the idea that addressing explicitly identified conservation needs and treatments is key to success. Choice of analytical technique can substantially affect the final form and value of a spatial model (Wilson et al. 2005, Thompson and Millspaugh 2009), and we present observations on factors that can, in our experience, affect the usefulness of a model.

Models can be classified in many different ways, but for the purposes of this chapter we place models into three general categories: conceptual or expert opinion models, statistical or empirical models, and black boxes. Conceptual or expert opinion models can provide useful guidance based on biological relationships that are identified and related to spatial data (Clevenger et al. 2002, Sanderson et al. 2002, Johnson et al. 2010). These models have the advantage of making use of experience and knowledge of species experts, and results can be almost identical to those of statistical models (Niemuth et al. 2005). Involving experts and other stakeholders in model development can be politically powerful, and conceptual models can provide useful conservation guidance and a better understanding of factors that should be considered if data are collected to develop a statistical model. However, experts on species biology may not understand or agree upon

spatial issues and interrelation of data, scale, and conservation treatments, which can complicate development of useful models. Even in the absence of perfect agreement, we believe that conceptual models are superior to the common alternative of drawing focal area boundaries based on the experience or favorite areas of participants. Both approaches use similar methods, but conceptual models force the identification of relationships and characteristics that are important to the question or species of interest (Starfield 1997). These relationships can then be applied to spatial data to identify multiple potential conservation treatments across the entire landscape, whereas drawing focal areas only requires identification of geographic areas, conveys no understanding of mechanisms, and applies only to areas familiar to the experts.

Statistical or empirical approaches have the advantage that relationships are inferred from data, thereby avoiding biases associated with expert opinion. As such, these models can provide insights that might be new or contradict conventional wisdom. Because of their reliance on data, statistical models are generally considered more rigorous than conceptual models and have the additional advantage that estimates of uncertainty are provided for parameter estimates. Statistical models often require substantial amounts of data, which can be expensive and time consuming to collect. Considerable training and technical expertise may be required for proper development and application of some statistical models, particularly with complex data sets often associated with spatial analyses. Consequently, statistics can override ecological relationships (Austin 2002), which can affect interpretation and application of models with respect to conservation treatments. For these reasons, it is imperative that biologists and field staff be involved with the model development process and that statisticians work with an appropriate set of a priori models (Burnham and Anderson 2002, Millspaugh et al. 2009, Noon et al. 2009).

Our distinction between statistical models and

black boxes relates more to the process than the mechanisms of the analysis; in fact, black boxes may well be based on sound statistical procedures, but the inputs, assumptions, and process are often unclear, even though they might greatly affect model output (Merow et al. 2013). Black box approaches may include occurrence or richness data sets that obviate the need for analysis (US Fish and Wildlife Service 2019a, Western Association of Fish and Wildlife Agencies 2019) and software and websites that offer simple means of developing spatial models (NatureServe 2019), sometimes at the expense of process transparency. In many of these cases the process, including scale and analytical framework, may be determined by the software rather than need, purpose, or biology, with little or no link to explicit conservation treatments. In other cases, users may be able to subjectively weight different components of the analysis or rate data quality, with no understanding of how those values are used or affect the outcome, all of which reduce transparency and interpretability. We do not argue that analyses conducted within black boxes are wrong but suggest that users or administrators might be swayed by ease of use rather than go through the effort required to obtain answers pertinent to the question or treatment of interest. Most important, black box approaches can take the thought out of what should be a thought-intensive process (Starfield 1997).

No matter what type of analytical technique is used, the limitations of process, data, and models should be explicitly acknowledged, as should the importance of people in the field who use the models. Data and models are not reality, and spatial models provide only a view of the world at a fairly coarse scale. Wildlife require habitat at fine scales as well as coarse, and field staff are essential to ensure that all elements of habitat requirements are met and provide feedback about models based on their knowledge of treatments and local conditions.

Programmatic and Staffing Considerations

Finally, even the most clearly articulated and demonstrated conservation needs will go unmet if a project to develop spatial tools lacks programmatic support and qualified staff. Conservation practitioners may have to adopt mindsets and approaches differing from their previous training and experience, and educate administrators as to why new approaches or data are necessary. Researchers should consider management issues and constraints, which may entail citing or publishing in the management literature or planning documents such as joint venture implementation plans (Fields 2017, Connelly and Conway, Chap. 10, this volume). Researchers generally prefer to publish in high-impact journals that are often theoretical or conceptual in nature (Aarssen et al. 2008); however, gray literature may contain specific information that is pertinent to conservation actions that is not available in the mainstream scientific literature (Corlett 2011). Decision makers should rely on science, decision support tools, and field staff to help make decisions and focus on the end goal of conservation rather than program development. The most effective way to foster development and implementation of useful spatial tools will be to identify a need, then develop useful tools that meet the need. People will seek out and use those tools that aid them in making decisions, just as consumers in a free market economy purchase products that meet their needs, whereas products in a command or centrally planned economy often languish because they do not meet a need.

The purpose of a spatial decision-support tool is to increase the efficiency of conservation by providing insights and helping guide decisions over spatial extents that exceed the knowledge of local experts (Guisan and Thuiller 2005, Franklin 2013). Models can increase the cost-effectiveness of conservation (Haight and Gobster 2009), but if there is no conservation action taking place or if the amount of action is minimal, the time and expense associated with developing a spatial tool may exceed the benefits received from the tool, potentially decreasing the amount of conservation taking place. Too often, people want to do monitoring or research, both of which are expensive in terms of time and money, without considering the return on investment that the final product should provide or ensuring that the data will be useful (Legg and Nagy 2006). Instituting a new program will do little good unless the information gathered informs future decisions.

Creating rigorous, useful spatial models and decision-support tools is much more complicated than simply having someone with GIS skills stack multiple spatial data layers or intersect species observation data with county polygons. The process requires capable people with a diversity of analytical skills, biological background, awareness of conservation issues and treatments, and significant experience with data processing and development of spatial tools. The people and process involved with developing a spatial model are critical to its success. Noss (2003) identified eight standards for conservation planning, each with specific sub-criteria: staff qualifications, choice of conservation target, methodological comprehensiveness and rigor, replicability, analytic rigor, peer review, overall quality of scholarship, and iterative improvement. These are simple concepts, but they are too often cast aside for the expedience of using simple methodology, existing staff, poor data, or the desire for a simple, universal spatial tool. Rather than working toward a single spatial tool that meets all needs, conservation practitioners would be better off developing a conservation toolbox, where accurate data and multiple rigorous spatial models serve as components from which decision-support tools can be developed that are appropriate to the needs and treatments under consideration. Such a toolbox and approach will facilitate the development of "small, simple models that focus relentlessly on the problem to be solved" (Starfield 1997:261).

Administrators must recognize that development of scientifically sound spatial tools over broad spatial and temporal scales requires a significant investment of time and money, which reinforces the necessity of clearly identified needs and treatments. Gains in efficiency and effectiveness that come with spatial decision-support tools must be weighed against the cost required to develop those tools. If little money is being spent on the ground, efficiency gained from development of a spatial model may be insufficient to offset the cost of data collection and/or model development. The time and financial investments necessary to develop spatial tools, along with an absence of in-house expertise, may necessitate collaboration with multiple agencies, organizations, and researchers. In these cases, all partners must be cognizant and supportive of the appropriate spatial and temporal scales and characteristics of the conservation program necessary to succeed.

Good science associated with solid data and rigorous spatial models and decision-support tools can provide transparency and accountability as well as increased efficiency for conservation delivery. But the definition of good science often varies among people depending on their personal values or their organization's position on a topic. Politics, money, and logistics are valid reasons for making decisions but should never be disguised as science. Decision makers also must learn to critically evaluate scientific products, as many spatial products are heavily promoted but lack the explicit purpose, appropriate scale, biological linkages, or data quality necessary to be useful for conservation delivery. Similarly, decision makers must realize that scenarios and hypothetical landscapes used in many publications might have heuristic value, but hypothetical landscapes are of less value to people who must make decisions based on the reality of risk of habitat conversion, land prices, and variability in resources.

Summary

Properly developed and implemented, spatial models and decision-support tools can substantially increase efficiency and cost-effectiveness of on-the-ground conservation delivery, but this does not mean that the output of spatial models should be followed blindly. For all the advantages that a model-based approach provides, a model is simply a useful abstraction that cannot characterize all the dynamics of a system (Fig. 12.7). When local knowledge is available that reliably surpasses the general relationships described in a landscape model, the local knowledge should be used. Model developers should work with biologists and field staff who deliver conservation to develop a strategy of continuous feedback that helps refine the model and improve delivery (US Fish and Wildlife Service 2008). This process might provide one of the greatest benefits of a model-based approach to conservation delivery, as it helps people understand the strengths, needs, and limitations of their conservation efforts.

The value of spatial tools goes beyond delivery. Spatial models can be used to demonstrate the biological benefits of conservation programs, provide spatially explicit population estimates, and evaluate potential stressors, to name a few uses (Reynolds et al. 2006, Niemuth et al. 2018). A rigorous program that uses spatial models to target conservation delivery and demonstrate conservation benefits is also highly attractive to administrators and potential funders; for example, a variant of the waterfowl thunderstorm map (Reynolds et al. 2006) was used to support the acquisition and guide the expenditure of $590,000 (USA) for ruddy duck (*Oxyura jamaicensis*) conservation in North Dakota and South Dakota following the April 2000 Chalk Point oil spill off the coast of Maryland, USA. Similarly, spatial models developed for waterbirds (Niemuth et al. 2009) were used to justify the award of $6,000,000 (USA) for black tern (*Chlidonias niger*) conservation in North Dakota and South Dakota following the April 2010 Deepwater Horizon oil spill in the Gulf of Mexico.

Benefits such as these, however, are unlikely to be realized if spatial models and decision-support tools suffer from lack of purpose, poor resolution, inaccurate data, inappropriate analysis, or are not closely linked to implementation. A model that purports to

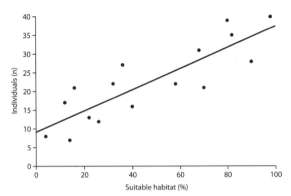

Figure 12.7. Spatial models and decision-support tools will include error, whether from sampling error, unexplained variation in data, or misclassification of landcover data, but the general patterns they portray will be useful for guiding decisions. In this hypothetical example predicting number of individuals as a function of percentage of the landscape in a suitable habitat class, the model (solid line) does not exactly fit any of the observations (black dots) and is substantially wrong for some, but it is useful for characterizing the overall relationship between landscape composition and number of individuals. Similarly, local knowledge that reliably surpasses a statistical model can and should be used when appropriate; if possible, such knowledge should then be incorporated into future analyses to improve the value of the model.

identify the best places for conservation but does not specify what those places are best for will have little real-world value. Virtually every publication describing a species distribution model contains an obligatory sentence or paragraph indicating that the model can be used to guide conservation, but the reality is that, in our experience, the value of many published models for conservation delivery is limited, as they simply do not have an explicit link to delivery. Development of useful spatial tools requires a substantial investment of time, money, and, most importantly, thought about how the tools will be developed, who the users will be, and how the output will be used to deliver conservation. Many of the problems we have described can easily be avoided if conservation practitioners work in the context of a conservation issue, do their homework, maintain a strong biological foundation, and remember that the end goal of applied landscape ecological research is not program development or publication but conservation.

LITERATURE CITED

Aarssen, L. W., T. Tregenza, A. E. Budden, C. J. Lortie, J. Koricheva, and R. Leimu. 2008. Bang for your buck: rejection rates and impact factors in ecological journals. Open Ecology 1:14–19.

Abrahms, B., S. C. Sawyer, N. R. Jordan, J. W. McNutt, A. M. Wilson, and J. S. Brashares. 2017. Does wildlife resource selection accurately inform corridor conservation? Journal of Applied Ecology 54:412–422.

AMEC Environment and Infrastructure. 2014. Peer review of technical guidance on selecting species for landscape scale conservation. US Fish and Wildlife Service, Denver, Colorado, USA. http://www.fws.gov/science/pdf/Final-Summary-Report-Complete-Technical-Guidance-on-Selecting-Species-for-Landscape-Scale-Conservation.pdf. Accessed 2 Aug 2019.

Anderson, D. R. 2001. The need to get the basics right in wildlife field studies. Wildlife Society Bulletin 29:1294–1297.

Anselin, L. 2006. How (not) to lie with spatial statistics. American Journal of Preventive Medicine 30(2):S3–S6.

Anteau, M. J. 2012. Do interactions of land use and climate affect productivity of waterbirds and prairie-pothole wetlands? Wetlands 32:1–9.

Askins, R. A. 1993. Population trends in grassland, shrubland, and forest birds in eastern North America. Current Ornithology 11:1–34.

Austin, M. P. 2002. Spatial prediction of species distribution: an interface between ecological theory and statistical modelling. Ecological Modelling 157:101–118.

Baker, N. T., and W. W. Stone. 2015. Estimated annual agricultural pesticide use for counties of the conterminous United States, 2008–12. US Geological Survey Data Series 907, Washington, DC, USA.

Batt, B. D. J., M. G. Anderson, C. D. Anderson, and F. D. Caswell. 1989. Use of prairie potholes by North American ducks. Pages 204–227 in A. van der Valk, editor. Northern prairie wetlands. Iowa State University, Ames, Iowa, USA.

Bertas33ello, L. E., P. S. C. Rao, J. W. Jawitz, G. Botter, P. V. V. Le, P. Kumar, and A. F. Aubeneau. 2018. Wet-landscape fractal topography. Geophysical Research Letters 45:6983–6991.

Beyersbergen, G. W., N. D. Niemuth, and M. R. Norton. 2004. Northern Prairie and Parkland Waterbird Conservation Plan. Prairie Pothole Joint Venture, Denver, Colorado, USA.

Boryan, C., Z. Yang, R. Mueller, and M. Craig. 2011. Monitoring US agriculture: the US Department of Agriculture, National Agricultural Statistics Service, Cropland Data Layer Program. Geocarto International 26:341–358.

Bouffard, S. H., and M. A. Hanson. 1997. Fish in waterfowl marshes: waterfowl managers' perspective. Wildlife Society Bulletin 25:146–157.

Boyce, M. S. 2006. Scale for resource selection functions. Diversity and Distributions 12:269–276.

Boyce, M. S., and L. L. McDonald. 1999. Relating populations to habitats using resource selection functions. Trends in Ecology and Evolution 14:268–272.

Bureau of Indian Affairs. 2019. Great Plains region: overview. https://www.bia.gov/regional-offices/great-plains/. Accessed 12 Aug 2019.

Burnham, K. P., and D. R. Anderson. 2002. Model selection and multimodel inference: a practical information-theoretic approach. Springer, New York, New York, USA.

Carlisle, J. D., D. A. Keinath, S. E. Albeke, and A. D. Chalfoun. 2018. Identifying holes in the greater sage-grouse conservation umbrella. Journal of Wildlife Management 82:948–957.

Che-Castaldo, J. P., and M. C. Neel. 2012. Testing surrogacy assumptions: can threatened and endangered plants be grouped by biological similarity and abundances? PLOS One 7(12):e51659.

Clevenger, A. P., J. Wierzchowski, B. Chruszcz, and K. Gunson. 2002. GIS-generated, expert-based models for identifying habitat linkages and planning mitigation passages. Conservation Biology 16:503–514.

Connor, E. F., and E. D. McCoy. 1979. The statistics and biology of the species-area relationship. American Naturalist 113:791–833.

Conroy, M. J., and B. R. Noon. 1996. Mapping of species richness for conservation of biological diversity: conceptual and methodological issues. Ecological Applications 6:763–773.

Corlett, R. T. 2011. Trouble with the gray literature. Biotropica 43:3–5.

Cowardin, L. M., D. S. Gilmer, and C. W. Shaiffer. 1985. Mallard recruitment in the agricultural environment of North Dakota. Wildlife Monographs 92:1–37.

Cowardin, L. M., T. L. Shaffer, and P. M. Arnold. 1995. Evaluation of duck habitat and estimation of duck population sizes with a remote-sensing based system. National Biological Service, Biological Science Report 2, Jamestown, North Dakota, USA.

Cox, R. R., M. A. Hanson, C. C. Roy, N. H. Euliss, D. H. Johnson, and M. G. Butler. 1998. Mallard duckling growth and survival in relation to aquatic invertebrates. Journal of Wildlife Management 82:124–133.

Dahl, T. E. 1990. Wetland losses in the United States 1780s

to 1980s. US Department of the Interior, Fish and Wildlife Service, Washington, DC, USA.

Deryabina, T. G., S. V. Kuchmel, L. L. Nagorskaya, T. G. Hinton, J. C. Beasley, A. Lerebours, and J. T. Smith. 2015. Long-term census data reveal abundant wildlife populations at Chernobyl. Current Biology 25:R824–R826.

Dieni, J. S., and S. L. Jones. 2002. A field test of the area search method for measuring breeding bird populations. Journal of Field Ornithology 73:253–257.

Distler, T., J. G. Schuetz, J. Velásquez-Tibatá, and G. M. Langham. 2015. Stacked species distribution models and macroecological models provide congruent projections of avian species richness under climate change. Journal of Biogeography 42:976–988.

Elith, J., C. H. Graham, R. P. Anderson, M. Dudík, S. Ferrier, A. Guisan, R. J. Hijmans, F. Huettmann, J. R. Leathwick, A. Lehmann, et al. 2006. Novel methods improve prediction of species' distributions from occurrence data. Ecography 29:129–151.

Farhat, Y. A., W. M. Janousek, J. P. McCarty, N. Rider, and L. L. Wolfenbarger. 2014. Comparison of butterfly communities and abundances between marginal grasslands and conservation lands in the eastern Great Plains. Journal of Insect Conservation 18:245–256.

Fields, S. P., editor. 2017. Prairie Pothole Joint Venture implementation plan. US Fish and Wildlife Service, Denver, Colorado, USA.

Fischer, J., D. B. Lindenmayer, and A. Cowling. 2004. The challenge of managing multiple species at multiple scales: reptiles in an Australian grazing landscape. Journal of Applied Ecology 41:32–44.

Fleishman, E., R. F. Nos, and B. R. Noon. 2006. Utility and limitations of species richness metrics for conservation planning. Ecological Indicators 6:543–553.

Forman, R. T. T. 1995. Land mosaics: the ecology of landscapes and regions. Cambridge University Press, Cambridge, UK.

Franklin, J. 2013. Species distribution models in conservation biogeography: developments and challenges. Diversity and Distributions 19:1217–1223.

Galatowitsch, S. M., and A. G. van der Valk. 1996. Vegetation and environmental conditions in recently restored wetlands in the prairie pothole region of the USA. Vegetatio 126:89–99.

Gallant, A. L. 2009. What you should know about land-cover data. Journal of Wildlife Management 73:796–805.

Game, E. T., P. Kareiva, and H. Possingham. 2013. Six common mistakes in conservation priority setting. Conservation Biology 27:480–485.

Godfray, H. C., T. Blacquière, L. M. Field, R. S. Hails, G. Petrokofsky, S. G. Potts, N., E. Raine, A. J. Vanbergen, and A. R. McLean. 2014. A restatement of the natural science evidence base concerning neonicotinoid insecticides and insect pollinators. Proceedings of the Royal Society B 281:20140558.

Greenwood, R. J., A. B. Sargeant, D. H. Johnson, L. M. Cowardin, and T. L. Shaffer. 1995. Factor associated with duck nest success in the Prairie Pothole Region of Canada. Wildlife Monographs 128:1–57.

Guisan, A., and W. Thuiller. 2005. Predicting species distribution: offering more than simple habitat models. Ecology Letters 8:993–1009.

Haight, R. G., and P. H. Gobster. 2009. Social and economic considerations for planning wildlife conservation in large landscapes. Pages 123–152 in J. J. Millspaugh and F. R. Thompson III, editors. Models for planning wildlife conservation in large landscapes. Elsevier, Burlington, Massachusetts, USA.

Harris, S. W., and W. H. Marshall. 1963. Ecology of water-level manipulations on a northern marsh. Ecology 44:331–343.

Hoekstra, J. M., T. M. Boucher, T. H. Ricketts, and C. Roberts. 2005. Confronting a biome crisis: global disparities of habitat loss and protection. Ecology Letters 8:23–29.

Homer, C. G., J. A. Dewitz, L. Yang, S. Jin, P. Danielson, G. Xian, J. Coulston, N. D. Herold, J. D. Wickham, and K. Megown. 2015. Completion of the 2011 National Land Cover Database for the conterminous United States—representing a decade of land cover change information. Photogrammetric Engineering and Remote Sensing 81:345–354.

Johnson, D. H. 2001. Validating and evaluating models. Pages 105–119 in T. M. Shenk and A. B. Franklin, editors. Modeling in natural resource management. Island Press, Washington, DC, USA.

Johnson, R. R., M. E. Estey, D. A. Granfors, R. E. Reynolds, and N. D. Niemuth. 2010. Delineating grassland bird conservation areas in the U.S. Prairie Pothole Joint Venture. Journal of Fish and Wildlife Management 1:38–42.

Jones-Farrand, D. T., T. M. Fearer, W. E. Thogmartin, F. R. Thompson III, M. D. Nelson, and J. M. Tirpak. 2011. Comparison of statistical and theoretical habitat models for conservation planning: the benefit of ensemble prediction. Ecological Applications 21:2269–2282.

Kantrud, H. A., G. L. Krapu, and G. A. Swanson. 1989. Prairie basin wetlands of the Dakotas: a community profile. US Fish and Wildlife Service Biological Report 85(7.28), Washington, DC, USA.

Kantrud, H. A., and R. E. Stewart. 1984. Ecological distribution and crude density of breeding birds on prairie wetlands. Journal of Wildlife Management 48:426–437.

Kim, K. C. 1997. Preserving biodiversity in Korea's demilitarized zone. Science 278:242–243.

Klett, A. T., T. L. Shaffer, and D. H. Johnson. 1988. Duck nest success in the Prairie Pothole Region. Journal of Wildlife Management 52:431–440.

Knopf, F. L. 1994. Avian assemblages on altered grasslands. Studies in Avian Biology 15:247–257.

Koh, I., E. V. Lonsdorf, N. M. Williams, C. Brittain, R. Isaacs, J. Gibbs, and T. H. Ricketts. 2016. Modeling the status, trends, and impacts of wild bee abundance in the United States. Proceedings of the National Academy of Sciences 113:140–145.

Krebs, C. J. 1989. Ecological methodology. Harper & Row, New York, New York, USA.

Krygier, J., and D. Wood. 2016. Making maps: a visual guide to map design for GIS. Third edition. Guilford Press, New York, New York, USA.

Lark, T. J., J. M. Salmon, and H. K. Gibbs. 2015. Cropland expansion outpaces agricultural and biofuel policies in the United States. Environmental Research Letters 10:044003.

Legg, C. J., and L. Nagy. 2006. Why most conservation monitoring is, but need not be, a waste of time. Journal of Environmental Management 78:194–199.

Lozier, J. D., P. Aniello, and M. J. Hickerson. 2009. Predicting the distribution of Sasquatch in western North America: anything goes with ecological niche modelling. Journal of Biogeography 36:1623–1627.

MacArthur, R. H., and E. O. Wilson. 1967. The theory of island biogeography. Princeton University Press, Princeton, New Jersey, USA.

Mayor, S. J., D. C. Schneider, J. A. Schaefer, and S. P. Mahoney. 2009. Habitat selection at multiple scales. Ecoscience 16:238–247.

McKelvey, K. S., K. B. Aubry, and M. K. Schwartz. 2008. Using anecdotal occurrence data for rare or elusive species: the illusion of reality and a call for evidentiary standards. BioScience 58:549–555.

Merow, C., M. J. Smith, and J. A. Silander. 2013. A practical guide to MaxEnt for modeling species' distributions: what it does, and why inputs and settings matter. Ecography 36:1058–1069.

Millspaugh, J. J., R. A. Gitzen, D. R. Larsen, M. A. Larson, and F. R. Thompson III. 2009. General principles for developing landscape models for wildlife conservation. Pages 1–32 in J. J. Millspaugh and F. R. Thompson III, editors. Models for planning wildlife conservation in large landscapes. Elsevier, Burlington, Massachusetts, USA.

Millspaugh, J. J., and F. R. Thompson III, editors. 2009. Models for planning wildlife conservation in large landscapes. Elsevier, Burlington, Massachusetts, USA.

Monmonier, M. 1996. How to lie with maps. University of Chicago Press, Chicago, Illinois, USA.

Morris, L. R., K. M. Proffitt, and J. K. Blackburn. 2016. Mapping resource selection functions in wildlife studies: concerns and recommendations. Applied Geography 76:173–183.

Murkin, H. R., E. J. Murkin, and J. P. Ball. 1997. Avian habitat selection and prairie wetland dynamics: a 10-year experiment. Ecological Applications 7:1144–1159.

Naidoo, R., A. Balmford, P. J. Ferraro, S. Polasky, T. H. Ricketts, and M. Rouget. 2006. Integrating economic costs into conservation planning. Trends in Ecology and Evolution 21:681–687.

National Agricultural Statistics Service. 2015. Land values: 2015 summary. US Department of Agriculture, Washington, DC, USA.

NatureServe. 2019. NatureServe Vista. www.natureserve .org/conservation-tools/natureserve-vista. Accessed 12 Aug 2019.

Naugle, D. E., R. R. Johnson, M. E. Estey, and K. F. Higgins. 2001. A landscape approach to conserving wetland bird habitat in the Prairie Pothole Region of eastern South Dakota. Wetlands 21:1–17.

Niemuth, N. D., M. E. Estey, S. P. Fields, B. Wangler, A. A. Bishop, P. J. Moore, R. C. Grosse, and A. J. Ryba. 2017. Developing spatial models to guide conservation of grassland birds in the U.S. Northern Great Plains. The Condor 119:506–525.

Niemuth, N. D., M. E. Estey, and C. R. Loesch. 2005. Developing spatially explicit habitat models for grassland bird conservation planning in the Prairie Pothole Region of North Dakota. Pages 469–477 in C. J. Ralph and T. D. Rich, editors. Bird conservation implementation and integration in the Americas: proceedings of the Third International Partners in Flight Conference 2002. USDA Forest Service Report PSW-GTR-191, Albany, California, USA.

Niemuth, N. D., M. E. Estey, and R. E. Reynolds. 2009. Data for developing spatial models: criteria for effective conservation. Pages 396–411 in Proceedings of the Fourth International Partners in Flight Conference, 13–16 February 2008, T. D. Rich, C. D. Thompson, D. Demarest, et al., editors. Partners in Flight, McAllen, Texas, USA.

Niemuth, N. D., M. E. Estey, R. E. Reynolds, C. R. Loesch, and W. A. Meeks. 2006. Use of wetlands by spring-migrant shorebirds in agricultural landscapes of North Dakota's Drift Prairie. Wetlands 26:30–39.

Niemuth, N. D., R. E. Reynolds, D. A. Granfors, R. R. Johnson, B. Wangler, and M. E. Estey. 2008. Landscape-level planning for conservation of wetland birds in the U.S. Prairie Pothole Region. Pages 533–560 in J. J.

Millspaugh and F. R. Thompson III, editors. Models for planning wildlife conservation in large landscapes. Elsevier, Burlington, Massachusetts, USA.

Niemuth, N. D., A. J. Ryba, A. T. Pearse, S. M. Kvas, D. A. Brandt, B. Wangler, J. E. Austin, and M. J. Carlisle. 2018. Opportunistically collected data reveal habitat selection by migrating whooping cranes in the U.S. Northern Plains. The Condor 120:343–356.

Noon, B. R., K. S. McKelvey, and B. G. Dickson. 2009. Multispecies conservation planning on U.S. federal lands. Pages 51–83 in J. J. Millspaugh and F. R. Thompson III, editors. Models for planning wildlife conservation in large landscapes. Elsevier, Burlington, Massachusetts, USA.

North American Waterfowl Management Plan Committee. 1986. North American Waterfowl Management Plan. US Fish and Wildlife Service, Washington, DC, USA.

Noss, R. F. 1987. Do we really want diversity? Whole Earth Review 1987(2):126–128.

Noss, R. F. 2003. A checklist for wildlands network designs. Conservation Biology 17:1270–1275.

Olimb, S. K., and B. Robinson. 2019. Grass to grain: probabilistic modeling of agricultural conversion in the North American Great Plains. Ecological Indicators 102:237–245.

Oslund, F. T., R. R. Johnson, and D. R. Hertel. 2010. Assessing wetland changes in the Prairie Pothole Region of Minnesota from 1980 to 2007. Journal of Fish and Wildlife Management 1:131–135.

Pecenka, J. R., and J. G. Lundgren. 2015. Non-target effects of clothianidin on monarch butterflies. Science of Nature 102:19.

Peterjohn, B. G., and J. R. Sauer. 1999. Population status of North American grassland birds from the North American Breeding Bird Survey, 1966–1996. Studies in Avian Biology 19:27–44.

Phillips, S. J., R. P. Anderson, M. Dudik, R. E. Schapire, and M. E. Blair. 2017. Opening the black box: an open-source release of Maxent. Ecography 40:887–893.

Potts, S. G., J. C. Biesmeijer, C. Kremen, P. Neumann, O. Schweiger, and W. E. Kunin. 2010. Global pollinator declines: trends, impacts and drivers. Trends in Ecology and Evolution 25:345–353.

Rashford, B. S., J. A. Walker, and C. T. Bastian. 2011. Economics of grassland conversion to cropland in the Prairie Pothole Region. Conservation Biology 25:276–284.

Reynolds, R. E., T. L. Shaffer, C. R. Loesch, and R. R. Cox Jr. 2006. The Farm Bill and duck production in the Prairie Pothole Region: increasing the benefits. Wildlife Society Bulletin 34:963–974.

Reynolds, R. E., T. L. Shaffer, R. W. Renner, W. E. Newton, and B. D. J. Batt. 2001. Impact of the Conservation Reserve Program on duck recruitment in the U.S. Prairie Pothole Region. Journal of Wildlife Management 65:765–780.

Ribic, C. A., R. R. Koford, J. R. Herkert, D. H. Johnson, R. B. Renfrew, N. D. Niemuth, D. E. Naugle, and K. K. Bakker. 2009. Area sensitivity in North American grassland birds: patterns, processes, and research needs. Auk 126:233–244.

Robbins, C. S. 1979. Effect of forest fragmentation on bird populations. Pages 198–212 in Proceedings of the workshop on management of north-central and northeastern forests for nongame birds. R. M. DeGraaf and N. Tilghman, editors. US Forest Service General Technical Report NC-51, Washington, DC, USA.

Rodda, G. H. 1993. How to lie with biodiversity. Conservation Biology 7:959–960.

Sanderson, E. W., K. H. Redford, A. Vedder, P. B. Coppolillo, and S. E. Ward. 2002. A conceptual model for conservation planning based on landscape species requirements. Landscape and Urban Planning 58:41–56.

Shenk, T. M., and A. B. Franklin, editors. 2001. Modeling in natural resource management. Island Press, Washington, DC, USA.

Sherry, T. W., and R. T. Holmes. 1988. Habitat selection by breeding American redstarts in response to a dominant competitor, the least flycatcher. The Auk 105:350–364.

Starfield, A. M. 1997. A pragmatic approach to modeling for wildlife management. Journal of Wildlife Management 61:261–270.

Stauffer, D. F. 2002. Linking populations and habitats: where have we been? where are we going? Pages 53–62 in J. M. Scott, P. J. Heglund, M. L. Morrison, et al., editors. Predicting species occurrences: issues of accuracy and scale. Island Press, Washington, DC, USA.

Stephens, S. E., J. J. Rotella, M. S. Lindberg, M. L. Taper, and J. K. Ringelman. 2005. Duck nest survival in the Missouri Coteau of North Dakota: landscape effects at multiple spatial scales. Ecological Applications 15:2137–2149.

Stephens, S. E., J. A. Walker, D. R. Blunck, A. Jayaraman, D. E. Naugle, J. K. Ringelman, and A. J. Smith. 2008. Predicting risk of habitat conversion in native temperate grasslands. Conservation Biology 22:1320–1330.

Stohlgren, T. J., D. Barnett, C. Flather, P. Fuller, B. Peterjohn, J. Katesz, and L. L. Master. 2006. Species richness and patterns of invasion in plants, birds, and fishes in the United States. Biological Invasions 8:427–447.

Tacha, M., A. Bishop, and J. Brei. 2010. Development of the whooping crane tracking project geographic information system. Proceedings of the Eleventh North American Crane Workshop 11:98–104.

Thogmartin, W. E., A. L. Gallant, M. G. Knutson, T. J. Fox,

and M. J. Suárez. 2004. Commentary: a cautionary tale regarding use of the National Land Cover Dataset 1992. Wildlife Society Bulletin 32:970–978.

Thompson, F. R., III, and J. J. Millspaugh. 2009. A decision framework for choosing models in large-scale wildlife conservation planning. Pages 661–674 *in* J. J. Millspaugh and F. R. Thompson III, editors. Models for planning wildlife conservation in large landscapes. Elsevier, Burlington, Massachusetts, USA.

US Fish and Wildlife Service. 2008. Strategic habitat conservation handbook: a guide to implementing the technical elements of strategic habitat conservation. US Department of the Interior, Washington, DC, USA.

US Fish and Wildlife Service. 2019*a*. Information for planning and consultation. https://ecos.fws.gov/ipac/. Accessed 12 Aug 2019.

US Fish and Wildlife Service. 2019*b*. ECOS species profile for whooping crane (*Grus americana*). https://ecos.fws.gov/ecp0/profile/speciesProfile?spcode=B003>. Accessed 15 Aug 2019.

van der Valk, A. G., and C. B. Davis. 1978. The role of seed banks in the vegetation dynamics of prairie glacial marshes. Ecology 59:322–335.

Van Horne, B. 2002. Approaches to habitat modeling: the tensions between patterns and process and between specificity and generality. Pages 63–72 *in* J. M. Scott, P. J. Heglund, M. L. Morrison, et al., editors. Predicting species occurrences: issues of accuracy and scale. Island Press, Washington, DC, USA.

Wassenaar, L. I., and K. A. Hobson. 1998. Natal origins of migratory monarch butterflies at wintering colonies in Mexico: new isotopic evidence. Proceedings of the National Academy of Sciences 95:15436–15439.

Western Association of Fish and Wildlife Agencies. 2019. West-wide Crucial Habitat Assessment Tool. http://www.wafwachat.org/pages/west-wide-chat. Accessed 12 Aug 2019.

Whited, D., S. Galatowitsch, J. R. Tester, K. Schik, R. Lehtinen, and J. Husveth. 2000. The importance of local and regional factors in predicting effective conservation planning strategies for wetland bird communities in agricultural and urban landscapes. Landscape and Urban Planning 49:49–65.

Wiens, J. A. 1989. Spatial scaling in ecology. Functional Ecology 3:385–397.

Wiens, J. A. 2002. Predicting species occurrences: progress, problems, and prospects. Pages 739–749 *in* J. M. Scott, P. J. Heglund, M. L. Morrison, et al., editors. Predicting species occurrences: issues of accuracy and scale. Island Press, Washington, DC, USA.

Wilson, K. A., M. I. Westphal, H. P. Possingham, and J. Elith. 2005. Sensitivity of conservation planning to different approaches to using predicted species distribution data. Biological Conservation 22:99–112.

Wood, E. M., A. M. Pidgeon, V. C. Radeloff, D. Helmers, P. D. Culbert, N. S. Keuler, and C. H. Flather. 2014. Housing development erodes avian community structure in U.S. protected areas. Ecological Applications 24:1445–1462.

Zuur, A. F., E. N. Ieono, N. J. Walker, A. A. Saveliev, and G. M. Smith. 2009. Mixed effects models and extensions in ecology with R. Springer, New York, New York, USA.

13 — Managing Landscapes and the Importance of Conservation Incentive Programs

Mark J. Witecha and
Todd R. Bogenschutz

Introduction

Agriculture is a dominant land use across many regions, encompassing ~370 million ha throughout the United States (US Department of Agriculture 2012). Agricultural landscapes are home to an array of plant and wildlife species, including many imperiled or declining species (Murphy 2003, Potts et al. 2010, Pleasants and Oberhauser 2012). The declining plant and wildlife communities in these highly disturbed, pervasive landscapes have made agricultural lands the focus of many conservation efforts.

Despite intensive efforts, landscape-level conservation (i.e., a broadscale, holistic approach to conservation that focuses on balancing social, economic, and environmental factors, oftentimes spanning ownership and political boundaries) is challenging in agriculture-dominated areas (Atkinson et al. 2011, Reimer and Prokopy 2014). The profitability and reduced risk associated with farming during the early part of the 21st century have contributed to extensive habitat loss. Agricultural intensification has perhaps been most detrimental to species that inhabit open natural communities within agricultural landscapes, such as wetlands and grasslands (Wright and Wimberly 2013). Wetlands and grasslands tend to contain productive soils and are easily converted to cropland, making them susceptible to loss. By the mid-1980s, more than half of the wetlands in the lower 48 states had been lost (Dahl and Johnson 1991). Today, prairie grasslands comprise <3% of their original area in the contiguous United States (Bachand 2001).

In agricultural landscapes where public landownership is lacking, engaging private landowners is essential to conserving grasslands, wetlands, and forests, and the species that inhabit those communities. Financial incentives are often used to generate interest in adopting conservation practices on private lands. To be successful at a landscape level, conservation incentive programs must be economically competitive with crop, livestock, or timber production, or the conservation practices being implemented have to help increase production, reduce expenses, or provide some other tangible or intangible benefit that the landowner is interested in.

Providing competitive incentives to many producers, especially when commodity prices are high, is costly. Private lands conservation incentive programs also require extensive input of time and resources to coordinate. Conservation initiatives on public lands may only require collaboration between a few government agencies that possess technical expertise in natural resource management, whereas initiatives on private lands involve participation from many private landowners that do not necessarily have a strong ecological background. Government agencies are given the significant task of providing technical assistance to private landowners, administering contracts, issuing payments, ensuring compliance with formal standards, and marketing for conservation incentive programs.

The objectives of this chapter are to provide readers with a basic understanding of the various types of conservation incentive programs available to private landowners, challenges to implementing these programs, and keys to future success. Emphasis is placed on US Farm Bill conservation incentive programs given their significant funding and comprehensive breadth. Understanding the role of conservation incentive programs allows landscape ecologists to take a more holistic and inclusive approach to managing landscapes by engaging both public land managers and private landowners through voluntary conservation measures.

Farm Bill Primer

Farm Bill is a term used to describe an omnibus piece of legislation that requires reauthorization by Congress approximately every 5 years. The primary purpose of the Farm Bill is to provide stable sources of food and fiber for the United States and economic stability for private agricultural producers. The Farm Bill is broken down into provisions (i.e., titles) that cover a variety of topics such as conservation, agricultural commodities, nutrition and food stamps, trade, crop insurance, farm loans, agricultural research, and bioenergy. Certain titles and programs contained within the Farm Bill stir much discussion and debate; however, the omnibus nature of the Farm Bill also supports collaboration and negotiation among congressional delegates and stakeholder groups, as all parties involved have a vested interest

in seeing a Farm Bill passed despite their sometimes-conflicting viewpoints on specific components of the bill.

Since the Great Depression, Farm Bills have traditionally focused on subsidizing a few of the main agricultural commodities in the United States, including corn, soybeans, and wheat. The first Farm Bill, the Agriculture Adjustment Act of 1933, was part of Franklin Delano Roosevelt's New Deal and was designed to provide a safety net for producers to protect against market volatility (Rasmussen et al. 1976). The 1933 Farm Bill also created a program that paid farmers to leave land idle to help control the supply of major US commodities, a precursor to some modern conservation incentive programs.

The Food Security Act of 1985 was the first Farm Bill to include a Conservation Title and to create significant and enduring conservation provisions such as the Conservation Reserve Program (CRP; Gray and Teels 2006). Also included in the 1985 Farm Bill was the conservation compliance provision, which requires producers to meet basic standards when farming environmentally sensitive ground, specifically wetlands and highly erodible lands, in order to receive federal subsidies. The 1990 Farm Bill further improved these conservation provisions by specifying wildlife as a purpose of the programs. Farm Bill conservation programs are considered by many wildlife professionals to be the only programs capable of creating wildlife habitat at large scales on private agricultural lands.

Since the 1985 Farm Bill, the Conservation Title has undergone many changes to become what it is today and continues to evolve with every reauthorization. Although the conservation programs included in the Farm Bill are regularly created, merged, eliminated, or altered, this suite of programs remains a continuing and essential tool for landscape-level conservation. In 2015, the budget for Farm Bill conservation programs was $6.2 billion (USA; US Department of Agriculture 2015), exceeding all other federal sources of conservation funding.

Farm Bill Conservation Incentive Programs

Conservation incentive programs require reauthorization in each Farm Bill. Program funding levels are determined through two different processes: authorization and appropriations. When the Farm Bill is passed by Congress and signed by the president, it authorizes certain programs and sets their initial funding levels. The authorization process occurs approximately every 5 years. In contrast, the appropriations process is conducted annually by the Senate and House Appropriations Committees. The appropriations process determines the actual amount that can be spent on each conservation program within the fiscal year. Oftentimes, the funds appropriated for Farm Bill conservation programs fall well below authorized spending levels.

Farm Bill conservation programs are administered by the US Department of Agriculture (USDA) through the Natural Resources Conservation Service (NRCS) or Farm Service Agency (FSA). Each agency administers their own programs; however, NRCS and FSA oftentimes share offices within a county or multicounty work unit and collaborate on conservation programs. For all Farm Bill conservation programs, NRCS provides technical assistance for conservation planning.

Participation by landowners and producers in USDA conservation incentive programs is voluntary. To be eligible for Farm Bill conservation programs, applicants must meet certain criteria. For example, an applicant must own or control the land they want to enroll. Producers must follow highly erodible land and wetland conservation compliance requirements. Also, a producer's annual income must be below a specified level to be eligible to receive financial incentives (Farm Service Agency 2019).

Farm Bill conservation programs fall within one of five different categories: reserves, easements, enhancements, partnerships, and grants. Programs are divided into these five categories based on the types of incentives provided and the conservation objectives of the program. This suite of voluntary programs provides private landowners with diverse options for addressing various resource concerns on their land (Natural Resources Conservation Service 2019a).

Reserve Programs

Reserve programs remove environmentally sensitive lands (e.g., highly erodible, wetlands, stream corridors) from agricultural production for a designated period of time. Fields or portions of fields are generally planted to perennial cover such as grass or trees. The perennial cover provides numerous ecological benefits by preventing erosion and nutrient runoff, providing cover and food for wildlife, adding organic matter back into the soil, and sequestering greenhouse gases (Farm Service Agency 2017).

The most successful and pervasive reserve program in the Farm Bill is the CRP, which pays producers an annual per acre rate to plant cropland to perennial cover for 10- to 15-year periods. Landowners can enroll entire or partial fields into the CRP and in return receive cost share on establishment and management practices and the annual rental payment. The payment rate is based on soil productivity and the average price charged for leasing cropland within the county.

The CRP was initially created to help control the supply of agricultural commodities but has since evolved into a vitally important conservation program by incorporating diverse, native seed mixes and requiring periodic management. At its peak in 2007, more than 14 million ha were enrolled in the CRP. There are numerous wildlife benefits provided by the CRP. From 1992 to 2004, CRP contributed to a 30% increase in waterfowl production in the Prairie Pothole Region, USA (Reynolds et al. 2007). Nielson et al. (2008) reported that ring-necked pheasant (pheasant; *Phasianus colchicus*) populations increased an average of 22% for each 4% increase in CRP acreage along breeding bird survey routes in the core US pheasant range. Without CRP grasslands in the Prairie Pothole Region, certain grassland bird species could decrease by 2% to 52%, depending on the species (Niemuth et al. 2007).

Easement Programs

Farm Bill easement programs provide producers with options for long-term or permanent protection for their land. Easement programs use Farm Bill funds to purchase development or other rights, ensuring the land stays in the desired cover type. Easements are generally permanent or are held for 30 years, with permanent easements receiving higher payment rates than 30-year easements.

Easements can be purchased for agricultural lands, grasslands, and wetlands through the Agricultural Conservation Easement Program. The Agricultural Conservation Easement Program has two components: agricultural land easements to protect agricultural croplands and grasslands, and wetland reserve easements to protect wetlands. Agricultural easements provide numerous benefits, such as preventing the loss of agricultural lands in rapidly developing urban landscapes. Agricultural easements also protect hay lands, pasturelands, and rangelands, ensuring grasslands will not be developed or converted to cropland. For wetlands easements, the federal government purchases cropping and development rights and also pays for all or most of the wetland restoration and management expenses depending on the term length of the easement (Natural Resources Conservation Service 2019b).

Enhancement Programs

Farm Bill enhancement programs provide producers with technical and financial assistance for adopting or installing conservation measures. Enhancement programs largely target lands in agricultural production, but they do allow for enhancements to nonworking lands, including forests, wetlands, and grasslands, if intended to provide wildlife habitat. Enhancement programs are flexible and diverse, being used broadly to improve general agricultural practices and for targeted initiatives that focus on specific conservation issues. Examples of such initiatives include creating or enhancing habitat for wildlife species of concern, protecting water quality in priority watersheds, converting to organic farming, and protecting priority landscapes such as the Prairie Pothole Region.

The Environmental Quality Incentives Program and Conservation Stewardship Program are the largest Farm Bill enhancement programs. The Environmental Quality Incentives Program provides cost share for establishing or adopting a conservation practice, with USDA typically paying 50% to 90% of the cost. The Conservation Stewardship Program provides producers with annual incentive payments for adopting or maintaining conservation practices and will also provide supplemental payments for adopting an improved crop rotation. Diverse conservation practices are covered under enhancement programs, including nutrient management, conservation grazing, filter strips to protect waterbodies, and creation of wildlife habitat (Natural Resources Conservation Service 2019a).

Partnership Programs

Partnership programs use Farm Bill funds to subsidize projects proposed by nonfederal entities, including local or state government, Native American tribes, agricultural groups, universities, and nonprofit organizations. These collaborative projects between USDA and partners generally take a targeted approach to address specific conservation issues at a local, state, national, or priority area level. Partners are typically required to provide some level of match with partner programs. Partnership programs provide flexible guidelines in terms of scale of projects, conservation issues being addressed, and partners involved, allowing for innovation on how conservation practices are delivered on private lands. The largest partner programs that USDA offers are the Regional Conservation Partnership Program and the Conservation Reserve Enhancement Program. Both programs target soil, water, or wildlife priorities and require at least 25% matching funds by partners.

Grant Programs

The Farm Bill also gives USDA the authority to provide competitive grants at the national and state level. The USDA's largest conservation grant programs are the Conservation Innovation Grants Program and the Voluntary Public Access and Habitat Improvement Program. The Conservation Innovation Grants Program emphasizes innovation in agricultural conservation practices and technologies and provides grants for a diverse collection of conservation and outreach efforts. The Voluntary Public Access and Habitat Improvement Program provides grants to states and other entities to encourage public hunting or recreational access and wildlife habitat improvement on private lands. This program is often paired with other state habitat or access programs.

Other Conservation Incentive Programs

Individual states or private organizations also offer conservation incentive programs geared toward private landowners. Voluntary state incentive programs often function similarly to Farm Bill conservation programs, with state agencies using program funding to purchase easements or provide incentive payments and cost share to private landowners. State incentive programs, much like Farm Bill programs, can also be used as part of a focus area approach to address state conservation issues such as water quality or declining wildlife populations on private lands.

One way state incentive programs can differ from Farm Bill conservation programs is to focus on providing public access for hunting and other recreation to private lands. Farm Bill programs, with the exception of the Voluntary Public Access and Habitat Improvement Program, do not require landowners to allow public access (Natural Resources Conserva-

tion Service 2019c). States may offer cost share on habitat management practices or provide an incentive payment to private landowners who are willing to allow public access on their land. A stronger emphasis may be placed on public access programs in states with limited public land. An example would be Kansas, USA, which contains little public land; however, Kansas is able to provide ample opportunities for hunters and outdoor recreationists through its Walk-In Hunting Access Program. In 2004, more than 400,000 ha were enrolled in the Walk-In Hunting Access Program, helping to compensate for the lack of public lands.

While conservation grants are available through Farm Bill conservation programs, state and nongovernmental organization (NGO) grant programs can take different approaches to conservation on private lands. These nonfederal conservation grant programs and their objectives are as diverse as the hosting organizations, but they oftentimes focus on improving wildlife habitat for game, nongame (species of greatest conservation need), or threatened and endangered species. Grants can also focus on research issues, acquisition of priority habitats, or expanded access for hunting, fishing, trapping, and other wildlife-associated recreation on private land. The grant organization typically defines the eligible entities, eligible landscapes, match requirements, and reporting requirements for grant funds. Many grant programs have no match requirement if located in a priority landscape. An example of a state grant program would be the Iowa Department of Natural Resources' Habitat Management Grants Program, which awards small grants to organizations focused on restoring critical habitat or species identified in the Iowa Wildlife Action Plan (Iowa Department of Natural Resources 2016a). Nationally, the best-known wildlife and fisheries habitat grant program is sponsored by the National Fish and Wildlife Foundation (2016). The foundation typically has more than 50 different grant programs in any given year focused primarily on critical habitats identified in federal, state, or regional wildlife plans. Related to the Farm Bill, the foundation offers the Conservation Partners Program grant, which provides awardees with funding to provide staff and technical assistance to private landowners in regions where some of the nation's most crucial conservation issues can be addressed through Farm Bill programs (National Fish and Wildlife Foundation 2019).

Tax incentives are another way that state agencies promote habitat conservation on private lands. Tax incentives usually come in one of two forms: a property tax credit or a state income tax credit. A property tax credit provides the landowner with a complete or partial waiver of their annual property tax on enrolled lands. A state income tax credit provides landowners with a credit waiver that can be applied to their state income tax on enrolled acres. State resource agencies along with state assessors set the eligibility rules for these programs. Wisconsin, USA, offers a property tax credit to private landowners enrolled in the Managed Forest Law Program, which promotes sustainable forest management on private forestlands. Landowners can enroll their forestland into the program for 25 or 50 years, and they are required to follow a management plan that incorporates landowner objectives, sustainable forestry practices, wildlife management, water quality, and public recreation opportunities. In return, enrolled landowners pay an area share tax instead of regular property tax, greatly reducing their annual tax bill. Landowners willing to open their enrolled lands to public access receive additional savings (Wisconsin Department of Natural Resources 2019).

Some states also have statues or regulations on land use than can provide wildlife habitat on private lands. One example is the buffer law passed by the Minnesota legislature in 2015 (Minnesota Board of Soil and Water Resources 2016), which mandates that all public waterways in Minnesota, USA, have a perennial buffer to filter nutrients and reduce soil erosion. Depending upon the waterbody, buffers can vary from 5 to 15 m. Native vegetation is not required but is encouraged. Some limited use of the buffers (e.g., haying, grazing) is allowed. While such regulations are directed at water quality and not wildlife specifically, they can have a significant effect on the amount of perennial habitat available to wildlife in a larger landscape.

Current Challenges to Implementing Conservation Incentive Programs

Some of the greatest challenges associated with landscape-level management for wildlife relate to land ownership. The vast majority of lands within the United States are privately owned, with the exception being western states with large federal land holdings. Land owned by state agencies averages <5% of land area in the lower 48 states (Natural Resources Council of Maine 1995). Of these public lands, both state and federal, only a portion is managed primarily for wildlife. Thus one has to consider private landowners when attempting to manage wildlife across a landscape, as private landowners control the majority of land and can provide connecting corridors between tracts of public land. In thinking about landscape-level management, conservation professionals need to consider how land ownership, current habitat conditions, and available conservation incentive programs affect their strategy for targeted conservation on private lands. In states like Texas, USA, it may be possible to work with a single owner who controls access to tens to hundreds of thousands of hectares; however, in the more suburban East, a landscape management plan may involve hundreds of landowners with small properties of 10 to several hundred hectares. From the perspective of a wildlife manager, how many small landowners need to participate in a landscape management plan to be effective in conserving focal species or guilds depends on the species and the amount of habitat needed to reach the desired end result.

In many cases, private landowners derive income from timber production, livestock grazing, agricultural commodities, or other uses of their land. These land uses may be the primary source of income for the landowner. On working lands, landowners are sometimes hesitant to take land out of production and enter into conservation incentive programs, as doing so could affect their income. Landowners also make investments in their lands, such as roads, fencing, watering facilities, terraces, and drainage systems. Landowners looking at their return on investment know that idling land or altering farm management practices can significantly affect their income. These substantial economic factors must be considered when implementing landscape-level conservation incentive programs.

Commodity prices can also be a deterrent to enrolling in conservation incentive programs. Commodity prices have long- and short-term effects on wildlife habitat. Over the past several decades, certain row crops (e.g., corn, soybeans, cotton) have had higher returns per hectare than other grassland crops (e.g., small grains, hay, pasture; Farris et al. 1977). As a result, there has been a long-term decline in grassland crops, with significant effects on grassland wildlife (Farris et al. 1977).

In the short term, timber, livestock, and agricultural commodity prices can be highly volatile (Peters et al. 2009). Landowners generally look to maximize production when prices are high and use every available hectare for profit, as they know a down cycle will occur and profits may be low or even negative for several years. Clean farming, removal of fence lines and hedgerows to improve efficiency, and herbicide spraying of crop and fallow fields to reduce weeds are common practices to maximize income when commodity prices are high but are usually detrimental to wildlife (Hill 1985, Brennan and Kuvlesky 2005, Coates et al. 2017). High agricultural commodity prices can also lead to the rapid conversion of grassland cover (hay, pasture, range, CRP) to agricultural production (Wright and Wimberly 2013, Lark et al. 2015). Similarly, high timber prices can lead to significant loss of mature timber in a region. Rapid conversion of certain habitats can be problematic for landscape planning depending upon priority wildlife species and habitat needs. Implementing

conservation programs under such markets requires equally high incentive payments to encourage participation.

Conversely, with down markets, conservation incentive programs are sometimes easier to deliver, as landowners are looking for ways to maintain a positive cash flow. Historically, downturns in timber, livestock, and agricultural commodity markets led to increased interest in conservation incentive programs to minimize losses and maximize return on investment; however, federally subsidized crop insurance and revenue protection programs diminish the risk in markets, and landowners may be less inclined to consider conservation incentive programs. Highly erodible lands, wet areas, and other marginally profitable cropped lands are usually the first areas landowners consider enrolling in conservation programs when crop prices are low. These insurance and revenue products may be changing this dynamic. Claassen et al. (2011) suggest that federal subsidy programs may be encouraging the conversion of grassland cover (hay, pasture, rangelands, and small grains) to row crop production because of reduced risk and government payments.

Lastly, private landowners are busy people with limited time and budgets and are increasingly absentee owners who do not live on their property (Arbuckle 2010). Increasing absenteeism on private land complicates efforts to target incentive programs, as land stewardship responsibilities can remain with the owner or be assigned to the tenant or lessee (Zhang et al. 2018, Arbuckle 2019). Conservation planners need to address all of these issues when designing conservation initiatives focused on private land. Most landowners or tenants prefer simple conservation incentive programs, whereas complicated forms and rules discourage participation (Lute et al. 2018). When developing incentive programs for private land, conservation planners should strive to keep programs simple for landowners or tenants to understand. Programs should also offer competitive economic incentives and provide a service to the landowner, such as planning and locating contractors. Lastly, incentive programs should provide landowners the flexibility to achieve their own objectives.

Keys to Success and Future Outlook

The future of implementing conservation incentive programs across landscapes is a daunting one. Human populations continue to expand in the United States and worldwide, creating a demand for more food and fiber from the land. Historically, the land provided areas for livestock grazing, production of agricultural commodities, and timber for building, while marginal or less productive lands generally benefitted wildlife. With increasing human populations and decline in traditional energy resources such as coal and crude oil, more and more countries and companies are looking toward the future and the use of bioenergy to augment traditional sources of energy. Bioenergy products are deemed sustainable, renewable, and environmentally greener (e.g., reduce greenhouse gases) than traditional energy sources (US Department of Energy 2016). Some of the more promising technologies include biomass and wind energy. Biomass (e.g., wood, grass, algal) can be harvested and burned directly or converted to liquid fuels. Wind technologies use wind-powered turbines to convert wind into electricity. Both technologies are promoted as being good uses of marginal or less productive lands and thus will not compete for land already in agricultural or timber production. Marginal lands, however, are usually the best remaining candidates for enrolling in conservation incentive programs and may contain much of the remaining wildlife habitat in an area (Rupp et al. 2012, Winder et al. 2015). Converting them to bioenergy uses may reduce or eliminate their ability to support wildlife populations.

Greater demands on land use will require more targeted and focused delivery of conservation programs than in the past. Many conservation programs (e.g., federal, state, or NGO funded) have historically taken a shotgun approach. With federal conservation incentive programs, including Farm Bill programs, the shotgun approach is mostly an administrative artifact, as these programs are funded with federal tax dollars and therefore all taxpayers should have the opportunity to participate. For other state and NGO programs the shotgun approach is also easy to administer and market, as there are minimal eligibility requirements and fewer rules to explain. With limited budgets and more demand for land resources, however, a more holistic and targeted approach is needed to achieve measurable wildlife results. The keys to successful delivery of landscape-level conservation incentive programs to private landowners will be organizational partnerships, economic incentives, education, outreach, and keeping programs simple for the landowner.

For many landowners with agricultural or forested lands, their local USDA service center, soil and water conservation district, state forester, state wildlife biologist, or university extension specialist is their first contact for information on conservation incentive programs. Within USDA, both NRCS and FSA provide conservation incentive programs to promote forest, wetland, and grassland conservation (Stubbs 2018). Most federal and state wildlife and forestry agencies and state soil and water conservation districts also have funding for private landowner conservation efforts, as do NGOs (e.g., Ducks Unlimited, Pheasants Forever, Nature Conservancy). Unfortunately, in many states, these agencies and organizations are not jointly located; thus landowners must first find out who to contact to get information about what incentive programs they are eligible for. The USDA has implemented policy to co-locate its agencies within a county, and many state soil and water conservation districts are integrated within these local county USDA offices (Robertson 2000). Many state wildlife and forestry agencies have developed shared positions with USDA and co-locate these staff in USDA offices, as have NGOs like Ducks

Unlimited and Pheasants Forever (Iowa Department of Natural Resources 2016b). Successful landscape planning for wildlife should incorporate the suite of conservation programs currently available to private landowners whether they are federal, state, or NGO administered or funded. Co-location of these different organizations could improve overall coordination and communication of conservation incentive programs and create stronger partnerships between the various organizations. This would lead to a more holistic approach in delivering conservation incentive programs to private landowners.

Conservation incentive programs generally have identified priorities to reduce soil erosion, improve water quality, or improve wildlife habitat. Coordinated partnerships with a holistic approach can consider all incentives in landscape planning. Planting diverse native grassland species to reduce soil erosion or improve water quality will have wildlife benefits even if the priority focus for the practice is not wildlife. A holistic approach to using the various conservation incentive programs increases the suite of conservation practices and incentives available to the landowner, which in turn can lead to a great adoption of practices within priority landscapes.

Minimizing losses and inputs, and maximizing yields and profits, is a common goal with private landowners, whether on agricultural or forested lands, and is the driving force behind a new precision farming management effort (Schimmelpfennig 2016). Precision management looks at each hectare of land and determines whether the land is returning a profit or losing money. If the return on investment is negative, other options can be explored to turn a profit or reduce losses. For example, in the agricultural Midwest, a field may have a tile-drained wetland, but the area may still flood periodically. The average row crop return on these flood-prone areas is lower than the rest of the field. Adding additional tile to such areas is cost-prohibitive, so what other options exist? A farm consultant may suggest converting the area to hay or another crop more tolerant to flooding or suggest a conservation incentive pro-

gram to restore the wetland. Some conservation and wildlife NGOs are beginning to work with agricultural consultants and groups to promote precision agriculture as a means to more wildlife habitat. Increasingly, within agricultural landscapes, precision farming efforts may encourage additional conservation actions by landowners influenced primarily by profit and return on investment.

Close collaboration with partner agencies and organizations also fosters broader outreach to landowners and stimulates educational opportunities for partners and landowners. Often, state and federal agencies are not aware of what programs other divisions within the government offer. Good partnerships lead to better communication among organizations and thus better outreach to private landowners. Landowners typically get information from many different sources, including agricultural publications, agricultural and forestry cooperatives, governmental agencies, local media, and in many cases neighbors. A good partnership among organizations and programs increases outreach to landowners, as organizations typically market products through different media outlets, providing multiple avenues for a conservation message to reach landowners or tenants. This outreach fosters a stronger land ethic with landowners because the conservation message will be broader than just wildlife and include other quality-of-life issues, such as cleaner drinking water and reduced pollution. Public opinion polls have demonstrated that quality-of-life issues (e.g., water quality) garner more interest from a larger segment of the population than more general issues like wildlife habitat (Responsive Management 2016). Marketing to the suite of issues important to citizens and private landowners will improve participation in a conservation incentive programs over marketing efforts that cater just to wildlife benefits.

Successful landscape plans for wildlife require the delivery of conservation incentives to a threshold value of private landowners in a focal landscape. The landscape plan should identify the priority wildlife species and focal landscape (Hudson et al. 2017), the necessary priority habitats to restore or preserve, and what landscape metrics, juxtaposition, and configuration of habitat elements are needed for priority species. A holistic management approach with knowledgeable staff, competitive economic incentives, and marketing outreach and education will provide a one-stop shop for landowners within focal landscapes and keep the process simple for landowners. Landscape planning efforts that consider all these elements have a strong foundation to achieve their stated objectives.

How Do Landscape Ecologists Fit into the Future?

Perhaps most importantly, landscape ecologists play a pivotal role in identifying the habitats and landscape metrics necessary to restore and maintain priority wildlife species (Reiley and Benson 2019). Without these basics it is difficult to build effective conservation incentive programs. The understanding of these relationships also allows the development of focal landscapes where habitat improvements have the greatest likelihood of improving or maintaining populations, thus targeting conservation programs to where they can be most effective. Without this basic ecological information, wildlife managers are simply taking a shotgun approach to wildlife management with the hope that if enough practices are done, it will benefit the species. Nusser et al. (2004) demonstrated that pheasant response to shotgun approach of CRP enrollment in Iowa varied with geographic location such that the landscape context affected pheasant population response. Within a typical 1-km home range, Schmitz and Clark (1999) and Clark et al. (1999) reported that landscape metrics—such as mean grassland patch size, landscape core area, landscape shape index, and edge metrics—were predictive of female pheasant survival and nest success, whereas local cover type was not. Landscape models such as these could be used to predict the effect of agricultural conservation incentive programs like CRP for pheasants or nongame grassland birds

and demonstrate how much more effective a new incentive program could be if targeted spatially with other habitats in focal landscapes.

Nebraska Game and Parks Commission conducted a multiscale landscape assessment of wild pheasant populations (Jorgensen et al. 2014). Their results showed CRP and other grasslands best explained pheasant abundance at the local management scale of 1 km. At the landscape scale (5-km radius) the response of pheasants to local habitat was positively influenced, to a threshold, by the proportion of row crops and small grains and negatively influenced by the presence of trees. Pheasants responded positively to local management like CRP, but the landscape context surrounding these areas can have drastic ramifications on the outcome of local management. The authors note that even 15% tree cover in the larger landscape can significantly reduce the benefits of local management actions, while landscapes containing 15% small grains significantly enhanced local management (Jorgensen et al. 2014). This information was used to develop a predictive landscape model to better target limited pheasant habitat management dollars to regions of the state where they would most benefit pheasant populations. The effect of analyses like this, by landscape ecologists, is threefold. It improves knowledge and understanding regarding habitat management prescriptions for wildlife managers, it provides information necessary to effectively target programs to focal landscapes, and it provides information to increase landowner understanding of complex wildlife–habitat relationships.

Using pheasant as an example, wildlife managers know the amount of habitat locally is an important consideration to improve populations, but research also suggests that so are patch size, edge metrics, and patch configuration at the local scale. Not all local habitats are created equal from a landscape perspective, however, and landscape location does matter (e.g., proportion of trees, small grains in the surrounding landscape). The "if you build it, they will come" approach to habitat management has significant caveats. Wildlife managers cannot improve habitat prescriptions for wildlife without landscape ecologists sharing this kind of information with wildlife managers. Knowledge of the importance of the juxtaposition of different habitats, edge effects, and other landscape metrics is necessary for wildlife managers and administrators to develop more effective habitat conservation incentive programs.

Conservation incentive programs are most effective when linked with locally important habitats and targeted to the proper location in the landscape. Landscape ecologists can play a fundamental role in development focal landscapes. How large does a focal landscape need to be? What connectivity and configuration are needed between different habitats? What is the critical mass of habitat necessary to see a response from the species of interest? What proportion of landowners within a focal landscape must participate in an incentive program to provide meaningful benefits to the priority species? Landscape ecologists have the training and skills necessary to answer these types of complicated questions. Answers to these questions will also aid in developing partnerships and outreach materials. Having a better understanding of how a particular wildlife species interacts with its habitat and the importance of surrounding landscape features is powerful information for partners and landowners. Partners can get a better sense of how they fit into the larger picture. Their particular incentive program may not necessarily have wildlife as an objective, but in the larger landscape context the connectivity provided by, say, a water-quality grass buffer to nearby wildlife habitat may be significant. Information can also be used to educate landowners (e.g., why certain conservation incentives are available or not, why populations are responding in some areas and not others, why the special focus on their land).

Opportunities also exist for landscape ecologists to expand outside traditional wildlife boundaries. Is there an opportunity for analysis of some conservation incentive programs? Perhaps analysis to determine whether a conservation incentive program is providing critical connecting habitat across a focal landscape compared to other programs like land acquisition. Can other priorities like water quality and soil erosion be included in landscape analyses to form a more holistic approach to the delivery of conservation incentive programs (Rowe 2014)? Answering such questions helps to determine what the greatest needs are, not just for conservation incentive programs but in the field of natural resource management as a whole, remaining curious, innovative, and adaptive when managing landscapes.

Summary

Engaging private landowners is a complex but necessary component of landscape-scale conservation in many regions of the United States. Diverse economic, social, and environmental factors must be considered at all levels (i.e., policy, program funding, planning, implementation) to garner and maintain interest in conservation incentive programs at a meaningful scale, which often entails working with large numbers of landowners. Currently, the US Farm Bill provides the most comprehensive suite of programs available to private landowners and is an essential tool in many current landscape-scale conservation initiatives. Recent research suggests that coordinated and targeted implementation of Farm Bill conservation programs on private lands can improve wildlife habitat and wildlife populations (Reiley et al. 2019). Moving forward, emphasis should be placed on policy issues that greatly influence the implementation and effectiveness of conservation incentive programs (e.g., finding creative solutions to minimize risk for agricultural producers while not encouraging conversion of natural cover types). At a finer scale the entities administering conservation incentive programs can maximize effectiveness by employing holistic and targeted approaches and better sharing information and resources through partnerships.

LITERATURE CITED

Arbuckle, J. G. 2010. Rented land in Iowa: social and environmental dimensions. Iowa State University Extension PMR 1006, Ames, Iowa, USA. http://store.extension.iastate.edu/Product/Rented-Land-in-Iowa-Social-and-Environmental-Dimensions.

Arbuckle, J. G. 2019. Iowa farm and rural life poll: 2018 summary report. Iowa State University Extension SOC3090, Ames, Iowa, USA. https://store.extension.iastate.edu/product/15687.

Atkinson, L. M., R. J. Romsdahl, and M. J. Hill. 2011. Future participation in the Conservation Reserve Program in North Dakota. Great Plains Research 21:203–214.

Bachand, R. R. 2001. The American prairie: going, going, gone? National Wildlife Federation, Rocky Mountain Natural Resource Center, Boulder, Colorado, USA.

Brennan, L. A., and W. P. Kuvlesky Jr. 2005. North American grassland birds: an unfolding conservation crisis? Journal of Wildlife Management 69:1–13.

Claassen, R., F. Carriazo, J. C. Cooper, D. Hellerstein, and K. Ueda. 2011. Grassland to cropland conversion in the northern plains: the role of crop insurance, commodity, and disaster programs. US Department of Agriculture Economic Research Report No. ERR-120, Washington, DC, USA. https://www.ers.usda.gov/webdocs/publications/44876/7477_err120.pdf?v=0.

Clark, W. R., R. A. Schmitz, and T. R. Bogenschutz. 1999. Site selection and nest success of ring-necked pheasants as a function of location in Iowa landscapes. Journal of Wildlife Management 63:976–989.

Coates, P. S., B. E. Brussee, K. B. Howe, J. P. Fleskes, I. A. Dwight, D. P. Connelly, M. G. Meshriy, and S. C. Gardner. 2017. Long-term and widespread changes in agricultural practices influence ring-necked pheasant abundance in California. Ecology and Evolution 7:2546–2559.

Dahl, T. E., and C. E. Johnson. 1991. Wetlands: status and trends in the conterminous United States, mid-1970's to mid-1980's. US Fish and Wildlife Service, Washington, DC, USA.

Farm Service Agency. 2017. Environmental benefits of the Conservation Reserve Program, Prairie Pothole Region. https://www.fsa.usda.gov/Assets/USDA-FSA-Public/usdafiles/EPAS/natural-resouces-analysis/nra-landing-index/2017-files/Environmental_Benefits_of_the_Prairie_Pothole_CRP_2017_draft.pdf. Accessed 3 July 2019.

Farm Service Agency. 2019. Adjusted gross income. https://www.fsa.usda.gov/programs-and-services/payment-eligibility/adjusted-gross-income/index. Accessed 9 May 2019.

Farris, A. L., E. D. Klonglan, and R. C. Nomsen. 1977. The ring-necked pheasant in Iowa. Iowa Conservation Committee, Des Moines, Iowa, USA.

Gray, R. L., and B. M. Teels. 2006. Wildlife and fish conservation through the Farm Bill. Wildlife Society Bulletin 34(4):906–913.

Hill, D. A. 1985. The feeding ecology and survival of pheasant chicks on arable farmland. Journal of Applied Ecology 22:645–654.

Hudson, M. R., C. M. Francis, K. J. Campbell, C. M. Downes, A. C. Smith, and K. L. Pardieck. 2017. The role of the North American Breeding Bird Survey in conservation. Condor 119:526–545.

Iowa Department of Natural Resources. 2016a. Wildlife stewardship, grants and funding availability. https://www.iowadnr.gov/Conservation/Iowas-Wildlife/Wildlife-Diversity-Program/Diversity-Projects/State-Wildlife-Grants. Accessed 12 Aug 2019.

Iowa Department of Natural Resources. 2016b. Wildlife landowner assistance. http://www.iowadnr.gov/Conservation/Wildlife-Landowner-Assistance. Accessed 31 May 2016.

Jorgensen, C. F., L. A. Powell, J. J. Lusk, A. A. Bishop, and J. J. Fontaine. 2014. Assessing landscape constraints on species abundance: does the neighborhood limit species response to local habitat conservation programs? PLOS One 9(6):e99339.

Lark, T. J., J. M. Salmon, and H. K. Gibbs. 2015. Cropland expansion outpaces agricultural and biofuel policies in the United States. Environmental Research Letters 10:044003.

Lute, M. L., C. R. Gillespie, D. R. Martin, and J. J. Fontaine. 2018. Landowner and practitioner perspectives on private land conservation programs. Society and Natural Resources 31:218–231.

Minnesota Board of Soil and Water Resources. 2016. Minnesota Buffer Law. https://bwsr.state.mn.us/minnesota-buffer-law. Accessed 31 May 2016.

Murphy, M. T. 2003. Avian population trends within the evolving agricultural landscape of eastern and central United States. The Auk 120(1):20–34.

National Fish and Wildlife Foundation. 2016. National Fish and Wildlife Foundation grants. http://www.nfwf.org/whatwedo/grants/Pages/home.aspx. Accessed 31 May 2016.

National Fish and Wildlife Foundation. 2019. Conservation partners program. https://www.nfwf.org/conservation-partners/Pages/home.aspx. Accessed 12 Aug 2019.

Natural Resources Conservation Service. 2019a. Conservation practices. https://www.nrcs.usda.gov/wps/portal /nrcs/detailfull/national/technical/cp/ncps/?cid=nrcs 143_026849. Accessed 9 Aug 2019.

Natural Resources Conservation Service. 2019b. Wetland reserve easements. https://www.nrcs.usda.gov/wps /portal/nrcs/detailfull/nj/programs/easements/acep /?cid=stelprdb1248941. Accessed 12 Aug 2019.

Natural Resources Conservation Service. 2019c. Voluntary public access and habitat incentive program. https:// www.nrcs.usda.gov/wps/portal/nrcs/detail/national /programs/farmbill/?cid=stelprdb1242739. Accessed 12 Aug 2019.

Natural Resources Council of Maine. 1995. Public land ownership by state. http://www.nrcm.org/documents /publiclandownership.pdf. Accessed 31 May 2016.

Nielson, R. M., L. L. McDonald, J. P. Sullivan, C. Burgess, D. S. Johnson, D. H. Johnson, S. Bucholtz, S. Hyberg, and S. Howlin. 2008. Estimating the response of ring-necked pheasants (Phasianus colchicus) to the Conserva-tion Reserve Program. The Auk 125(2):434–444.

Niemuth, N. D., F. R. Quamen, D. E. Naugle, R. E. Reyn-olds, M. E. Estey, and T. L. Shaffer. 2007. Benefits of the Conservation Reserve Program to grassland bird popula-tion in the Prairie Pothole Region of North Dakota and South Dakota. US Fish and Wildlife Service, Bismarck, North Dakota, USA.

Nusser, S. M., W. R. Clark, J. Wang, and T. R. Bogenschutz. 2004. Combining data from state and national monitor-ing surveys to assess large-scale impacts of agricultural policy. Journal of Agricultural, Biological, and Environ-mental Statistics 9:381–397.

Peters, M., S. Langley, and P. Westcott. 2009. Agricultural commodity price spikes in the 1970s and 1990s: valu-able lessons for today. Amber Waves, Washington, DC, USA. http://www.ers.usda.gov/amber-waves/2009 /march/agricultural-commodity-price-spikes-in-the -1970s-and-1990s-valuable-lessons-for-today. Accessed 31 May 2016.

Pleasants, J. M., and K. S. Oberhauser. 2012. Milkweed loss in agricultural fields because of herbicide use: effect on the monarch butterfly population. Insect Conservation and Diversity 6(2):135–144.

Potts, S. G., J. C. Biesmeijer, C. Kremen, P. Neumann, O. Schweiger, and W. E. Kunin. 2010. Global pollinator declines: trends, impacts, and drivers. Trends in Ecology and Evolution 25(6):345–353.

Rasmussen, W. D., G. L. Baker, and J. S. Ward. 1976. A short history of agricultural adjustment, 1933–75. Economic Research Service Bulletin 391, Washington, DC, USA.

Reiley, B. M., and T. J. Benson. 2019. Differential effects of landscape composition and patch size on avian habitat use of restored fields in agriculturally fragmented landscapes. Agriculture Ecosystems and Environment 274:41–51.

Reiley, B. M., K. W. Stodola, and T. J. Benson. 2019. Are avian population targets achievable through programs that restore habitat on private-lands? Ecosphere 10(1):e02574.

Reimer, A. P., and L. S. Prokopy. 2014. Farm participation in U.S. Farm Bill conservation programs. Environmental Management 53:318–332.

Responsive Management. 2016. Water resources. https:// responsivemanagement.com/research-topics/water -resources/. Accessed 31 May 2016.

Reynolds, R. E., C. R. Loesch, B. Wangler, and T. L. Shaffer. 2007. Waterfowl response to the Conservation Reserve Program and swampbuster provisions in the Prairie Pot-hole Region, 1994–2004. US Fish and Wildlife Service, Bismarck, North Dakota, USA.

Robertson, R. E. 2000. U.S. Department of Agriculture: state office collocation. General Accounting Office Report GAO/RCED-00-208R, Washington, DC, USA. https://www.gao.gov/assets/90/89954.pdf.

Rowe, M. A. 2014. Collaborative landscape conservation in the Southwest Wisconsin Grassland and Stream Conservation Area. Page 9 in A. Glaser, editor. Ameri-ca's grasslands conference: the future of grasslands in a changing landscape. Proceedings of the 2nd biennial conference on the conservation of America's grasslands. National Wildlife Federation and Kansas State Univer-sity, Manhattan, Kansas, USA.

Rupp, S. P., L. Bies, A. Glazer, C. Kowaleski, T. McCoy, T. Rentz, S. Riffell, J. Sibbing, J. Verschuyl, and T. Wigley. 2012. Effects of bioenergy production on wildlife and wildlife habitat. Technical Review 12–03, The Wildlife Society, Bethesda, Maryland, USA.

Schimmelpfennig, D. 2016. Farm profits and adoption of precision agriculture. Economic Research Service Report ERR-217, Washington, DC, USA. https://www.ers .usda.gov/webdocs/publications/80326/err-217.pdf?v=0.

Schmitz, R. A., and W. R. Clark. 1999. Survival of ring-necked pheasant hens during spring in relation to landscape features. Journal of Wildlife Management 63:147–154.

Stubbs, M. 2018. Agricultural conservation: a guide to pro-grams. Congressional Research Service Report R40763, Washington, DC, USA. https://fas.org/sgp/crs/misc /R40763.pdf.

US Department of Agriculture. 2012. 2012 census of agri-culture: United States summary and state data. Volume 1. Geographic Area Series Part 51. National Agriculture

Statistics Service, Washington, DC, USA. http://www
.nass.usda.gov/AgCensus/.

US Department of Agriculture. 2015. Fiscal Year 2015
budget summary and annual performance plan. Office
of Budget and Program Analysis, Washington, DC, USA.
http://www.obpa.usda.gov/budsum/FY15budsum.pdf.

US Department of Energy. 2016. 2016 billion-ton report:
advancing domestic resources for a thriving bioeco-
nomy. Volume 1. Economic availability of feedstocks.
M. H. Langholtz, B. J. Stokes, and L. M. Eaton (Leads).
ORNL/TM-2016/160, Oak Ridge National Laboratory,
Oak Ridge, Tennessee, USA. https://info.ornl.gov/sites
/publications/Files/Pub62368.pdf.

Winder, V. L., A. J. Gregory, L. B. McNew, and B. K. Sander-
cock. 2015. Responses of male greater prairie-chicken to
wind energy development. The Condor 117(2):284–296.

Wisconsin Department of Natural Resources. 2019. Managed forest law. https://dnr.wi.gov/topic/forestland
owners/mfl/. Accessed 12 Aug 2019.

Wright, C. K., and M. C. Wimberly. 2013. Recent land use
change in the Western Corn Belt threatens grasslands
and wetlands. Proceedings National Academy of Sciences 110:4134–4139.

Zhang, W., A. Plastina, and W. Sawadgo. 2018. Iowa
farmland ownership and tenure survey 1982–2017: a
thirty-five year perspective. FM 1893, Iowa State University Extension and Outreach, Ames, Iowa, USA. https://
store.extension.iastate.edu/product/6492.

14

Davᴵᴅ M. Wᴵʟʟᴵᴀᴍs

Part III Synthesis

Establishing a Wildlife Management Foundation for Landscape Ecologists

It has been stated many times by managers and academics that wildlife management is not rocket science; it is much harder. I first heard those words at the beginning of a course on wildlife habitat management and have since used them in many courses and conversations. They have been used to communicate many complexities of wildlife management, but most commonly they express the challenges of implementing management actions based on science across a diverse landscape of people, landowners, values, and agencies and their directives, all within a political context. The nature of the foundations of wildlife management necessary for landscape ecologists to work effectively with managers differs substantively from that discussed in Part II of this volume. Although wildlife managers benefit most from understanding how landscape ecologists approach and quantify the relationships between landscape patterns and the ecological processes of wildlife (Part II), landscape ecologists benefit from a foundational understanding of the practical complexities to implementing management across large landscapes (Part III). Landscape ecologists also benefit from understanding existing policy and management systems that may facilitate accomplishing wildlife management across those landscapes. Chapters 10–13 address the challenges to managing wildlife across broad landscapes, what constitutes useful spatial models or decision tools for wildlife management, and how improving communication and leveraging conservation incentive programs might improve the potential for landscape ecologists and wildlife managers to work effectively together.

In Chapter 10, Connelly and Conway provide an overview of the practical challenges and constraints to managing wildlife at landscape scales. Challenges are described as difficult but able to be overcome, whereas constraints are barriers that cannot be overcome without changes to policy, legislation, or the administrative structure of agencies. For the landscape ecologist it is helpful to recognize and accept limitations to the application of landscape ecology to real landscapes, to be able to distinguish between challenges and constraints, and to be equipped to engage the wildlife management process most effectively. They conclude the chapter with a case study of sage-grouse (*Centrocercus urophasianus*) management, highlighting how those challenges, constraints, and opportunities are demonstrated in a real wildlife management system.

One of the primary challenges to applying landscape ecology to wildlife populations within real landscapes is a spatial mismatch between the landscape scales of population processes and the spatial extents of management. Wildlife management generally occurs at politically determined extents (e.g., wildlife management unit, county, state) despite knowledge that population processes are often indifferent to those arbitrary boundaries. Furthermore, within those boundaries there may be multiple agencies, organizations, and private landowners managing wildlife habitat at different spatial scales. Managers in one agency may have little or no influence on the actions of managers in another organization with different mandates, goals, and objectives. Connelly and Conway highlight the potential for such spatial and organizational mismatches to adversely affect wildlife populations even when individual agencies or management units are working for population improvements. They highlight how complicated these challenges can become when a single wildlife population exists across multiple states (e.g., sage-grouse). Although these challenges are complex and daunting, they can be overcome through a landscape approach to research and management that includes proper planning, better coordination and collaboration between management entities, and flexibility and compromise from agencies and landowners.

While those challenges to implementing landscape ecology to wildlife management can be addressed, there are constraints that cannot be overcome so easily. These constraints can occur because of legislation, policy, and the administrative structures of agencies. Connelly and Conway focus their discussion of these barriers to landscape-scale management of wildlife on areas where migration and dispersal are essential components of population dynamics. They emphasize that good management of ranges can be wasted if connecting habitat is compromised. Thus identifying connecting corridors and protecting critical land areas are essential to effective landscape-scale management. The authors suggest that landscape ecologists can help agencies prioritize key areas for management and identify the importance of protected areas relative to connecting areas, particularly if they are working together with wildlife managers from early in the process.

In Chapter 11, Vierling et al. further develop the importance of communication and provide recommendations for improving communication between landscape ecologists and wildlife managers. They use three case studies to demonstrate effective communication strategies among interdisciplinary scientists. In the first study they identify two challenges. First, there is a disconnect between landscape ecologists and wildlife managers related to access to the most recent and applicable research, particularly research that is presented in a way that is understandable to managers who may be unfamiliar with the jargon and approaches of landscape ecologists. Many agencies may have limited access to scientific journals because of costly subscription fees. Thus landscape ecologists should consider publishing in open access journals and use graphical abstracts that simply and effectively communicate the primary messages of the manuscript. Second, communication between landscape ecologists and wildlife managers could be improved by providing opportunities for personal interactions within agencies, professional organizations, working groups, and technical workshops. Those opportunities can assist wildlife managers and administrators in better understanding the tools and perspectives of the landscape ecologist.

Using the other two case studies, Vierling et al. propose additional recommendations to improve communication and overcome challenges faced when landscape ecologists and managers work together. They suggest that when maps are used with multiple stakeholder groups, the landscape ecologist should explain the concepts of spatial and temporal scale along with how those aspects affected map production. Landscape ecologists could benefit by limiting the use of jargon and engaging the local expertise of managers to identify appropriate levels of spatial variation in spatial models. That engagement is most effective through person-to-person meetings that happen early and often.

In addition to developing effective communication with wildlife managers, landscape ecologists need to be producing models and decision tools that are useful and practicable for the wildlife manager. In Chapter 12, Niemuth et al. tackle the diverse array of spatial models, provide considerations for what makes such models useful to the manager, and discuss how to avoid pitfalls in the development of those models. Of primary importance to whether a spatial model will be useful for decision makers is whether the tool was developed with specific management treatments in mind. Failure to align spatial models with goals, either from the outset or when goals change, will limit their utility in the future and may result in misidentification of priority areas for management. Likewise, the choice of response metric can also affect where management actions occur (e.g., a focus on species richness may divert resources away from individual species of interest and even result in reduced habitat quality). When multispecies management is a goal, landscape ecologists should work with managers to identify key treatment areas that benefit similar species rather than produce models that aggregate many disparate species.

Landscape ecologists can help wildlife managers better understand the quality and characteristics of spatial data, what analytical aspects need to be considered, and how they might engage the analytical process. Wildlife managers would benefit from assistance with choosing landscape scales (extent and grain) that are appropriate for the species and ecological processes in question, particularly when there may be multiscale influences to consider. Spatial predictor data such as landcover or climate data are the result of other models. The uncertainty of those models is seldom considered when studying relationships between those predictors and the ecological processes of wildlife. Whatever data are used, whether response or explanatory, landscape ecologists and wildlife managers should work together to identify questions that can be responsibly addressed and practically implemented. One of the powerful aspects of empirical and statistical models is their capacity to avoid biases and provide a perspective that changes conventional wisdom. Likewise, inappro-

priately applied models could incorrectly override orthodox understandings of ecology.

Niemuth et al. suggest that engaging managers early in the process can help target model goals toward a unified implementable purpose. Too often, research is initiated without considering how it will explicitly benefit management actions that are practicable. Engagement between the landscape ecologist and wildlife manager has benefits even when aspects of a model might be inaccessible or black-boxed. Rather than develop single-question or single-output models, Neimuth et al. suggest that landscape ecologists help develop a toolbox of potential models that are purposeful and inform decision-making tools. Ultimately, the quality of the data and the analyses are only as valuable as the degree to which they can be implemented by managers.

In Chapter 13, Witecha and Bogenschutz provide an overview of one of the important frameworks through which conservation of landscapes is accomplished in the United States: conservation incentive programs. These programs are essential for management of landscapes because management of wildlife habitat often occurs in conflict with other land use, particularly private agricultural practices. Wildlife management of landscapes needs to be both ecologically relevant and economically competitive with other potential land use. Some of the greatest challenges to landscape-level management of wildlife habitats are rooted in land ownership.

The Farm Bill contains provisions for a number of programs that may benefit the landscape-level management of wildlife habitat by making managed and conserved lands economically competitive with surrounding lands, and potentially expanding or connecting lands already managed by state and federal agencies. The conservation programs in the Farm Bill are one of the few opportunities for creating wildlife habitat at large scales on private lands.

Of particular interest is how landscape ecologists might improve the effect of conservation incentive programs in the future. Not only can they help identify the essential habitats and improvements needed to maintain and increase wildlife populations, but they can also help conservation programs identify areas for improvement that will most benefit wildlife species of interest. Landscape ecologists can assist the development of more rigorous and effective habitat conservation programs by incorporating their knowledge of the effects of habitat composition and configuration, edge effects, and other landscape metrics on wildlife populations.

Even the most rigorous management efforts cannot hope to be successful in isolation. Landing a man on the moon required more than rocket science. Overcoming the significant hurdles of that new frontier required a diverse set of interest groups and abilities, working together to develop programs that had never been imagined before. Likewise, effective management of wildlife populations across broad landscapes is more than incorporating some landscape ecology theory and metrics into wildlife research or management decisions. Landscape ecologists and wildlife managers must work together to understand the constraints to effective management, to tailor approaches to practicable goals and challenges of the system, and to engage a broad array of stakeholders and interest groups. Part IV will explore ways that wildlife managers and landscape ecologists are partnering with others to translate landscape ecology into management.

PART IV • Translating Landscape Ecology to Management

15

Age, Size, Configuration, and Context

JEFFREY K. KELLER

Keys to Habitat Management at All Scales

Introduction

Successful species management depends on identifying (the values of) a suite of landscape attributes (patch composition, size, structure, spatial arrangement, insularity, connectivity) most associated with species occurrence and abundance. Today, biologists have many tools to accomplish this. However, the use of powerful modern technologies to apply ecological theory to management is only beneficial when we consider the underlying biology of the systems we are attempting to manage, including the resolution and scale(s) at which species perceive and respond to that portion of the landscape they inhabit (Keller and Smith 2014). Otherwise, as Wiens (1989a:390) commented on interpreting species-habitat relationships at inappropriate scales, "we may think we understand the system, when we have not even observed it correctly." The main goal of this chapter is to heed Wiens's caution and ensure we observe the system correctly so that we can manage it more successfully.

Developing and testing a predictive habitat model for a particular region, however, are only part of the process of wildlife management. Implementing model-based management involves identifying a network of one or more sets of contiguous stakeholder properties with the potential to create or build upon existing areas of habitat with appropriate age, size, and configuration within the regional landscape context of those properties. Developing these partnerships can be every bit as challenging as developing predictive models, requiring managers to be biologists, wildlife ambassadors, educators, marketers, team builders, and negotiators. In this chapter, I review the influence of four major landscape attributes (age, size, configuration, and context) on species occurrence and explore the use of variables that quantify these four attributes as a basis for assessing and managing habitat at any scale. To illustrate the importance of stakeholder involvement even at the most local scales, I include a case study summation of preservation and alternative timber management for a 240-ha forested watershed in north-central New Jersey, USA, originally slated for single-tree harvest and development. Subsequent chapters discuss case studies that exemplify particular implementation approaches and further explore the process of stakeholder building.

Habitat Quantification: Linking Ecological Theory to Management

Wildlife biologists have been interested in how to better manage landscapes for species of interest since at least the time when Leopold (1933:fig. 9) recognized that the quality of quail habitat depended on configuration of the cover types used. With the development of concepts such as island biogeography theory (MacArthur and Wilson 1967) and the associated rise of the field of landscape ecology (Forman and Godron 1986), our understanding of species distribution patterns has become increasingly more informed. The parallel development of geographic information systems (GISs) and associated landscape metrics packages (e.g., FRAGSTATS; McGarigal and Marks 1995), coupled with advances in remotely sensed technologies (e.g., airborne light detection and ranging [lidar]; Zimble et al. 2003), has enhanced quantitative analysis of how landscape composition and arrangement influence species occurrence and abundance.

Equally important has been the recognition by ecologists that different organisms perceive their surroundings at different resolutions and respond to perceived landscape features at different scales, often related to their body size (Wiens and Milne 1989, Morrison et al. 1998, Keller and Smith 2014). Thus, although conservation biologists and wildlife managers routinely categorize patch configurations as representing corridors (i.e., connections between patches), islands (i.e., patches separated from other patches of similar composition), or core habitat (i.e., the interior of a patch minus a buffer of predefined width extending inward from the patch perimeter; Vogt et al. 2007), these terms are relative and entirely dependent on the resolution and scale of habitat use exhibited by individual species.

For example, what constitutes a corridor for a deer may be core habitat to a shrew. While the deer may respond to a forested connection between two larger patches of forest at the scale of the entire corridor, the shrew more likely responds to components that make up the corridor such as individual trees, shrubs, logs, boulders, small bodies of water, and edges between them (Huston 2002). Kotliar and Wiens (1990) defined these smallest landscape components to which a species responds as the *grain*. The contrast between the deer and shrew suggests that habitat management will be most effective when it closely reflects the grain and scale of landscape composition and structure to which the focal taxon responds, regardless of the geographic extent of any proposed management program.

In a GIS, which is increasingly the basis for many habitat models and most management planning, the

---

Box 15.1 Landscape Ecology Concepts

Scale

In ecology, *scale* refers to spatial extent (i.e., an area). Landscape attributes to which a species may respond are typically measured within a circular sample of a particular size (scale) centered on the species' known location to determine the level of response to the attributes at that scale (Jackson and Fahrig 2015). Researchers frequently measure landscape variables at multiple scales to determine the scale that yields the strongest species-landscape relationship. This is termed the *scale of effect*.

Grain

The smallest landscape component to which an organism responds is the *grain*. Kotliar and Wiens (1990) noted that at a scale smaller than the grain, the organism functionally perceives its environment as homogeneous. In GIS-based analysis, the ability to identify landscape elements that represent the grain is determined by the *spatial resolution* of the underlying imagery. Image resolution is measured as ground resolvable distance. *Ground resolvable distance* is the minimum distance between two objects that allows the two objects to be identified as separate entities.

ability to identify landscape elements that represent the grain is a function of the image resolution, minimum mapping unit, and classification system chosen by the researcher or wildlife manager (Gallant 2009, McDermid et al. 2009, Keller and Smith 2014). These three parameters, in turn, influence the classification accuracy of GIS imagery (i.e., how accurately it represents what is actually on the ground; Franklin et al. 2000, Fleming et al. 2004, Thogmartin et al. 2004, Hines et al. 2005) and the values of calculated landscape metrics (e.g., patch size, edge density) based on that imagery (Trani [Griep] 2002, Wagner and Fortin 2005, Keller and Smith 2014). Recent improvements in the ability to precisely locate study animals using radiotelemetry and global positioning system monitoring further emphasize the importance of employing appropriately resolved and accurately classified imagery in any subsequent habitat analysis. Ultimately, GIS resolution, the minimum size of mapped elements, and their classification accuracy strongly influence our interpretations of the scale(s) at which species appear to select habitat (Keller and Smith 2014). Understanding the influence of these image attributes on interpretations of species-habitat associations is important if biologists and managers are to successfully apply principles of landscape ecology to species management, particularly at the local scale of the individual tax parcels that make up a regional landscape of management interest.

The other major GIS-related influence on our interpretation of species-habitat associations is the use of landscape metrics (e.g., FRAGSTATS) to link species occurrence to habitat information contained in the selected imagery. Metrics describing landscape attributes such as patch size, configuration (patch shape and arrangement of patches), insularity (degree of isolation), productivity (often related to plant community age), and context (i.e., location within the regional landscape) have been used for three decades to model the habitat associations of individual species (Keller and Smith 2014). From a manage-

ment perspective, this research also has provided insight as to which landscape attributes may be limiting species occurrence even when appropriate habitat composition or structure appears to be present (Keller 1986, With and Crist 1995, Mönkkönen and Reunanen 1999, Lindenmayer et al. 2005, Bakermans and Rodewald 2006).

Additional Concepts
Patch

Patch has several definitions in ecology (MacArthur et al. 1962). Wiens (1976:82) proposed defining patch as "a surface area differing from its surroundings in nature and appearance." This definition, which lends itself readily to GIS applications, has largely been adopted as the standard interpretation of the term patch in ecology. In landscape ecology, patch is most frequently used to refer to a discrete and internally homogeneous entity such as a land use or cover type, a plant community, or a biotope in the landscape (Forman and Godron 1981, Wiens 1989*b*).

Keller and Smith (2014:13–15) argued that, from a wildlife perspective, Wiens's (1976) definition of patch as representing an area could also apply to the combinations of landscape components, such as shrubs adjacent to grass, used by edge species such as the song sparrow (*Melospiza melodia*) and eastern cottontail (*Sylvilagus floridanus*). As a result, they suggested definitions (Keller 1986, 1990, Keller et al. 2003, Keller and Smith 2014) to separate what they considered to be two major patch associations of wildlife discernible in a GIS landscape image: solid and edge.

Solid and Edge

Solid patches are clusters (i.e., a polygon in vector-based GIS, or contiguous pixels or grid cells in raster-based GIS; Keller et al. 1980, Turner 1989) of identical or structurally similar landscape components (e.g., open grass, open water, sawtimber trees [i.e., a forest], emergent marsh) associated with a particular species or assemblage. *Edge* patches are any combination of interfaces between adjacent structurally dissimilar landscape components (e.g., shrub-grass, deciduous tree-grass, open water-emergent marsh) associated with a particular species or assemblage.

Interpreting the concept of patch to include both solid and edge types discernible in a GIS is useful in understanding species-habitat relationships and furthering efforts to conserve or enhance habitat for species of interest. First, recognizing solid versus edge habitat associations aids in understanding differences in the observed occurrence of individual species on the landscape (Keller and Smith 2014; Diamond and Elliott, Chap. 6, this volume) and species richness (biodiversity) between one area and another. Second, differentiating between solid and edge patch types is useful in designing and evaluating the effects of management options to create or enhance habitat for individual species or promote biodiversity in general.

ADDITIONAL CONSIDERATIONS FOR EDGE

The concept of edge has various interpretations. In wildlife biology it has been used since Leopold (1933) to describe vegetation associations of many game species and is defined most typically in ecology as the boundary between two plant communities (e.g., forest and agriculture boundaries; Fig. 15.1). This definition, known as an ecotone (Clements 1905, Odum 1971), has become the default concept of edge in landscape ecology and wildlife management. In contrast, Keller and Smith (2014) concurred with Risser (1987) and Wiens (1989b) that myriad edges of numerous types are present at multiple grains and scales, both between and frequently within biotopes (Fig. 15.2). This alternate view also suggests that there are many edge types that, viewed at the reso-

Figure 15.1. A 1:40000 view of the Connecticut Hill Wildlife Management Area in the southern Finger Lakes region of central New York, USA, illustrating edge at the scale of ecotones, the major edges between different plant communities or land use types. Conservation planning is done most often at scales such as this, which, although useful, may miss key landscape features important to smaller species of wildlife. From Google Earth, 1 May 2007, New York, New York, USA.

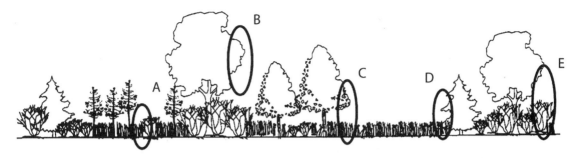

Figure 15.2. A perspective view of a mid-stage succes-sional oldfield illustrating 5 different types of edge to which various species of wildlife respond: (*A*) shrub-grass edge, (*B*) deciduous canopy-open air edge, (*C*) deciduous sapling and poletimber-grass edge, (*D*) coniferous sapling and poletimber-grass edge, and (*E*) deciduous canopy-deciduous shrub edge. Although each of these edge types has meaning to different species of wildlife, edges viewed at this scale are often not considered in wildlife management planning. Reproduced from Keller and Smith (2014:fig. 1.2).

lution perceived by the organism, are recognized by various species of wildlife as appropriate habitat. Ex-amples include the deciduous forest-talus slope edge used by the Alleghany woodrat (*Neotoma magister*), the emergent marsh-open water edge used by marsh birds (Rallidae and Ardeidae; Rehm and Baldassarre 2007), and the canopy-open air edge used by fly-catchers (Tyrannidae; labeled B in Fig. 15.2).

As a result, Keller and Smith (2014) emphasized the need to employ remotely sensed imagery at a res-olution appropriate for the species of interest when evaluating the potential effects of any proposed management options for wildlife, including those associated with edge habitat types. They noted that most GIS databases rely on Landsat or other lower-resolution imagery that cannot adequately depict local-scale edges (Fig. 15.2) associated with some smaller species of conservation interest (e.g., blue-winged warbler [*Vermivora cyanoptera*]). They then suggested a less arbitrary method of determining appropriate image resolution based on the body size of the organism of interest (Keller and Smith 2014). In general, they recommended using the highest res-olution imagery available to evaluate the presence of habitat for a particular species or assemblage. Otherwise, landscape components important to the species of interest may be indiscernible from the average conditions of the landscape (Huston 2002),

and habitat may be missed or potential habitat ma-nipulation effects misinterpreted.

Biotope

Looijen (1998:165) recommended the use of the term *biotope* to describe the physical place of a com-munity as "an area [topographic unit] characterized by distinct, more or less uniform, biotic and/or abi-otic conditions." Under this definition, biotope is essentially the equivalent of the current landscape ecology and GIS interpretation of Wiens's (1976:82) definition of a patch as "a surface area differing from its surroundings in nature and appearance."

Use of the term biotope appears to be largely ab-sent from the more recent North American ecologi-cal literature. Keller and Smith (2014) suggested that such omissions are unfortunate because the term groups such disparate but hierarchically similar en-tities as a coral reef, a stream channel, and all types of plant communities (as containing habitats for an-imals) under one organizational umbrella. They in-cluded it in their review of GIS-based wildlife–habi-tat analysis for its value in landscape ecology and GIS applications as a relatively unambiguous descriptor of the habitat of a community. For these same rea-sons, it is included here.

Effects of Resolution and Scale on Habitat Interpretation
Habitat Selection Scale

Lower resolutions of remotely sensed imagery are associated with decreased image interpretability and increased misclassification (Franklin et al. 2000, Fleming et al. 2004, Thogmartin et al. 2004, Hines et al. 2005). These limitations are problematic when one considers the fundamental differences in the grain and scale at which different wildlife species select habitat (Wiens and Milne 1989, Pearson and Gardner 1997).

Wiens (1989a:fig. 4) considered scale dependency in ecological systems theoretically and argued that within the spectrum of potential analytical scales there are domains (i.e., subsets of the scale continuum) of particular ecological phenomena within which process-pattern relationships are consistent, regardless of the scales of observation within that domain. Wiens (1989a) noted that proper analysis requires that the scale of researcher measurements and that of the organism's responses fall within the same domain. Keller and Smith (2014) suggested that, viewed in this way, domain represents the range of image resolution and level of image classification at which GIS-based habitat analysis and interpretation of species-habitat relationships should occur.

Scales of Edge Detection, Classification, and Quantification at Low Resolution

Although germane to both solid-patch and edge-patch species, not all of the problems of low-resolution imagery and classification accuracy discussed above apply to all species. For example, for solid-patch species associated with forest interiors or grasslands, lower resolution imagery, which is ideal for examining large geographic areas, is useful for considering issues such as fragmentation at the scale of land use or landcover types. The ecotone-scale edges typically associated with forest fragmentation produce quantifiable impacts to some forest interior species such as the barred owl (*Strix varia*) but are responded to positively by some larger-bodied edge species with extensive home ranges such as white-tailed deer (*Odocoileus virginianus*; Fig. 15.1).

Many early successional species of conservation concern, however, such as the indigo bunting (*Passerina cyanea*), field sparrow (*Spizella pusilla*), and blue-winged warbler, are associated with finer-grained edges (e.g., individual shrubs or saplings adjacent to small grassy openings; Fig. 15.2) than can be detected on Landsat or similarly resolved imagery (Keller and Smith 2014). This is also the case for a number of what are now described in the literature as forest interior-edge species (i.e., species associated with disturbances in the forest canopy) such as the cerulean warbler (*Setophaga cerulea*; Perkins and Wood 2014), hooded warbler (*Setophaga citrina*; Moorman et al. 2002), and black-throated blue warbler (*Setophaga caerulescens*; Keller 1986, Goetz et al. 2010, Keller and Smith 2014). As a result of the inability to detect these fine-grained edges at low image resolutions with concomitantly lower classification accuracy, the edge species above often have been categorized as associated with more general forested (i.e., solid) cover or land use types. This has led to reduced correlations of mapped habitat with the species actual occurrence, particularly for forest interior-edge species, because they use specific edge-associated subsets within more extensive forested areas. Beyond having low predictive capability, resulting models have the potential to produce misinformed management strategies and prescriptions (Keller and Smith 2014).

Additionally, because early successional stages and small canopy gaps provide habitat for the species that use them for only short periods of time before they progress to an older, unsuitable stage of forest regeneration (Keller et al. 2003, Schlossberg and King 2009; Fig. 15.3), species associated with these successional stages appear more adapted to locating

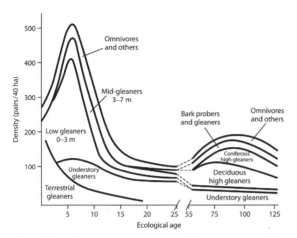

Figure 15.3. Density (mean pairs per 40 ha) of avian guilds following clear-cutting at the Connecticut Hill Wildlife Management Area in central New York, USA. Reproduced from Keller et al. (2003:fig. 7).

much smaller and more ephemerally available parcels than do most forest interior species (Roberts and King 2017). Therefore investigators or managers must employ a somewhat different approach to adequately assess habitat availability for these species, which are not subject to the same degree of ecotone-scale edge effects that influence the distribution and reproductive success of forest interior species (Brittingham and Temple 1983, Donovan et al. 1997).

All of these observations regarding habitat associations (solid vs. edge), scales of habitat use, image resolution, and classification accuracy suggest that caution should be used in overinterpreting or over-relying on habitat distribution information mapped using low-resolution databases such as Landsat or any databases with minimum mapping units that are as large or larger than the home ranges of species of interest. This appears especially true for smaller species with smaller home ranges, which as a group tend to select habitat at higher levels of resolution than are used in some mapping of wildlife habitat (Tattoni et al. 2012, Farrell et al. 2013, Vogeler et al. 2013, Keller and Smith 2014). Note, however, that Fleming et al. (2004) and Gaston et al. (2017) demonstrated the same problems of habitat misinterpreta-

tion when lower-resolution imagery was applied to habitat selection modeling for two large mammals, the white-tailed deer and brown bear (*Ursus arctos*), respectively.

Alternative High-Resolution Analysis and Assessment

High-resolution models exist for some species of management interest (Goetz et al. 2010, Farrell et al. 2013, Vogeler et al. 2013) and can be tested for applicability on areas not used in model development. For smaller species for which existing models relied on low-resolution satellite- or small-scale aerial photography-based GIS databases, however, the problem of unmeasurable habitat at low image resolutions remains. In these cases, development of updated models using higher-resolution imagery may be required to better inform management decisions.

I offer one possible approach to this issue with four steps. First, use an existing database to identify the location of known populations of smaller species of conservation interest that, natural history and expert opinion suggest, are either edge-associated or associated with landcover types with higher-than-average misclassification rates (e.g., shrublands, young cutover areas; Smith et al. 2001). Second, combine high-resolution imagery (e.g., lidar, high-resolution aerial photography) of these known locations of occurrence with expert opinion to identify and refine the taxon-specific landscape composition of suspected solid-patch or edge-patch associations. Use these suspected associations to identify a suite of predictor variables and develop a taxon-specific habitat model at a taxon-specific image resolution (Keller and Smith 2014:chap. 6; Diamond and Elliott, Chap. 6, this volume). Next, test predictive ability of the occupancy model on known locations of species occurrence not used in model development. Finally, if model tests validate the model's predictive ability, apply the model to areas that were previously classified as habitat using lower-resolution imagery but for which actual occurrence is unknown. Conduct surveys of those areas for the species in question, and test the resulting data against the higher-resolution habitat model to determine whether the model provides significantly higher prediction of species occurrence. Conduct sufficient detection surveys to apply occupancy modeling analysis (MacKenzie et al. 2002, MacKenzie and Royle 2005).

Based on the results of such a process, consider higher-resolution mapping for species associated with frequently misclassified landcover types or considered to be fine-grained edge species. Among many possible species, this approach would apply to early successional edge species and to species now recognized as forest interior-edge species, such as cerulean warbler (Perkins and Wood 2014) and golden-winged warbler (*Vermivora chrysoptera*; J. L. Larkin, Indiana University of Pennsylvania, personal communication 2016).

Principles of Conservation Design and Management

Researchers have identified a suite of landscape attributes that are repeatedly correlated with the presence of individual species and, at higher organizational levels such as assemblages, with biodiversity (MacArthur and MacArthur 1961, MacArthur and Wilson 1967, McClintock et al. 1977, Rabenold 1978, Bakermans and Rodewald 2006). To better understand the biological basis for the influence of these attributes on species-habitat relationships and apply them to species management, I grouped them into four general categories: age, size, configuration, and context (ASCC).

Age (Ecological)

Natural disturbance has always been an influential factor in ecosystems. More recently, anthropogenic disturbance has been superimposed on this regime.

In the case of terrestrial systems, responses to disturbance vary depending on geographic location but typically include immediate and subsequent changes in species composition, which for plants are most influenced by underlying soil types and moisture availability. Although such changes have been observed to be relatively predictable in more mesic environments (a process referred to as succession by Odum 1971), semiarid or arid locations often do not follow predictable patterns of return to the original vegetative community (Fulbright 1996, Fulbright et al. 2008). Whether predictable or stochastic, it is this temporal dynamic that produces the ever-changing spatial mosaics we observe across landscapes. As a result, habitat management includes a temporal component that is equally important to the spatial component most often represented in species-habitat models.

As noted above, in more mesic environments such as the eastern deciduous forest of North America, the process known as ecological succession is commonly defined as the more or less predictable and orderly changes that occur in the structure and composition of vegetation and associated animal communities over time. Two types of succession are generally recognized, and although applied most precisely in reference to natural processes, both types are often used to characterize silvacultural practices. The first type (even-aged succession) usually follows a catastrophic event such as a blowdown or fire where all regeneration starts at the same time immediately following the disturbance and all propagules of the disturbed community are already in place. For example, following a harvest practice such as clear-cutting, stump sprouts and root suckers of the harvested trees begin to grow the following season along with seedlings from seeds already in place in the soil prior to the harvest.

The second type of succession (uneven aged) typifies agricultural abandonment, a frequent occurrence in the Northeast during the last century (Stanton and Bills 1996). More rarely, it is associated with the primary succession (plant colonization) that occurs on newly available substrates such as a recent lava flow, an exposed beach, or the previously inundated substrates exposed following abandonment of a beaver (*Castor canadensis*) pond. Uneven-aged succession occurs where propagules of various plant species reach the site over time and different ages of woody vegetation are present simultaneously. The importance of these different successional types to animals is in the structure and productivity of the plant communities that result. These differences are important to plants because of the various growing conditions created due to levels of shading or local moisture availability.

The effects of vegetation succession on individual species and broader assemblages of animals have been studied most frequently for birds, which show marked changes in species composition and abundance during the course of forest succession (Fig. 15.3; Conner and Adkisson 1975, Titterington et al. 1979, DeGraaf 1991, Keller et al. 2003, Schlossberg and King 2009). Keller et al. (2003) and others (Smith and Shugart 1987, Holmes et al. 1996, Goetz et al. 2010) have demonstrated that these patterns are strongly correlated with the distribution and density of leaves at various heights during succession, and with related abundance and distribution of prey items, primarily insects. Young clear-cuts and other early successional stages appear to support the highest density and richness of birds and other animals (Kirkland 1977) during forest succession in the Northeast owing to this high productivity and concentration of leaf area at or near ground level (Keller et al. 2003).

Knowing where a plant community of interest falls along a successional continuum (Fig. 15.3) allows prediction of current and future habitat availability to various members of the avian assemblage. Similar successional relationships have been documented for the more subtle changes that occur over time within grasslands (Bollinger 1995, Mitchell et al. 2000), and the relationships of various game

species to successional changes are well known (Leopold 1933, Bump et al. 1947). Again, such patterns may not occur in all environments but can be very useful in management where they are observed.

Size

Following decades of observation of patterns of species richness on oceanic islands, MacArthur and Wilson (1967) offered an explanation (i.e., island biogeography theory) of these patterns based on colonization and extinction rates related to island distance from the mainland and island size, respectively. In short, islands closer to mainland sources of species are colonized more frequently and thus tend to support higher numbers of species. Larger islands support larger populations and hence are less likely to have those populations go extinct as a result of chance events. Thus larger islands also tend to support higher numbers of species. Conversely, small islands, especially those more distant from the mainland, support fewer species. This combination of island size and distance to the mainland interacts to produce what MacArthur and Wilson (1967) called the equilibrium number of species that can be anticipated to occur on a given island of a size and distance from the mainland or from some alternate source of colonists.

Subsequent research showed that these principles of island size and distance also applied to patches or biotopes on the mainland. For example, Forman et al. (1976) reported that bird species richness in forest islands increased as woodlot size increased in central New Jersey farmland. Despite ensuing debates over the influence of island size on species richness and the suggestion of alternative explanations such as heterogeneity, island shape, landscape context, and total patch amount (Osman 1977, Litwin and Smith 1992, Rodewald and Yahner 2001a, Pearman 2002, Fahrig 2013), several underlying principles related to the importance of size alone are consistent on habitat islands and oceanic islands.

Figure 15.4. Relative territory sizes of (*left to right*) downy (*Picoides pubescens*), hairy (*Leuconotopicus villosus*), and pileated (*Dryocopus pileatus*) woodpeckers.

First, larger species occur only in larger biotopes (habitat islands; Keller 1986). This is a direct result of the relationship between increasing body size and increasing territory size (Schoener 1968), as illustrated by three common species of woodpeckers (downy [*Picoides pubescens*], hairy [*Picoides villosus*], pileated [*Dryocopus pileatus*]) in the Northeast (Fig. 15.4). Second, larger patches, solid or edge, of appropriate habitat should contain more guild members because a patch that is suitable for a larger member of a guild is more likely suitable for a smaller member of that guild also (Fig. 15.5; Keller 1986).

Keller and Smith (2014) suggested that one of the reasons that patch size is not consistently correlated with species richness is that its measurement usually fails to simultaneously consider shape. They further argued that the clustered arrangements of habitat often associated with threshold occurrences of species on the landscape (Hiebeler 2000) are related to the energetic efficiency of such arrangements (Covich 1976). Covich (1976) argued that, in an energetics context, a circle is the optimal shape in horizontal space for an all-purpose territory. This efficiency is primarily the result of two properties of circles. A circle has the minimum perimeter to area ratio; therefore it has the minimum amount of border per unit area to defend against conspecifics. It also has the shortest average distance between points within it, which optimizes foraging. Keller (1986) and Keller and Smith (2014) discussed the use of this concept, termed the maximum diameter circle (MDC), to measure the functional size of solid-patch types such as closed-canopy forest, open water, and grasslands observable on remotely sensed imagery (Figs. 15.6 and 15.7).

Among examples of unrecognized but similar applications of this approach, Keller et al. (1993), Darveau et al. (1995), Confer and Pascoe (2003),

Figure 15.5. The relationship of patch size to species richness. A patch large enough for the largest member of a guild should be large enough to support smaller guild members.

and King et al. (2009) (all bird assemblages), Stoddard and Hayes (2005) (salamanders), and Kubel and Yahner (2008) (golden-winged warbler) reported species richness correlations or threshold occurrence relationships with the width (equivalent to MDC) of various corridor types (e.g., powerline right-of-way, fencerow, riparian corridor). Davis (2004) reported that among similarly sized native grassland patches, those with a lower perimeter (edge) to patch area ratio (i.e., a larger MDC) supported greater richness and abundance of area-sensitive grassland birds.

MEASURING THE FUNCTIONAL SIZE OF EDGE PATCHES
To measure the size of edge patches (e.g., shrub-open grass edge, shrub-water edge) in a manner comparable to MDC, Keller (1986) also developed an edge-scanning algorithm. The algorithm locates areas with the highest density of edges (meters per square meter) for a given structural type, such as shrubs adjacent to grass (e.g., song sparrow [*Melospiza melodia*] habitat), within a series of progressively larger circular samples on the GIS map (Keller and Smith 2014:figs. 3.4, 3.5). Keller (1986) called these measurements of edge density within circles the diameter of the equivalent area circle (DEAC) because they were an equivalent measure of functional patch size (for edge species) to the MDC for solid-patch species (Fig. 15.6; Keller 1986, Keller and Smith 2014). The key to application of this variable is to have a reasonable understanding of the edge type(s) used by a particular species or group of species and use remotely sensed imagery with sufficient resolution to identify and quantify those edges. This requires some understanding of the species' natural history and its functional role within its community (Keller and Smith 2014).

Configuration

As noted above, patch size and shape are important to consider when attempting to assess patch quality (e.g., species richness or density of individuals supported) or estimate potential effects of landscape management. Researchers have reported that clustered arrangements (optimally circles) of the landscape components or patch types important to particular species are correlated with the occurrence of those species (Andren 1994, Hiebeler 2000, Villard and Metzger 2014).

Typically, landscape metrics packages such as Spatial Analyst (Environmental Systems Research Institute, Redlands, California, USA) or FRAGSTATS, offered with GIS applications, do not include variables that measure clustered arrangements or the circles described above, although the "girth" variable (Center for Land Use Education and Research 2019) and "landscape composition" variable (Collier et al. 2012) are surrogates. Instead, metrics packages include variables that measure primarily proportions or indices. Keller et al. (1980) and Keller and Smith (2014) noted that the main problem with proportions and index-type variables, the latter of which includes general measures of patch shape, is that they reduce information to such a degree that many possible configurations of landscape components or cover types can produce the same or similar values (Fig. 15.8).

This lack of mathematical and associated biological discrimination among proportion- or index-type variables greatly reduces the explanatory and, ultimately, predictive utility of such measures (Cale

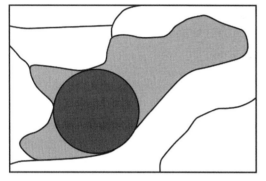

Figure 15.6. The maximum diameter circle (MDC) representing the largest optimally shaped territory that fits within a particular plant community of interest (i.e., a solid-patch type) identifiable on remotely sensed imagery. The MDC represents the functional size of the patch. Reproduced from Keller and Smith (2014:fig. 3.2).

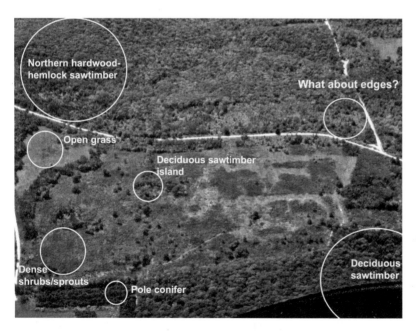

Figure 15.7. Maximum diameter circles (MDCs) for different patch types. Note the heterogeneity in the oldfield at the upper right, which is inconsistent with the concept of MDC as pertaining to solid-patch types. Reproduced from Keller and Smith (2014:fig. 3.3).

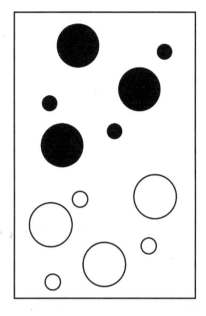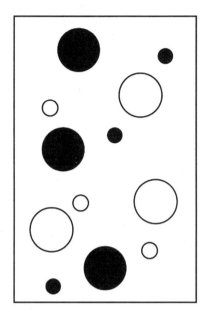

Figure 15.8. Hypothetical distributions of identical amounts of two different landscape component types on two plots. Though representing identical proportions of component types, these different spatial arrangements can represent different habitats to different species. Reproduced from Keller et al. (1980:fig. 5).

and Hobbs 1994, Neel et al. 2004, Vogt et al. 2007). Simply knowing the proportion of an area in forest, for example, tells a manager little about the potential value of the area to forest interior species that might be of concern in such a setting, but see Fahrig (2013) for a discussion of the importance of total patch amount versus the size of individual patches in a local landscape. Although other measures of configuration such as gradient analysis exist (Keller and Smith 2014:fig. 3.6), they are likely unduly complex for local management planning purposes but might be useful where gradients from more anthropogenically influenced areas to areas less so warrant formal quantification.

The MDC and DEAC are relatively simple measures that overcome the limitations of proportions and indices in quantifying the configuration of plant communities and other types of patches that include habitat for species of management or conservation interest. Measures of minimum inter-patch distances, patch connectivity, and the landscape composition variable (Collier et al. 2012) also add useful information about patch configuration and associated habitat quality. All of these spatial measures are potentially useful as comparative measures of existing versus proposed conditions when considering management options at any scale.

Context

Landscape context has been a somewhat recent topic of interest among ecologists. Context considers the effects of the surrounding landscape (i.e., matrix) on a particular species or assemblage's long-term viability at a given location (Bakermans and Rodewald 2006, Cornell and Donovan 2010, Kennedy et al. 2011, Thompson et al. 2012, Fahrig 2013). It is similar in many ways to the considerations of island biogeography. How different is the matrix surrounding the patch or habitat island occupied by the species? And, as a result, how difficult is it to cross the matrix to colonize the habitat island or, conversely, to immigrate from it? Additionally, how far is the habitat island from similar habitat?

Many researchers examined the influence of landscape composition on brood parasitism, nest depredation, and various measures of nest success (Donovan et al. 1997, Rodewald and Yahner 2001*b*, Thompson et al. 2002, Driscoll and Donovan 2004, Winter et al. 2006). Among these studies, the matrix size analyzed surrounding a site of interest varied widely from circles of tens of meters in diameter to ≥5 km. The scale at which context was found to influence local species occurrence or reproductive success, if at all, also varied depending on the focal taxon, the biotope examined, and the structure and composition of the surrounding land use (Prevedello and Vieira 2010, Kennedy et al. 2011). More recently, Jackson and Fahrig (2015) questioned whether the scales of analysis in many such studies are appropriate and suggested taxon-based criteria for selecting an appropriate range of scales.

Aerial views of Morris County in north central New Jersey (Fig. 15.9) and Central Park in Manhattan, New York, New York, USA (Fig. 15.10), illustrate the different contexts in which vegetation patches used by particular species can occur. In general, researchers reported that the most structurally different contexts, particularly anthropogenically influenced settings (e.g., suburbia surrounding a forest), have the greatest effect on the focal taxa (Bakermans and Rodewald 2006, Kennedy et al. 2011), particularly less mobile ones, such as amphibians (Gagne and Fahrig 2007, Pillsbury and Miller 2008, Sawatzky et al. 2018).

Understanding the life history of the organism in question is essential in evaluating the influence of landscape context on habitat within a particular

Figure 15.9. An aerial view of a portion of the Highlands Physiographic Province in north central New Jersey, USA, illustrating the interspersion of forests, agriculture, and development. The context (i.e., surrounding land use types and their extent) in which forests, hayfields, and other vegetated patch types occur influences the composition and population viability of wildlife species associated with each patch type. Generally, the higher the intensity of development surrounding a particular patch, the greater the negative effect on species of management or conservation interest within the patch. From Google Earth, 2014, New York, New York, USA.

Figure 15.10. An aerial view of Central Park in Manhattan, New York, New York, USA. While Manhattan is a true island surrounded by water, Central Park is a 337-ha habitat island surrounded by thousands of square kilometers of concrete and asphalt, in addition to water. The park's context thus limits immigration and emigration by animals incapable of flight or plants with seeds not dispersed by wind or birds. From Google Earth, 2010, New York, New York, USA.

patch or site (Van Horne 2002, Dunford and Freemark 2004, Bakermans and Rodewald 2006, Prevedello and Vieira 2010, Keller and Smith 2014). Although a patch's plant composition and structure may suggest that habitat exists, the degree of isolation of the patch may prevent sustained occupation of the site by a species otherwise associated with that plant composition and structure (e.g., smaller isolated agricultural fields; Fig. 15.9). Alternatively, connection of an otherwise isolated patch to a larger patch of similar habitat (i.e., a source area) may facilitate access and sustained occupation of the first patch (Whitcomb et al. 1981).

Perhaps one of the best examples of the importance of context on the East Coast is the position within Manhattan of Central Park (Fig. 15.10), a 337-ha green space surrounded by thousands of square kilometers of largely concrete, asphalt, and water, all of which limit colonization of the park mostly to species capable of flight (e.g., birds, bats, flying insects). The park's vegetation composition (i.e., its patch types) and the size and arrangement of those patches then determine whether sufficient habitat exists to support viable long-term populations of those colonist species able to reach the park.

Using Age, Size, Configuration, and Context to Assess Habitat and Manage Landscapes

Keller (1986) suggested that, collectively, the four attributes of landscapes discussed above (ASCC) serve as indicators of the threshold characteristics of patches of habitat (i.e., minimum size with optimal shape, degree of insularity [isolation], and productivity) that dictate the initial occurrence of a species within a landscape (Huggett 2005). For example, although a patch may appear to represent habitat for a species (i.e., be of the appropriate age [correct composition and structure]), it may be unoccupied because of deficiencies in one or more of the other three landscape attributes. It may be too small (size). It may be too linear and therefore inefficiently shaped energetically (configuration). Or it may be too isolated within a matrix of cover types that do not contribute to habitat (context).

In other cases a natural area may be large (size), contain an extensive core (shape defined by MDC), and exist within a favorable matrix of non-anthropogenic landscape (context) yet simply lack habitat (appropriate age) for the species of management interest. This scenario would describe many second-growth forests in the Northeast that, when unmanaged or only selectively cut, offer limited habitat for species associated with early succession (Rudnicky and Hunter 1993, King and DeGraaf 2000, King et al. 2001, Keller et al. 2003, King and Schlossberg 2014).

Threshold Patch Size

Different arrangements of solid and edge patches composed of deciduous high canopy (Fig. 15.11) illustrate the influence of age, size, configuration, and context landscape attributes on the general pattern of a species appearance within a landscape. For the red-eyed vireo (*Vireo olivaceus*), a bird generally accepted to be associated with deciduous high canopy in the Northeast, the occurrence of some minimum size of clustered (solid patch) high canopy (i.e., a woodlot) with the energetically optimal shape (a circle) fosters the occurrence of the first breeding pair of red-eyed vireos within a forest patch (Galli et al. 1976, Rusterholz and Howe 1979, Martin 1981). This is the threshold patch size (Fig. 15.11A) associated with the species occurrence on the landscape (Keller 1986, Huggett 2005).

As patches of appropriate habitat further increase in size, populations of species associated with the patch type also increase and eventually reach sustainable levels. The generally contiguous, extensive forest (Fig. 15.9) represents this latter type of patch and as such serves as a source area for many forest interior species within the region. Similarly, as the patch increases in size, larger species with similar habitat requirements colonize the patch as it reaches their threshold size for occupation. In this example the scarlet tanager (*Piranga olivacea*) is also a high-canopy foliage-gleaning bird like the red-eyed vireo. Because of its larger body size, it requires a larger area of forest in which to successfully reproduce and thus rarely occurs in forests that do not also support the red-eyed vireo (Keller 1986). This situation is similar to the guild of woodpeckers of differing body sizes referenced above (Fig. 15.5).

Fragmentation

In contrast to the addition of species as patch size increases, habitat loss—often associated with fragmentation of large patches into smaller patches—is the major reason that long-term population levels decline to unsustainable levels. At small population

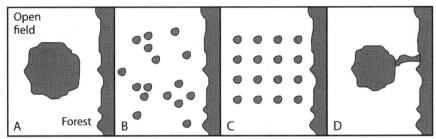

Figure 15.11. An example of the sequence of solid patch development for deciduous high canopy used by the red-eyed vireo (*Vireo olivaceus*). (*A*) Threshold patch size at which the first breeding pair of the species associated with the patch type, in this case the red-eyed vireo with deciduous canopy, occupies the patch. (*B*) Random distribution of an equivalent number of trees, as in (*A*). (*C*) Regular distribution of an equivalent number of trees, as in (*A*). (*D*) Decreased threshold patch size due to the presence of a connecting corridor or, alternatively, without the connection, due to productivity. Adapted from Keller (1986).

sizes, species become vulnerable to local extinction (Fahrig 1997, 1998). Local extinctions and associated declines in population levels across a region are the underlying cause for listing species as threatened or endangered at the state level (D. Brauning, Pennsylvania Game Commission, personal communication 2012). Similar declines across all or major portions of a species range may lead to federal or international listing. Thus comparing existing patch attributes with those that would be obtained under alternative post-management conditions is critical to determining the potential effects of various management options on species or assemblages of concern.

Patch Configuration: Edge-Patch Species versus Solid-Patch Species

The concept of a set of threshold patch characteristics (i.e., minimum size with optimal shape of appropriate habitat) that determine the occurrence of solid-patch species such as the red-eyed vireo and scarlet tanager also applies to species associated with edge patches. For example, although a random distribution (i.e., a scattered configuration) of trees (Fig. 15.11*B*) would not represent habitat for a solid-patch species like the red-eyed vireo, it could potentially represent habitat for a high-canopy edge species like the Baltimore oriole (*Icterus galbula*). In this case, threshold patch size for the Baltimore oriole would be the diameter of a circle (i.e., the DEAC) with the

highest density (length per unit area) of canopy-opening edges within which its breeding requirements are met.

Despite the difference in configuration, the major landscape component used by the red-eyed vireo and the Baltimore oriole (i.e., sawtimber-sized deciduous trees) is the same. Thus it is largely the spatial arrangement (i.e., configuration) of the component on the landscape (Fig. 15.8) that dictates where these two birds occur. Regular distributions of trees (Fig. 15.11*C*) do not typically occur in nature and few species—usually only generalists like the American robin (*Turdus migratorius*)—are associated with this type of configuration.

Patch Connectivity and Isolation

In general, a species occurrence on the landscape is associated with sufficiently sized, clustered arrangements of the landscape components it uses (Fig. 15.11*A*) or, more broadly, the patch type that contains those elements (Fig. 15.11*A* can be interpreted more broadly as forest). The functional size of the patch is further influenced by variables such as its connectivity (Fig. 15.11*D*; Whitcomb et al. 1981) or, if unconnected, its degree of isolation (i.e., distance from a similar patch type; Fig. 15.11*A*). Connectivity can be assessed easily by examining remotely sensed imagery for links between particular plant communities, or between other biotopes of interest, and larger areas of the same type. Distance between patches is easily measured also, either manually or with GIS software.

One can better understand the influence of isolation simply by envisioning each of the four landscape snapshots (Fig. 15.11A-D) as representing a much larger geographic area where the single trees depicted (Fig. 15.11B, C) were each a patch of forest with its own small population of red-eyed vireos. Under this scenario, even if the most isolated forest patch (Fig. 15.11B) could support a small population of red-eyed vireos, it would be unlikely to maintain that population over time due to its size and degree of isolation from other red-eyed vireo populations. However, clustered arrangements of multiple patches relatively close to one another (e.g., the tightly grouped three- and four-patch subclusters; see Fig. 15.11B) can help maintain viable metapopulations in the absence of a single patch large enough to support such a population (Simberloff 1986, Fahrig 2013),where the presence of only one or two of the small patches (Fig. 15.11B) would be insufficient to do so. Thus clustering of landscape components is important at the scale of threshold species occurrence and at larger geographic scales necessary to support viable breeding populations of species of concern. This is especially true for less mobile species such as amphibians and smaller mammals (Pillsbury and Miller 2008, Prevedello and Vieira 2010, Fahrig 2013).

Other Patch Characteristics

Lastly, patch connectivity (Fig. 15.11D) or patch quality (e.g., in the form of higher insect productivity; Stenger 1958) can influence the threshold size of a patch for a particular species, guild, or assemblage. Connectivity was discussed above, and patch productivity can be inferred from the ecological age (Fig. 15.3) or latitude (Rabenold 1978) of the terrestrial plant communities being considered. Higher productivity, at least in the form of leaf area production and associated prey availability, has frequently been

linked to early successional communities such as regenerating forests and shrublands in the East and to riparian communities in the West (Rosenberg et al. 1982, Keller et al. 2003).

The Importance of Grain and Scale in Interpreting Patch Characteristics

The characteristics of patches described above (Fig. 15.11) can be applied in several ways. First, landscape attributes can be quantified and correlated with the distribution patterns of species or assemblages of interest to develop species-habitat models, which then can be tested to validate their use over broader geographic areas. Second, these relationships can be used to guide identification of habitat management alternatives. Third, manipulation alternatives can be quantified to evaluate their potential effects on focal and nontarget taxa.

To appropriately interpret potential associations or inferred occurrences of a particular species with mapped cover types, it is essential to consider the grain (resolution) and scale (spatial extent) at which the organism likely perceives and responds to the landscape (Wiens and Milne 1989, Huston 2002, Martin and Fahrig 2012, Keller and Smith 2014, Jackson and Fahrig 2015). This is the challenge of correctly observing the system and returns us to the shrew and white-tailed deer example noted above. In general, smaller species with generally smaller home ranges perceive landscape components (i.e., trees, shrubs, rocks, logs, grassy, canopy openings, water bodies) used in habitat selection at smaller scales and at higher resolutions commensurate with their size (Keller and Smith 2014). Concomitantly, how the species functions within the environment—its niche (e.g., flycatchers vs. foliage gleaners)—will further delineate that subset of the landscape it inhabits. Dispersal ability (walking vs. flying) additionally influences component perception and the scale of habitat use by the organism (Holland et al. 2004, Holland and Bennett 2009, Jackson and Fahrig 2015).

Thus, whether ground or GIS based, habitat variables need to describe landscape components that reflect a species' use of the landscape (Van Horne 2002). Historically, this generally has been true of ground-based variables (e.g., shrub density, height, leaf area) but has not always been true in GIS-based studies owing to data availability, cost of data acquisition, or cost of analysis (Keller and Smith 2014). As a result, low-resolution data such as Landsat (30-m pixel size) often have been used in habitat analyses for smaller species such as passerines, which has sometimes resulted in misinterpretations of species-habitat relationships (Keller and Smith 2014).

Understanding life history traits and function of the focal taxon enables more accurate identification of important aspects of habitat composition and structure within an appropriate land use or cover type (successional stage) and the desirable size of an optimally (circularly) shaped patch or patches of such a cover type within the constraints of its regional availability and setting (insularity, connectivity, context). For the researcher, this suggests the resolution and range of scales (Jackson and Fahrig 2015) at which modeling should take place. For the manager, this knowledge facilitates habitat manipulation methods and prioritization of tax parcel acquisition or landowner partnering to improve management or restoration outcomes.

Questions to Ask

As a practical extension of the above discussion, I suggest an approach to applying ASCC principles as part of the evaluation and decision-making process when developing a habitat management program. Due to the limited size or relative isolation of parcels controlled by government agencies or nongovernmental organizations in many areas of the eastern and midwestern United States, successful management of a species or assemblage of interest may be challenging. In these cases the first step may be to examine adjacent tax parcels for their potential to supplement the existing management area. For purposes of habitat assessment and management, consider asking ten questions:

1. What successional age or combination of ages represents habitat for the focal taxon? More generally, what is the composition and structure of habitat for the focal taxon? For example, a coral reef has a particular composition and structure but is not always considered from the standpoint of ecological age; as noted above, succession may not follow predictable paths in all environments.

2. Based on the scale at which it uses the landscape, would you characterize the focal taxon as associated with solid or edge habitat? Or, if foraging areas are separate from nesting areas, perhaps it is associated with both? Use the taxon's ecological function to help answer this question. Then use this assessment to identify landscape components for manipulation.

3. What is the home range size of the focal taxon? What is its typical dispersal distance?

4. How large is the MDC or DEAC for habitat of the focal taxon within the existing management area?

5. If too small now, how large should the expanded area of habitat be to cause a change in occupation by the focal taxon?

6. What is the context of the area to be managed? Is there additional area available for realistic expansion?

7. What cover types are in adjacent areas?

8. Do the cover types include any similar to the one to be managed or created, or can they be manipulated to create that type (e.g., through timber harvest, burning, or planting)?

   a. If yes, how large is the adjacent similar area?

   b. Will adding it to the existing management area increase the MDC or DEAC for the focal taxon? If yes, can additional habitat manipulation be effected to further increase the MDC and DEAC of the habitat of interest?

   c. If no, are there connecting corridors, or can they be created between the existing management area and habitat within adjacent parcels, or across adjacent parcels to existing or potential habitat on noncontiguous parcels?

9. Are there opportunities to create stakeholder partnerships with adjacent or nearby landowners to facilitate meaningful habitat management across multiple parcels?

10. How long will the manipulation effect last? Is there a required frequency of manipulation to maintain the desired habitat? Will potential stakeholders commit to a long-term management program on their parcels if one is required?

Partnering to Build Suitably Sized Parcels

As a brief example of the application of this process, imagine that data exist to determine the average MDC required to define a patch of habitat that can sustain a small population of a species of management interest and that the MDC (Fig. 15.6) on a current management parcel falls a bit below that threshold. Perhaps the parcel containing that MDC contains the correct vegetation composition and structure to provide habitat for the species, but based on the current MDC and regional landscape context the parcel is only occasionally colonized and occupied for one or a few years. Enlarging this habitat enough to sustain regular occupancy would contribute to the maintenance of a regional (meta) population of the species. The use of MDC (or DEAC for edge species) allows a manager to quickly and quantitatively evaluate the potential gains in habitat that could be achieved by establishing partnerships with adjacent parcel owners and thus prioritize partnering efforts or even parcel acquisition.

For example, a nearly 60% increase in MDC area accrues when even one of the adjacent parcels is included in the management program (Fig. 15.12), despite the mere 25% increase in MDC width over that in Figure 15.6. Also note that it is not essential to manage entire parcels to increase the amount or quality of habitat, only those portions that help to increase the MDC of habitat for the focal taxon. This is true until the area under consideration becomes

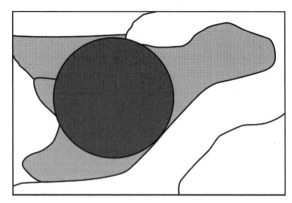

Figure 15.12. Use of an additional stakeholder parcel to create or add habitat to that already available on a currently managed property. Even a small addition, when well placed, can substantially increase the patch type's area. Maximum diameter circle (MDC), a measure of the functional size of the habitat, is a useful tool to assess the ecological value of this potential increase. Compare functional patch size here to that shown in Figure 15.6.

many multiples of home range size. At that point the area is large enough to support a metapopulation, and shape becomes less important (Simberloff 1986, Keller and Smith 2014).

A Case Study

Habitat management, enhancement, or restoration typically occurs as a result of one or more of three processes: regulatory compliance, profit motivation, or as an intended benefit to publicly owned resources (Keller 1991). Regardless of the motivation of such projects, it is useful to ask the ASCC questions described above to evaluate how proposed management or restoration alternatives can integrate with existing landscape to increase functional size or connectivity of species-specific habitat. The case study presented below provides an example of the application of ASCC principles and illustrates each type of project motivation (compliance, profit, public benefit). However, it represents only a fraction of the range of potential applications of ASCC principles and differs from habitat management often practiced by natural resource agencies or nongovernmental organizations in that it focuses on increasing biodiversity rather than creating or enhancing habitat for a particular species. Hence the emphasis is on plant community diversification rather than on creation of more taxon-specific habitat for game species or particular species of conservation interest.

Crown Tower: Timber Harvest Spatial Arrangement

A subdivision of more than 200 units was originally proposed on a parcel of more than 240 ha within a high-quality forested watershed in north central New Jersey. Among many forest interior species, the parcel was documented habitat for the state-listed barred owl (*Strix varia*). A consortium of local, state, federal, and nonprofit organizations eventually purchased the property for addition to an adjacent state park. As part of the sales agreement, however, the owner, who had obtained a state-approved forestry management plan, retained the right to selectively harvest more than one million board feet of timber before relinquishing the property. Such single-tree harvesting has recognized impacts on forest interior species, while creating single tree canopy openings that are too small to provide habitat for most early successional species associated with forest regeneration (Costello et al. 2000, Schlossberg and King 2007, Roberts and King 2017).

Final approval of the proposed harvest method, however, rested with the municipal planning board, which had the right to review the proposed forest management plan. The board used this opportunity to make it a condition of approval of the plan that a portion of the harvest be performed as two 6-ha seed tree cuts with scattered mature trees left standing and slash left evenly distributed across the cutover area to inhibit deer movements and access to seedlings, root suckers, and stump sprouts. This approach, which creates both larger contiguous areas of regenerating forest and greater structural diversity (Williamson 1970, Keller 1986, King and DeGraaf 2000, Schlossberg and King 2007), has been demonstrated to increase the richness of many early successional shrub-scrub species, including game species that use these stages such as ruffed grouse (*Bonasa umbellus*) and American woodcock (*Scolopax minor*; DeGraaf 1991, Costello et al. 2000, Keller et al. 2003, Schlossberg and King 2009).

As a result of this revised management plan, the harvest produced an early successional bird assemblage similar to that described by Keller et al. (2003). Concomitantly, it achieved the desired timber harvest and preserved a much larger portion of undisturbed core forest than would have obtained under the originally proposed management plan (J. K. Keller, Habitat by Design, unpublished data; Franklin et al. 2007). Thus all four landscape attributes (ASCC) were considered in designing an alternative harvest regime that both preserved habitat for forest interior species and increased overall species richness.

Summary

Wildlife management is based on an understanding of species-habitat relationships. To understand these relationships, we first need to observe the system correctly. Therefore, when developing species-habitat models to guide management, it is important at the outset that we accurately identify the resolution (grain) at which the species or assemblage of interest perceives and responds to landscape composition and structure.

Within the realm of landscape structure, it can be theoretically and managerially useful to group species as associated either with solid habitat or solid patch types such as closed-canopy forest or grasslands, or with edge types such as shrubs adjacent to grass or tree canopy adjacent to open air. In many geographic regions, such structure is a function of successional age, although this is less so in arid or more variable environments. Additionally important is our understanding of the scale (geographic extent) at which species respond to the landscape features with which they are associated. This response has been interpreted as a function of body size, which in turn influences territory size and dispersal capability.

Body size, territory size, and dispersal capability can be used to estimate the size of an area required to support reproductively successful individuals or sustainable metapopulations. Because the optimal shape of a territory is a circle, a simple measure of habitat size, the MDC, can be used to quantify the functional size of solid-patch types such as forest using remotely sensed imagery recorded at a taxon-appropriate resolution. A comparable measure of size for patches composed of edge types of interest, the DEAC, can be calculated for edge-related species. These measures combine patch size and shape (configuration) into a practical tool for evaluating existing habitat or identifying management strategies to improve or increase habitat area.

At larger geographic scales the landscape context in which the management parcel is located—measured as degree of patch isolation or connectivity, regional amount of available habitat, or intensity of anthropogenic presence—may influence species reproductive success and long-term viability. Incorporating questions of patch age, size, configuration, and context into habitat assessment and management, including identification of desirable partnering opportunities for expanding management areas, can improve decision making and provide a defensible process for wildlife managers.

LITERATURE CITED

Andren, H. 1994. Effects of habitat fragmentation on birds and mammals in landscapes with different proportions of suitable habitat: a review. Oikos 71:355–366.

Bakermans, M. H., and A. D. Rodewald. 2006. Scale-dependent habitat use of Acadian flycatcher (*Empidonax virescens*) in central Ohio. Auk 123:368–382.

Bollinger, E. K. 1995. Successional changes and habitat selection in hayfield bird communities. Auk 112:720–730.

Brittingham, M., and S. Temple. 1983. Have cowbirds caused forest songbirds to decline? BioScience 33:31–35.

Bump, G., R. W. Darrow, F. C. Edminster, W. F. Crissey, and F. Everett. 1947. The ruffed grouse: life history, propagation, management. New York Conservation Department, Albany, New York, USA.

Cale, P. G., and R. J. Hobbs. 1994. Landscape heterogeneity indices: problems of scale and applicability, with particular references to animal habitat description. Pacific Conservation Biology 1:183–193.

Center for Land Use Education and Research. 2019. Shape metric tool. http://clear.uconn.edu/tools/Shape_Metrics/method.htm. Accessed 2 Aug 2019.

Clements, F. E. 1905. Research methods in ecology. University Publishing, Lincoln, Nebraska, USA.

Collier, B. A., J. E. Groce, M. L. Morrison, J. C. Newnam, A. J. Campomizzi, S. L. Farrell, H. A. Mathewson, R. T. Snelgrove, R. J. Carroll, and R. N. Wilkins. 2012. Predicting patch occupancy in fragmented landscapes at the rangewide scale for an endangered species: an example of an American warbler. Diversity and Distributions 18:158–167.

Confer, J. L., and S. M. Pascoe. 2003. The avian community on utility rights-of-ways and other managed shrublands in northeastern United States. Forest Ecology and Management 85:193–206.

Conner, R. N., and C. S. Adkisson. 1975. Effects of clearcutting on the diversity of breeding birds. Journal of Forestry 73:781–785.

Cornell, K. L., and T. M. Donovan. 2010. Scale-dependent mechanisms of habitat selection for a migratory passerine: an experimental approach. Auk 127:899–908.

Costello, C. A., M. Yamasaki, P. J. Pekins, W. B. Leak, and C. D. Neefus. 2000. Songbird response to group selection harvests and clearcuts in a New Hampshire northern hardwood forest. Forest Ecology and Management 127:41–54.

Covich, A. P. 1976. Analyzing shapes of foraging areas: some ecological and economical theories. Annual Review of Ecology and Systematics 7:235–258.

Darveau, M., P. Beauchesne, L. Belanger, J. Huot, and P. LaRue. 1995. Riparian forest strips as habitat for breeding birds in boreal forest. Journal of Wildlife Management 59:67–78.

Davis, S. K. 2004. Area sensitivity in grassland passerines: effects of patch size, patch shape, and vegetation structure on bird abundance and occurrence in southern Saskatchewan. Auk 121:1130–1145.

DeGraaf, R. M. 1991. Breeding bird assemblages in managed northern hardwood forests in New England. Pages 153–171 in J. E. Rodiek and E. G. Bolen, editors. Wildlife and habitats in managed landscapes. Island Press, Washington, DC, USA.

Donovan, T. M., P. W. Jones, E. M. Annand, and F. R. Thompson III. 1997. Variation in local scale edge effects: mechanisms and landscape context. Ecology 78:2064–2075.

Driscoll, M. J. L., and T. M. Donovan. 2004. Landscape context moderates edge effects: nesting success of wood thrushes in central New York. Conservation Biology 18:1330–1338.

Dunford, W., and K. Freemark. 2004. Matrix matters: effects of surrounding land uses on forest birds near Ottawa, Canada. Landscape Ecology 20:497–511.

Fahrig, L. 1997. Relative effects of habitat loss and fragmentation on population extinction. Journal of Wildlife Management 61:603–610.

Fahrig, L. 1998. When does fragmentation of breeding habitat affect population survival? Ecological Modelling 105:273–292.

Fahrig, L. 2013. Rethinking patch size and isolation effects: the habitat amount hypothesis. Journal of Biogeography 40:1649–1663.

Farrell, S. L., B. A. Collier, K. L. Skow, A. M. Long, A. J. Campomizzi, M. L. Morrison, K. B. Hays, and R. N. Wilkins. 2013. Using LiDAR-derived vegetation metrics for high-resolution species distribution models for conservation planning. Ecosphere 4(3):42.

Fleming, K. K., K. A. Didier, B. R. Miranda, and W. F. Porter. 2004. Sensitivity of a white-tailed deer habitat-suitability index model to error in satellite land-cover data: implications for wildlife habitat-suitability studies. Wildlife Society Bulletin 32:158–168.

Forman, R. T., A. E. Galli, and C. Leck. 1976. Forest size and avian diversity in New Jersey woodlots with some land use implications. Oecologia 26:1–8.

Forman, R. T., and M. Godron. 1981. Patches and structural components for a landscape ecology. Bioscience 31:733–740.

Forman, R. T., and M. Godron. 1986. Landscape ecology. Wiley, New York, New York, USA.

Franklin, J., D. K. Simons, D. Beardsley, J. M. Rogan, and H. Gordan. 2000. Evaluating errors in a digital vegetation map with forest inventory data and accuracy assessment using fuzzy sets. Transactions in Geographic Information Systems 5:285–304.

Franklin, J. F., R. J. Mitchell, and B. J. Palik. 2007. Natural disturbance and stand development principles for ecological forestry. General Technical Report NRS-19, USDA Forest Service, Northern Research Station, Newtown Square, Pennsylvania, USA.

Fulbright, T. E. 1996. Viewpoint: a theoretical basis for planning brush management to maintain species diversity. Journal of Range Management 49:554–559.

Fulbright, T. E., J. A. Ortega-S., A. Rasmussen, and E. J. Redeker. 2008. Applying ecological theory to habitat management: the altering effects of climate. Pages 241–258 in T. E. Fulbright and D. G. Hewitt, editors. Wildlife science: linking ecological theory and management application. CRC Press, Boca Raton, Florida, USA.

Gagne, S. A., and L. Fahrig. 2007. Effect of landscape context on amphibian communities in breeding ponds. Landscape Ecology 22:205–215.

Gallant, A. L. 2009. What you should know about land-cover data. Journal of Wildlife Management 73:796–805.

Galli, A. E., C. Leck, and R. T. Forman. 1976. Avian distribution patterns in forest islands of New Jersey. Auk 93:356–364.

Gaston, A., C. Ciudad, M. C. Mateo-Sánchez, J. I. García-Vinas, C. López-Leiva, A. Fernández-Landa, M. Marchamalo, J. Cuevas, B. de la Fuente, M.-J. Fortin, et al. 2017. Species' habitat use inferred from environmental variables at multiple scales: how much we gain from high-resolution vegetation data? International Journal of Applied Earth Observation and Geoinformation 55:1–8.

Goetz, S. J., D. Steinberg, M. G. Betts, R. T. Holmes, P. J. Doran, R. Dubayah, and M. Hoften. 2010. Lidar remote sensing variables predict breeding habitat of a neotropical migrant bird. Ecology 91:1569–1582.

Hiebeler, D. 2000. Populations on fragmented landscapes with spatially structured heterogeneities: landscape generation and local dispersal. Ecology 81:1629–1641.

Hines, E. M., J. Franklin, and J. R. Stephenson. 2005. Estimating the effects of map error on habitat delineation for the California spotted owl in Southern California. Transactions in GIS 9:541–559.

Holland, G. J., and A. F. Bennett. 2009. Differing responses to landscape change: implications for small mammal assemblages in forest fragments. Biodiversity Conservation 18:2997–3016.

Holland, J. D., D. G. Bert, and L. Fahrig. 2004. Determining the spatial scale of species' response to habitat. Bioscience 54:227–233.

Holmes, R. T., P. P. Marra, and T. W. Sherry. 1996. Habitat-specific demography of breeding black-throated blue warblers (Dendroica caerulescens): implications for population dynamics. Journal of Animal Ecology 65:183–195.

Huggett, A. J. 2005. The concept and utility of "ecological thresholds" in biodiversity conservation. Biological Conservation 124:301–310.

Huston, M. A. 2002. Introductory essay: critical issues for improving predictions. Pages 7–21 in J. M. Scott, P. J. Heglund, and M. L. Morrison, editors. Predicting species occurrences: issues of accuracy and scale. Island Press, Washington, DC, USA.

Jackson, H. B., and L. Fahrig. 2015. Are ecologists conducting research at the optimal scale? Global Ecology and Biogeography 24:52–63.

Keller, C. M. E., C. S. Robbins, and J. S. Hatfield. 1993. Avian communities in riparian forests of different widths in Maryland and Delaware. Wetlands 13:137–144.

Keller, J. K. 1986. Predicting avian species richness by

assessing guild occupancy: the minimum critical patch hypothesis. Dissertation, Cornell University, Ithaca, New York, USA.

Keller, J. K. 1990. Using aerial photography to model species-habitat relationships: the importance of habitat size and shape. Pages 34–46 in R. S. Mitchell, C. J. Sheviak, and D. J. Leopold, editors. Ecosystem management: rare species and significant habitats. New York State Museum Bulletin 471, Albany, New York, USA.

Keller, J. K. 1991. Creation and restoration of wetlands in suburban landscapes. Pages 243–251 in D. J. Decker, M. E. Krasny, G. R. Goff, et al., editors. Challenges in the conservation of biological resources: a practitioner's guide. Westview Press, Boulder, Colorado, USA.

Keller, J. K., D. Heimbuch, and M. E. Richmond. 1980. Optimization of grid cell shape for the analysis of wildlife habitat. Pages 153–162 in F. Shahrokhi and T. Paludan, editors. Remote sensing of the earth resources. Volume 8. Earth Resources Conference, Tullahoma, Tennessee, USA.

Keller, J. K., M. E. Richmond, and C. R. Smith. 2003. An explanation of patterns of breeding bird species richness and density following clearcutting in northeastern USA forests. Journal of Forest Ecology and Management 174:541–564.

Keller, J. K., and C. R. Smith. 2014. Improving GIS-based wildlife–habitat analysis. Springer Briefs in Ecology. Springer, New York, New York, USA.

Kennedy, C. M., E. H. Campbell Grant, M. C. Neel, W. F. Fagan, and P. P. Marra. 2011. Landscape matrix mediates occupancy dynamics of neotropical avian insectivores. Ecological Applications 21:1837–1850.

King, D. I., R. B. Chandler, S. Schlossberg, and C. C. Chandler. 2009. Effects of width, edge and habitat on the abundance and nesting success of scrub-shrub birds in powerline corridors. Biological Conservation 142:2672–2680.

King, D. I., and R. M. DeGraaf. 2000. Bird species diversity and nesting success in mature, clearcut and shelterwood forest in Northern New Hampshire. Forest Ecology and Management 129:227–235.

King, D. I., R. M. DeGraaf, and C. R. Griffin. 2001. Productivity of early successional shrubland birds in clearcuts and groupcuts in an eastern deciduous forest. Journal of Wildlife Management 65:345–350.

King, D. I., and S. Schlossberg. 2014. Synthesis of conservation value of the early-successional stage in forests of eastern north America. Forest Ecology and Management 324:186–195.

Kirkland, G. L., Jr. 1977. Responses of small mammals to the clearcutting of northern Appalachian forests. Journal of Mammalogy 58:600–609.

Kotliar, N. B., and J. A. Wiens. 1990. Multiple scales of patchiness and patch structure: a hierarchical framework for the study of heterogeneity. Oikos 59:253–260.

Kubel, J. E., and R. H. Yahner. 2008. Quality of anthropogenic habitats for golden-winged warblers in central Pennsylvania. Wilson Journal of Ornithology 120:801–812.

Leopold, A. 1933. Game management. Charles Scribner's Sons, New York, New York, USA.

Lindenmayer, D. B., R. B. Cunningham, and J. Fischer. 2005. Vegetation cover thresholds and species responses. Biological Conservation 124:311–316.

Litwin, T. S., and C. R. Smith. 1992. Factors influencing the decline of neotropical migrants in a northeastern forest fragment: isolation, fragmentation, or mosaic effects? Pages 483–496 in J. M. Hagan and D. Johnston, editors. Ecology and conservation of neotropical migrant landbirds. Smithsonian Institution Press, Washington, DC, USA.

Looijen, R. C. 1998. Holism and reductionism in biology and ecology: the mutual dependence of higher and lower level research programmes. Dissertation, University of Groningen, Groningen, Netherlands.

MacArthur, R. H., and J. W. MacArthur. 1961. On bird species diversity. Ecology 42:594–598.

MacArthur, R. H., J. W. MacArthur, and J. Preer. 1962. On bird species diversity. II. Predictions of bird censuses from habitat measurements. American Naturalist 96:167–174.

MacArthur, R. H., and E. O. Wilson. 1967. The theory of island biogeography. Princeton University Press, Princeton, New Jersey, USA.

MacKenzie, D. I., J. D. Nichols, G. B. Lachman, S. Droege, J. A. Royle, and C. A. Langtimm. 2002. Estimating site occupancy rates when detection probabilities are less than one. Ecology 83:2248–2255.

MacKenzie, D. I., and J. A. Royle. 2005. Designing occupancy studies: general advice and allocating survey effort. Journal of Applied Ecology 42:1105–1114.

Martin, A. E., and L. Fahrig. 2012. Measuring and selecting scales of effect for landscape predictors in species-habitat models. Ecological Applications 22:2277–2292.

Martin, T. E. 1981. Limitation in small habitat islands: chance or competition. Auk 98:715–734.

McClintock, L., R. F. Whitcomb, and B. L. Whitcomb. 1977. Island biogeography and "habitat islands" of eastern forests. II. Evidence for the value of corridors and minimization of isolation in the preservation of biotic diversity. American Birds 31:6–16.

McDermid, G. J., R. J. Hall, G. A. Sanchez-Azofeifa, S. E. Franklin, G. B Stenhouse, T. Kobliuk, and E. F. LeDrew. 2009. Remote sensing and forest inventory for wildlife habitat assessment. Forest Ecology and Management 257:2262–2269.

McGarigal, K., and B. J. Marks. 1995. FRAGSTATS: spatial pattern analysis program for quantifying landscape structure. US Department of Agriculture Forest Service General Technical Report PNW-351, Portland, Oregon, USA.

Mitchell, L. R., C. R. Smith, and R. A. Malecki. 2000. Ecology of grassland breeding birds in the northeastern United States—a literature review with recommendations for management. New York Cooperative Fish and Wildlife Research Unit, Department of Natural Resources, Cornell University, Ithaca, New York, USA.

Mönkkönen, M., and P. Reunanen. 1999. On critical thresholds in landscape connectivity: a management perspective. Oikos 84:302–305.

Moorman, C. E., D. C. Guynn Jr., and J. C. Kilgo. 2002. Hooded warbler nesting success adjacent to group-selection and clearcut edges in a southeastern forest. Condor 104:366–377.

Morrison, M. L., B. G. Marcot, and R. W. Mannan 1998. Wildlife–habitat relationships: concepts and applications. Second edition. University of Wisconsin Press, Madison, Wisconsin, USA.

Neel, M. C., K. McGarigal, and S. A. Cushman. 2004. Behavior of class-level landscape metrics across gradients of class aggregation and area. Landscape Ecology 19:435–455.

Odum, E. P. 1971. Fundamentals of ecology. Third edition. Saunders, Philadelphia, Pennsylvania, USA.

Osman, R. W. 1977. The establishment and development of a marine epifaunal community. Ecological Monographs 47:37–63.

Pearman, P. P. 2002. The scale of community structure: habitat variation and avian guilds in tropical forest understory. Ecological Monographs 72:19–39.

Pearson, S. M., and R. H. Gardner. 1997. Neutral models: useful tools for understanding landscape patterns. Pages 215–230 in J. A. Bissonette, editor. Wildlife and landscape ecology. Springer-Verlag, New York, New York, USA.

Perkins, K. A., and P. B. Wood. 2014. Selection of forest canopy gaps by male cerulean warblers in West Virginia. Wilson Journal of Ornithology 126:288–297.

Pillsbury, F. C., and J. R. Miller. 2008. Habitat and landscape characteristics underlying anuran community structure along an urban-rural gradient. Ecological Applications 18:1107–1118.

Prevedello, J. A., and M. V. Vieira. 2010. Does the type of matrix matter? a quantitative review of the evidence. Biodiversity Conservation 19:1205–1223.

Rabenold, K. N. 1978. Foraging strategies, diversity, and seasonality in bird communities of Appalachian spruce-fir forests. Ecological Monographs 48:397–424.

Rehm, E. M., and G. A. Baldassarre. 2007. The influence of interspersion on marsh bird abundance in New York. Wilson Journal of Ornithology 119:648–654.

Risser, P. G. 1987. Landscape ecology: state of the art. Pages 1–14 in M. G. Turner, editor. Landscape heterogeneity and disturbance. Springer-Verlag, New York, New York, USA.

Roberts, H. P., and D. I. King. 2017. Area requirements and landscape-level factors influencing shrubland birds. Journal of Wildlife Management 81:1298–1307.

Rodewald, A. D., and R. H. Yahner. 2001a. Influence of landscape composition on avian community structure and associated mechanisms. Ecology 82:3493–3504.

Rodewald, A. D., and R. H. Yahner. 2001b. Avian nesting success in forested landscapes: influence of landscape composition, stand and nest-patch microhabitat, and biotic interactions. Auk 118:1018–1028.

Rosenberg, K. V., R. D. Ohmart, and B. W. Anderson. 1982. Community organization of riparian breeding birds: response to an annual resource peak. Auk 99:260–274.

Rudnicky, T. C., and M. L. Hunter Jr. 1993. Reversing the fragmentation perspective: effects of clearcut size on bird species richness in Maine. Ecological Applications 3:357–366.

Rusterholz, K. A., and R. W. Howe. 1979. Species-area relations of birds on small islands in a Minnesota lake. Evolution 33:468–477.

Sawatzky, M. E., A. E. Martin, and L. Fahrig. 2018. Landscape context is more important than wetland buffers for farmland amphibians. Agriculture, Ecosystems and Environment 269: 97–106.

Schlossberg, S. R., and D. I. King. 2007. Ecology and management of scrub-shrub birds in New England: a comprehensive review. Report submitted to Natural Resources Conservation Service, Resource Inventory and Assessment Division, Beltsville, Maryland, USA.

Schlossberg, S. R., and D. I. King. 2009. Postlogging succession and habitat usage of shrubland birds. Journal of Wildlife Management 73:226–231.

Schoener, T. W. 1968. Sizes of feeding territories among birds. Ecology 49:123–141.

Simberloff, D. S. 1986. Design of nature reserves. Pages 315–338 in M. B. Usher, editor. Wildlife conservation evaluation. Chapman and Hall, New York, New York, USA.

Smith, C. R., S. D. DeGloria, M. E. Richmond, S. K. Gregory, M. Laba, S. D. Smith, J. L. Braden, E. H. Fegraus, E. A. Hill, D. E. Ogureak, et al. 2001. The New York Gap Analysis Project final report. New York Cooperative Fish and Wildlife Research Unit, Cornell University, Ithaca, New York, USA.

Smith, T. M., and H. H. Shugart. 1987. Territory size variation in the ovenbird: the role of habitat structure. Ecology 68:695–704.

Stanton, B. F., and N. L. Bills. 1996. The return of agricultural lands to forest: changing land use in the twentieth century. College of Agriculture and Life Sciences Extension Bulletin 96-03, Cornell University, Ithaca, New York, USA.

Stenger, J. 1958. Food habits and available food to ovenbirds in relation to territory size. Auk 75:335–346.

Stoddard, M. A., and J. P. Hayes. 2005. The influence of forest management on headwater stream amphibians at multiple spatial scales. Ecological Applications 15:811–823.

Tattoni, C., F. Rizzolli, and P. Pedrini. 2012. Can LiDAR data improve bird habitat suitability models? Ecological Modelling 245:103–110.

Thogmartin, W. E., A. L. Gallant, T. J. Fox, M. G. Knutson, and M. J. Suarez. 2004. A cautionary tale regarding the use of the National Land Cover Dataset 1992. Wildlife Society Bulletin 32:970–978.

Thompson, F. R., T. M. Donovan, R. M. DeGraaf, J. Faaborg, and S. K. Robinson. 2002. A multi-scale perspective of the effects of forest fragmentation on birds in eastern forests. Pages 9–19 in T. L. George and D. S. Dobkin, editors. Effects of habitat fragmentation on birds in western landscapes: contrasts with paradigms from the eastern United States. Studies in Avian Biology 25. Allen Press, Lawrence, Kansas, USA.

Thompson, F. R., M. B. Robbins, and J. A. Fitzgerald. 2012. Landscape-level forest cover is a predictor of cerulean warbler abundance. Wilson Journal of Ornithology 124:721–727.

Titterington, R. W., H. S. Crawford, and B. N. Burgason. 1979. Songbird responses to commercial clear-cutting in Maine spruce-fir forests. Journal of Wildlife Management 43:602–609.

Trani (Griep), M. K. 2002. The influence of spatial scale on landscape pattern description and wildlife habitat assessment. Pages 141–155 in J. M. Scott, P. J. Heglund, and M. L. Morrison, editors. Predicting species occurrences: issues of accuracy and scale. Island Press, Washington, DC, USA.

Turner, M. G. 1989. Landscape ecology: the effect of pattern and process. Annual Review of Ecology and Systematics 20:171–197.

Van Horne, B. 2002. Approaches to habitat modeling: the tensions between pattern and process and between specificity and generality. Pages 63–72 in J. M. Scott, P. J. Heglund, and M. L. Morrison, editors. Predicting species occurrences: issues of accuracy and scale. Island Press, Washington, DC, USA.

Villard, M. A., and J. P. Metzger. 2014. Beyond the fragmentation debate: a conceptual model to predict when habitat configuration really matters. Journal of Applied Ecology 51:309–318.

Vogeler, J. C., A. T. Hudak, L. A. Vierling, and K. T. Vierling. 2013. Lidar-derived canopy architecture predicts brown creeper occupancy of two western coniferous forests. Condor 115:614–622.

Vogt, P., K. H. Riitters, C. Estreguil, J. Kozak, T. G. Wade, and J. D. Wickham. 2007. Mapping spatial patterns with morphological image processing. Landscape Ecology 22:171–177.

Wagner, H. H., and M. J. Fortin. 2005. Spatial analysis of landscapes: concepts and statistics. Ecology 86:1975–1987.

Whitcomb, R. F., C. S. Robbins, J. F. Lynch, B. L. Whitcomb, M. K. Klimiuewicz, and D. Bystrak. 1981. Effects of forest fragmentation on avifauna of the eastern deciduous forest. Pages 125–205 in R. L. Burgess and D. M. Sharpe, editors. Forest island dynamics in man-dominated landscapes. Springer-Verlag, New York, New York, USA.

Wiens, J. A. 1976. Population responses to patchy environments. Annual Review of Ecology and Systematics 7:81–120.

Wiens, J. A. 1989a. Spatial scaling in ecology. Functional Ecology 3:385–397.

Wiens, J. A. 1989b. The ecology of bird communities. Cambridge University Press, Aberdeen, United Kingdom.

Wiens, J. A., and B. T. Milne. 1989. Scaling of "landscapes" in landscape ecology, or, landscape ecology from a beetle's perspective. Landscape Ecology 3:87–96.

Williamson, K. 1970. Birds and modern forestry. Bird Study 17:167–176.

Winter, M., D. H. Johnson, J. A. Shaffer, T. M. Donovan, and W. D. Svedarsky. 2006. Patch size and landscape effects on density and nesting success of grassland birds. Journal of Wildlife Management 70:158–172.

With, K. A., and T. O. Crist. 1995. Critical thresholds in species' responses to landscape structure. Ecology 76:2446–2459.

Zimble, D. A., D. L. Evans, G. C. Carlson, R. C. Parker, S. C. Grado, and P. D. Gerard. 2003. Characterizing vertical forest structure using small-footprint airborne LiDAR. Remote Sensing of Environment 87:171–182.

16 — A Joint Venture Approach

Gregory J. Soulliere and
Mohammed A. Al-Saffar

Introduction

Bird habitat management is implemented at local scales, but effective conservation of migratory birds (e.g., habitat restoration and retention that results in sustaining target populations) involves an understanding and integration of population-level and landscape-scale priorities. In addition, because conservation requires political and financial support, social considerations are becoming a priority in landscape planning for birds. Bird habitat joint ventures (JVs) were established to implement the North American Waterfowl Management Plan (NAWMP), the first of several continental bird conservation strategies. Over the years, habitat JVs expanded to include planning for other bird taxa, and recently some have integrated human dimensions into landscape conservation design for birds. These facets of contemporary bird conservation are challenging JVs to be less opportunistic and more strategic with bird habitat decisions. Fortunately, the science and technical tools available to regional conservation planners have continued to advance. Like the NAWMP, joint venture regional plans are living documents, evolving with improved knowledge regarding bird populations, landscapes and spatial data, social and scale considerations, and key relationships influencing birds and people. In this chapter, we provide information regarding establishment and evolution of joint ventures and examples of how the Upper Mississippi / Great Lakes Joint Venture (UMGLJV) integrated principles of landscape ecology and other biological and social priorities into waterfowl habitat conservation planning.

History of Joint Ventures

The term *joint venture* originated in the private sector and was commonly used when referring to a temporary strategic alliance between business partners, typically a means to share risks and rewards through the partnership (Schermerhorn et al. 1991). For example, in the early 1980s, American automobile manufacturers began cooperating with foreign automakers to produce higher-quality cars for a lower price, enhancing participation and competition in the global marketplace. This collaboration—or sharing of information, technology, and parts—resulted in American-made autos with foreign components and foreign-made vehicles produced using American technologies. In addition to production, business joint ventures partnered on marketing initiatives, human resource management techniques, and many other aspects of our profit-based economy.

Recognizing potential value of strategic public-private partnerships, leaders within the US Fish and Wildlife Service (USFWS) established funding for administration of bird conservation JVs. The USFWS defined joint venture as "a self-directed partnership of agencies, organizations, corporations, tribes, or individuals that has formally accepted the responsibility of implementing national or international bird conservation plans within a specific geographic area or for a specific taxonomic group, and has received general acceptance in the bird conservation community for such responsibility" (US Fish and Wildlife Service 2005:1). Working collectively and independently, JV partners conduct activities in support of bird conservation goals cooperatively developed by the regional partnership. Five functional elements must be included in the operation of a JV to receive USFWS financial support: coordination; planning; project development and implementation; monitoring, evaluation, and applied research; and communications and outreach. The use of these elements to achieve established bird conservation goals is described in each JV's *implementation plan*, a document developed and approved by the partnership as their strategic road map to goal achievement.

The theme of JVs is one of relationships that build synergy, or a greater collective outcome than parties could realize individually. Partners with varied resources and expertise work together toward common goals, resulting in reduced overlap in effort and thus greater efficiency. The NAWMP was established in 1986 by governmental and nongovernmental partners with a common vision: to restore duck populations at a time when they were at historically low levels in much of North America. North American Waterfowl Management Plan (1986) developers recommended habitat joint ventures to coordinate plan implementation in key geographies, and by 1994 there were 10 JVs in the United States and 3 in Canada (North American Waterfowl Management Plan 1994). From this early effort grew a network of 22 regional habitat joint ventures (Fig. 16.1) currently blanketing most of the continent. Regional JV memberships are typically dominated by state and federal agencies involved with wildlife conservation and large nongovernmental conservation organizations (e.g., Ducks Unlimited, The Nature Conservancy, Pheasants Forever).

The end of a long dry period in the midcontinent prairie-pothole region (primary duck breeding area) helped continental waterfowl populations rebound in the mid-1990s. At the same time, NAWMP conservation accomplishments and political and financial support were expanding, largely because of regional JVs and their partnership networks. The NAWMP became recognized as a model for effective bird conservation. Subsequently, continental conservation plans were developed for additional bird taxa, including landbirds, shorebirds, and waterbirds (colonial nesting waterbirds and wading birds). The North American Bird Conservation Initiative (NABCI) also was established to help provide coordination between plans. Moreover, in the continued spirit of cooperation, the original NAWMP joint ventures expanded to all-bird joint ventures, embracing the goals of other bird conservation initiatives.

Like most wildlife conservation plans prior to 2000, JV implementation plans and their bird habitat objectives had a limited science foundation. Conservation objectives included protection, restoration, and enhancement of bird habitats, but quantities and locations of conservation activity were largely opportunistic. Habitat objectives were simplistic (e.g., increase habitat by 10%), lacked a landscape ecology context, and were waterfowl focused, which meant that conservation needs of other species (often species of higher conservation concern) were not well integrated in the conservation decision making.

Evolving Role in Landscape Conservation

By the early 2000s, NAWMP administrators and JV partner agencies realized the need for improved science to more effectively direct resources for waterfowl conservation. About this same time, many JVs

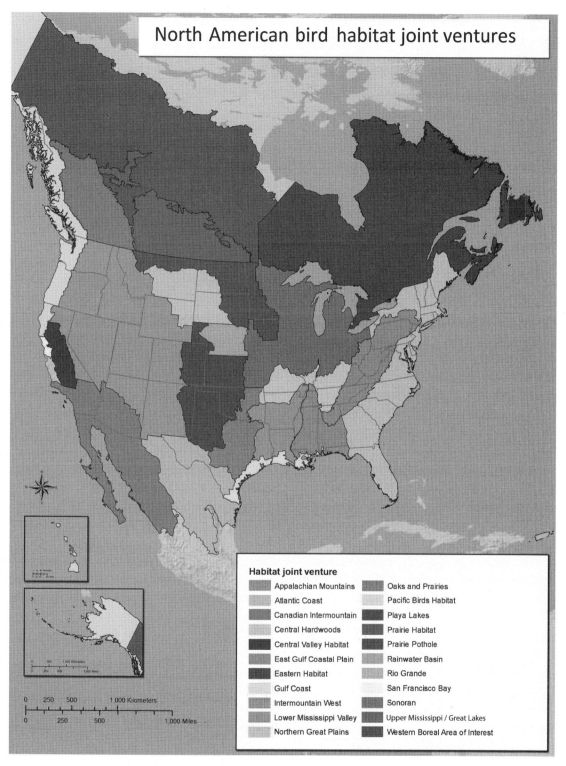

Figure 16.1. Boundaries of bird habitat joint ventures in North America. Joint ventures are self-directed partnerships of agencies, organizations, corporations, tribes, and individuals responsible for implementing national or international bird conservation plans within a specific geographic area.

were transitioning from a waterfowl focus to all-bird conservation. Bird habitat planning quickly became more complex, and the need to concentrate on principles of landscape ecology was increasingly evident. Joint venture management boards, typically composed of partner-organization administrators, supported hiring JV science coordinators, and they established technical committees to provide science-based guidance in bird habitat decisions. The paradigm shift was rapid, with newly established JV science teams improving understanding of ecological relationships and the use of more sophisticated planning tools to quantify habitat needs and target bird conservation. Joint ventures soon became leaders in developing decision support tools to strategically direct bird habitat work, with technical experts combining bird population distribution and abundance with digital cover-type data to establish spatial relationships between species and habitats. Predictive models and maps became the foundation for habitat restoration and retention decisions, and JV implementation plans increasingly included model-based habitat conservation recommendations directly linked to population objectives.

Through JVs, the bird conservation community had explicit population abundance objectives at multiple scales (e.g., continental, regional, state) and means to answer questions related to what, where, when, and how much habitat was needed to achieve these population objectives. Where scientists lacked biological monitoring data for model development, expert opinion was used. Information gaps and model assumptions were stated explicitly, and most JVs used a portion of their USFWS dedicated funds for evaluation, including the testing of planning assumptions and filling of information gaps. Like the NAWMP, joint venture implementation plans were iterative and regularly (5- to 10-year intervals) updated on the basis of new knowledge from research and monitoring by science partners.

Joint venture evolution from waterfowl to all-bird conservation was and remains a significant challenge, but progress has been steady and influenced mainly by the scientific community. Following the NAWMP, there was a surge in large-scale plans completed for other primary bird groups, including the North American Landbird Conservation Plan (Rich et al. 2004), the US Shorebird Conservation Plan (Brown et al. 2001), and the North American Waterbird Conservation Plan (Kushlan et al. 2002). In addition, NABCI was founded to help coordinate efforts among the four major bird plans. Early NABCI priorities included increasing the effectiveness of existing and new initiatives, fostering greater cooperation among the nations and peoples of the continent, and building on existing structures such as joint ventures, plus inspiring new JVs and supporting mechanisms as appropriate (North American Bird Conservation Initiative 2000a). Vision statements from each of the four continental bird-group plans were similar and collectively described by NABCI: "Healthy and abundant populations of North American birds are valued by future generations and sustained by habitats that benefit birds and people" (North American Bird Conservation Initiative 2017:2). The US NABCI Committee agreed to promote all-bird habitat conservation through existing and new JVs with a nationwide coverage, thus eliminating redundant partnership structures and separate biological planning processes for the various bird groups (North American Bird Conservation Initiative 2000a).

The 2004 NAWMP revision led to a conceptual shift in large-scale bird conservation, with planning, implementation, and evaluation viewed as integral components of management (North American Waterfowl Management Plan 2004). Accomplishing this shift required adopting principles of adaptive management, ideally practiced at multiple scales. Subsequently, the USFWS developed and promoted the adaptive management framework known as Strategic Habitat Conservation (SHC), a means to achieve greater long-term value from conservation investments (National Ecological Assessment Team 2006). This approach is also partner based, founded on the latest relevant science, and comprises an iterative planning cycle to assess costs and benefits of conservation techniques (return on investment). Understanding of species-habitat relationships improves through testing key planning assumptions and monitoring progress toward attaining population goals through habitat actions. The UMGLJV has embraced SHC as a conservation-planning model, but they expanded it to include biological and social objectives (Fig. 16.2) to better align with recent NAWMP revisions (North American Waterfowl Management Plan 2012, 2018). Below are descriptions and additional details of the SHC components most directly related to joint venture applications of landscape ecology. Also provided are specific examples regarding waterfowl habitat planning from the UMGLJV.

Biological and Social Planning

Planning establishes a foundation for effective bird habitat conservation by describing current conditions and trends, establishing species-habitat relationships, and identifying conservation goals. *Focal species* are selected as representatives of landcover and community types used by bird groups or guilds. They serve as a tool to focus biological planning to benefit multiple species using common habitats. Population objectives (e.g., abundance, relative density, duck use days) developed for focal species are then used to generate habitat objectives. Population response by focal species is the primary measure of progress toward biological objectives. Past JV measures of dollars (USA) spent and habitat area restored or enhanced do not adequately promote accountability for agencies and organizations with a primary charge of growing and sustaining wildlife populations. Clear descriptions of current and desired population and habitat conditions, use of effective management actions, and science-based monitoring to measure results are all needed to increase accountability and ultimately grow support for programs. Moreover, strategic placement of waterfowl habitat to maximize social values (e.g., hunting,

wildlife viewing, flood abatement, water filtration, groundwater recharge) theoretically also should result in increased political and financial support for conservation.

Focal Species and Habitat Associations

Joint ventures use focal species to simplify biological planning by reducing the number of models required to develop habitat objectives. By accommodating the habitat needs of focal species, JVs assume that habitat required by other species within a guild (occurring in the same community type) also can be provided. In a recently completed waterfowl habitat conservation strategy (Soulliere et al. 2017), the UMGLJV used five criteria for selecting breeding focal species. Criteria included relatively high abundance in the JV region; high importance of regional abundance to continental population size; representative of a complex of cover types that can be described by the National Wetland Inventory (NWI) and National Land Cover Database (NLCD) classification; factors limiting population growth are relatively well understood; and a system of population monitoring has been established. Nonbreeding focal species were also selected to represent species-habitat associations and monitor regional distribution and abundance. Population trends (e.g., relative abundance) for designated areas based on monitoring breeding and nonbreeding JV focal species are assumed to reflect habitat carrying capacity (related to quality and quantity) for the suite of species represented.

Focal species habitat objectives are typically expressed as quantity values, yet quality (i.e., habitat features resulting in relatively high recruitment and survival) is an equally important consideration. Habitat quality for wetland birds is often measured by site characteristics such as plant community diversity, water depth and interspersion, and food density, when demographic measures are not available. Predicting bird habitat quality and developing effective conservation prescriptions are complex,

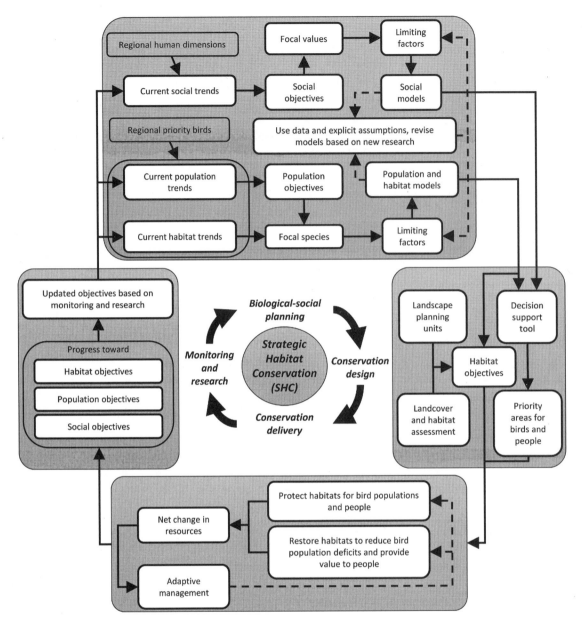

Figure 16.2. Planning model for bird habitat conservation by the Upper Mississippi / Great Lakes Joint Venture. Primary elements in this adaptive management framework include biological and social planning, conservation design, conservation delivery, and evaluation (monitoring and research). Feedback mechanisms are represented by dashed lines. Although the diagram suggests a looped process, multiple steps can occur at any point in time, and research to test planning assumptions and fill information gaps is continual.

as site-specific monitoring data must be combined with information regarding species-specific habitat requirements, landscape ecology, site land use history, and sound predictions of risk versus reward for various management treatments. Moreover, local-scale decision makers must consider site hydrology, soil types and nutrient richness, the dynamic nature of vegetation composition and structure with changing environmental conditions (e.g., precipitation in the near term, climate change in the long term), and other aspects influencing habitat quality. Ignoring these factors disregards landscape features critical to biological diversity and the potential carrying capacity of an area for birds.

Following focal species selection, the UMGLJV used the most recently updated spatial data available from the NWI (US Fish and Wildlife Service 2016), supplemented with NLCD (Homer et al. 2015), to broadly describe habitat associations required by waterfowl guilds during breeding and nonbreeding periods. Most species of waterfowl use areas with multiple wetland types (e.g., combinations of emergent, aquatic bed, unconsolidated or open water), and the juxtaposition of these wetland types, and specific upland cover, often determines habitat suitability. For spatial data analysis and habitat modeling, simple cover-type combinations comprising typical habitats for each JV focal species were identified. In addition, waterfowl and other waterbirds (e.g., wading and secretive marsh birds) have extensive overlap in habitat requirements (Soulliere et al. 2017, 2018). Thus waterbirds were included when developing species-habitat guilds to help ensure that conservation delivery complements rather than excludes species using common habitats. For instance, habitat generalists (e.g., mallard [*Anas platyrhynchos*]) typically benefit from management of species with more specialized habitat requirements (e.g., common gallinule [*Gallinula chloropus*], king rail [*Rallus elegans*]), but the opposite may not be true.

Several steps were required by the UMGLJV to establish habitat associations and divide species into planning guilds. First, primary wetland bird habitats were grouped into four NWI wetland classes (Table 16.1): emergent (including persistent and non-persistent herbaceous plants), forested (deciduous only), aquatic bed, and unconsolidated (including unconsolidated bottom and shore, which together represented open water communities). Spatial data at the NWI class level represented wetland area in terms of dominant vegetation and physical geography (Federal Geographic Data Committee 2013), which are important features of bird habitats

and useful for planning at a regional scale. Habitat associations for bird guilds were further refined by adding secondary attributes, which included both NWI wetland classes and NLCD landcover classes. The combination of wetland types and key upland features in proximity are essential in providing suitable habitats for many species. For example, blue-winged teal (*Spatula discors*) are associated with the NWI emergent wetland class during the breeding period, but habitat for this species must also include shallow aquatic bed or open water, plus surrounding areas of upland grassland and herbaceous cover for nesting. Similarly, cavity-nesting ducks are in the forested wetland guild, but they will use a variety of forested, emergent, and or scrub-shrub wetlands where adequate food and nest sites are available. For example, the wood duck (*Aix sponsa*) commonly uses emergent and aquatic bed wetlands when located near mature deciduous forest (wetland or upland forest). Both breeding waterfowl guilds within the emergent NWI class (Table 16.1) required at least some aquatic bed and or shallow unconsolidated (open water), and species within the aquatic bed guild typically use sites with an emergent wetland border.

POPULATION ASSESSMENT AND OBJECTIVES

Waterfowl comprise the bird group with the longest history of management at continental, national, and regional scales, and this group has some of the most comprehensive population survey data. In the UMGLJV region the Waterfowl Breeding Population and Habitat Survey and the North American Breeding Bird Survey were used to establish population abundance estimates. The UMGLJV used long-term average (LTA; 1991 to 2015) abundance estimates and 80th percentiles of LTA (80th%) to establish current breeding population estimates and population deficits (80th% − LTA = deficit). Whereas the 80th percentile of LTA abundance was used to establish a habitat carrying capacity target (population and habitat retention objectives), the deficit served as a basis for calculating habitat restoration

Table 16.1. Species-Habitat Associations for Wetland Bird Guilds Occurring in the Upper Mississippi / Great Lakes Joint Venture Region during Breeding and Nonbreeding Periods

Primary	Emergent		Forested	Aquatic Bed	Unconsolidated
Secondary	Aquatic bed or unconsolidated	Aquatic bed and grassland or herbaceous	Aquatic bed or emergent, or scrub-shrub and deciduous forest [a]	Emergent and unconsolidated	Aquatic bed or emergent, plus islands
Breeding waterfowl	**Mallard (Anas platyrhynchos)** Gadwall (Mareca strepera) Green-winged teal (Anas caroliniensis)	**Blue-winged teal (Spatula discors)** Northern shoveler (Spatula clypeata) Canada goose (Branta canadensis)	**Wood duck (Aix sponsa)** Common goldeneye (Bucephala clangula) Hooded merganser (Lophodytes cucullatus)	**Ring-necked duck (Aythya collaris)** American black duck (Anas rubripes) Redhead (Aythya americana) Trumpeter swan (Cygnus buccinator)	Common merganser (Mergus merganser) Red-breasted merganser (Mergus serrator)
Nonbreeding waterfowl	**Northern pintail (Anas acuta)** **Green-winged teal** Mallard Blue-winged teal Northern shoveler		**Wood duck** American black duck	Gadwall Canvasback (Aythya valisineria) American wigeon (Mareca americana) Redhead Ring-necked duck Ruddy duck (Oxyura jamaicensis) Snow goose (Anser caerulescens) Ross's goose (Anser rossii) Canada goose Trumpeter swan Tundra swan (Cygnus columbianus)	**Lesser scaup (Aythya affinis)** Greater scaup (Aythya marila) Surf scoter (Melanitta perspicillata) White-winged scoter (Melanitta fusca) Common scoter (Melanitta nigra) Long-tailed duck (Clangula hyemalis) Bufflehead (Bucephala albeola) Common goldeneye Hooded merganser Common merganser Red-breasted merganser
Breeding waterbirds	**American bittern (Botaurus lentiginosus)** Least bittern (Ixobrychus exilis) Common gallinule (Gallinula chloropus) American coot (Fulica americana)	**King rail (Rallus elegans)** **Sora (Porzana carolina)** **Yellow rail (Coturnicops noveboracensis)** Black rail (Laterallus jamaicensis) Virginia rail (Rallus limicola) Sandhill crane (Grus canadensis) Whooping crane (Grus americana)	**Black-crowned Night Heron (Nycticorax nycticorax)** Great blue heron (Ardea herodias) Great egret (Ardea alba) Snowy egret (Egretta thula) Little blue heron (Egretta caerulea) Cattle egret (Bubulcus ibis) Green heron (Butorides virescens) Yellow-crowned Night Heron (Nyctanassa violacea)	**Black tern (Chlidonias niger)** Pied-billed grebe (Podilymbus podiceps) Red-necked grebe (Podiceps grisegena) Forster's tern (Sterna forsteri)	**Common tern (Sterna hirundo)** **Common loon (Gavia immer)** Double-crested cormorant (Phalacrocorax auritus) American white pelican (Pelecanus erythrorhynchos) Ring-billed gull (Larus delawarensis) Herring gull (Larus argentatus) Great black-backed gull (Larus marinus) Caspian tern (Sterna caspia) Least tern (Sterna antillarum)

Table 16.1. continued

Primary	Emergent		Forested	Aquatic Bed	Unconsolidated
Nonbreeding waterbirds	**American bittern**	**Sora**	**Great blue heron**	**Pied-billed grebe**	**Common tern**
	Least bittern	**Sandhill crane**	Great egret	**American coot**	**Common loon**
		Cattle egret	Snowy egret	Red-necked grebe	Double-crested cormorant
		Yellow rail	Little blue heron	Common gallinule	
		Black rail	Green heron	Forster's tern	American white pelican
		King rail	Black-crowned Night Heron	Black tern	Ring-billed gull
		Virginia rail	Yellow-crowned Night Heron		Herring gull
					Great black-backed gull
					Caspian tern
					Least tern

Source: Adapted from Soulliere at al. (2017).

Note: Nonbreeding periods include migration and winter. Primary (National Wetland Inventory [NWI] wetland classes) and Secondary (NWI classes or National Land Cover Database [NLCD] upland cover classes) column headings reflect spatial data used in habitat modeling for each guild. Individual species regularly use multiple wetland types, and bird groupings were for general planning purposes. Names in boldface are joint venture (JV) focal species emphasized in planning. Multiple focal species were used for a single category to encompass larger geographic areas across the JV region and or foraging sub-guilds. Cover type categories were developed using NWI and NLCD classifications to enable conservation planning and landcover monitoring with spatial data available for a large geographic area.

[a]Species in forested wetland groups require upland or wetland deciduous forest for different purposes during breeding (e.g., waterbird rookeries, duck nest cavities) and nonbreeding (e.g., waterbird roosting) periods. Species in the forested wetland guild also readily use emergent, aquatic bed, and scrub-shrub wetlands for foraging so long as suitable deciduous forest is nearby for nesting and roosting.

objectives for each breeding focal species (Soulliere et al. 2017).

Joint ventures providing primarily migration or wintering areas during the annual life cycle of birds focus their conservation planning on the nonbreeding period. Various approaches are used to calculate nonbreeding waterfowl objectives (Petrie et al. 2011), but most common are the energy-based models consisting of three components: a regional population goal for each species (typically stepped down from the NAWMP), estimated energy demand per individual, and estimated energy supply per wetland unit area. Rather than focal species, habitat objective setting for the nonbreeding period requires estimating forage needs of species groups (nonbreeding guilds; Table 16.1), typically by means of wetland community types (e.g., aquatic bed, emergent marsh). Use day (or energy day) objectives are generated for each species on the basis of their estimated duration of stay in the JV region during migration staging and winter periods. Using the estimated food energy available in wetland types where waterfowl guilds most commonly occur, plus the estimated energy required (based on use days), habitat objectives to achieve carrying capacity targets are calculated.

MODEL-BASED DECISION TOOLS

Biological models provide a means for more effective conservation planning in the absence of complete knowledge. For example, JVs use models combining digital spatial data for landcover (potential habitats) and population survey data to predict the distribution of priority species across the region (Fig. 16.3). They also use models to target conservation and translate population objectives into habitat objectives. Behavior and habitat requirements, however, change with the seasons, and birds may use different areas for courtship, nesting, brood rearing, postbreeding molt, migration staging, and wintering. With wetland birds, availability of habitats can vary seasonally and among years depending on past and current wetland water budgets and weather severity. Large JV regions depend on readily available NLCD and NWI spatial data for modeling, but these data do not reflect the biological and environmental vari-

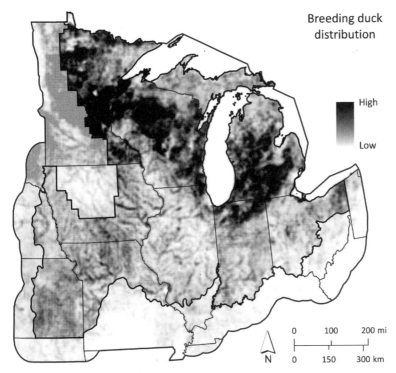

Breeding duck distribution

High

Low

Figure 16.3. Breeding duck distribution in the Upper Mississippi/Great Lakes Joint Venture region based on habitat suitability models developed from population survey data and spatial cover-type data (Soulliere et al. 2017). Population abundance data available for the northern half of the region were combined with landscape cover-type data characteristic of quality breeding habitats to predict occurrence in the whole region.

ation in natural systems. Thus JV decision models must be robust to account for temporal and spatial variation in species life requisites and the potential habitats where those requirements are fulfilled.

Conservation Design

Conservation design is a process (to design) and a product (a design) that helps achieve partner missions, mandates, and goals while ensuring sustainability of ecosystem services for current and future generations (Bartuszevige et al. 2016, Campellone et al. 2018). Conservation design for JVs is a means to quantify and broadly target regional objectives, while also providing a foundation for more refined *step-down* efforts developed for smaller geographic scales. The process involves combining geospatial data with biological and social information and models to create tools such as decision-support maps that prioritize landscapes to support JV objectives. Using spatial analysis, JVs assess landscape conditions and the characteristics that would need to change to

achieve specific desired outcomes. A primary step to complete this effort includes partner consensus around objectives and commitment to conservation implementation.

Landscape conservation design by JVs involves using the best available tools and information to efficiently target conservation and achieve objectives developed during biological and social planning. Understanding relationships between species populations and their habitats is essential to assess the ability of landscapes to support populations and determine the best strategies for attaining desired outcomes. Conservation design should result in a science-based, spatially explicit representation of the desired future landscape conditions needed to meet bird population requirements. Spatially explicit habitat objectives are based on contemporary understanding of species-habitat associations, factors limiting population growth, and characteristics of high-quality habitat (i.e., features resulting in enhanced breeding success and survival). The continually changing landscape of some regions owing to social, economic, and natural influences of landscape modification, however, add a challenging level of complexity to planning. Understanding ecological and social systems can help JV partners respond to system change and retain the most important bird habitats in the future.

Landscape Planning Units

One enduring accomplishment of the NABCI was establishment of Bird Conservation Regions (BCRs) as ecologically distinct planning units (Fig. 16.4). These regional landscapes have generally similar bird communities, habitats, and resource management issues (North American Bird Conservation Initiative 2000b). Bird Conservation Regions are scale-flexible, nested ecological units delineated in 1998 by the North American Commission for Environmental Cooperation (North American Bird Conservation Initiative 2000b). The original Commission for Environmental Cooperation framework comprised a hierarchy of three levels of ecoregions. At each step down in level, spatial resolution increases and ecoregions encompass areas that are progressively more similar in their biotic (e.g., flora, fauna) and abiotic (e.g., soils, drainage patterns, temperature, annual precipitation) characteristics. Joint ventures have embraced NABCI-sanctioned BCRs for landscape planning, and the boundaries of many JVs have gradually been adjusted from politically based to these ecologically based (BCR) boundaries. The use of BCRs and associated ecological information has also allowed partners to communicate with more universal terminology related to landscape ecology and planning. For example, JV technical teams, management boards, land managers, and administrative leaders (e.g., NAWMP Committee) now commonly speak in terms of BCRs when describing regional ecological units related to bird planning and habitat conservation.

Bird Conservation Regions may be partitioned into smaller units for finer-scale conservation plan-

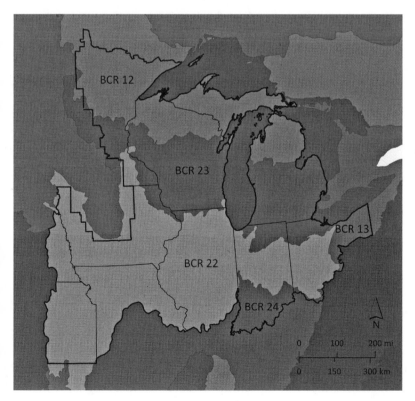

Figure 16.4. Boundaries of the Upper Mississippi/Great Lakes Joint Venture region and associated Bird Conservation Regions (BCRs) used for bird conservation planning: BCR 22 (Eastern Tallgrass Prairie), 23 (Prairie Hardwood Transition), 12 (Boreal Hardwood Transition), 24 (Central Hardwoods), and 13 (Lower Great Lakes and St. Lawrence Plain). Many joint ventures have adjusted their operational boundaries to more closely align with BCRs, whereas some continue to have governmental alignment or a mix of political and ecological boundaries.

ning or aggregated to facilitate larger conservation partnerships across a species range, recognizing that migratory birds normally use multiple BCRs throughout their annual life cycle. Bird Conservation Regions also facilitate domestic and international cooperation in bird conservation because they traverse state, provincial, and national borders (North American Bird Conservation Initiative 2000b). In summary, BCRs help the JV community facilitate communication among the multiple bird taxa initiatives, systematically and scientifically apportion North America into conservation planning units, facilitate a regional approach to bird conservation, promote new and expanded partnerships, and identify overlapping or conflicting conservation priorities (Commission for Environmental Cooperation 1998).

Landscape and Habitat Assessment

Following population assessment for focal species and objective setting, JVs evaluate their habitat base and trends important to conservation design. Some JVs have a wetland-bird focus, and others mainly emphasize landbird conservation. Landscape composition and associated bird habitats largely determine the role each JV partnership will play in the various bird-taxa conservation plans. Likewise, some regions are critical to priority species during the breeding period, whereas other locations serve to provide important migration and wintering habitats. Large and diverse regions like the area encompassed by the UMGLJV provide multiple bird taxa with critical habitats during multiple life cycle periods. Landcover composition, combined with soil characteristics and water chemistry, results in functional differences across BCRs within large JV regions. Understanding these functional differences among landscapes and trends in key cover types is important for effective bird conservation decisions.

The critical nature of understanding landscape ecology became apparent in a recent waterfowl conservation planning effort by the UMGLJV (Soulliere et al. 2017). Beginning with the 2011 NLCD, the JV

assessed current composition of primary land cover types and distribution of potential habitats important to wetland birds across the planning area (Fig. 16.5). Information from the NLCD was then combined with NWI data and US census data to quantify relevant characteristics of the individual BCRs (Table 16.2).

The assessment revealed relatively intact landscapes of natural cover types with abundant forest and wetland coverage across the northern portion of the UMGLJV region, especially BCR 12 (Figs. 16.4, 16.5). This subregion, with its generally nutrient-poor soils and short growing season, had limited cropland coverage, relatively few people, and lower breeding and nonbreeding duck densities per wetland area. Thus BCR 12 was considered lower priority for future conservation delivery. In contrast, the southern half of the UMGLJV region (BCR 22) had experienced extensive conversion to cropland and related wetland loss (cropland on hydric soils), had high duck densities per wetland area, and high human population, with much of the developed land concentrated in large urban centers. This portion of the UMGLJV region was considered a high priority target area with abundant opportunity for wetland-bird habitat restoration.

The landscape evaluation also included assessment of recent landcover trends using 2001 and 2011 NLCD. Although area size of various landscape cover types does not translate into quantities of bird habitats, an inventory of key landcovers important to birds provides information useful for planning at the regional scale. Evaluation of landcover amounts and trajectories also provides a basis to reconsider the feasibility of achieving bird population objectives established in earlier JV planning efforts. Knowledge of whether JV partners are gaining or losing priority bird habitats and where within their jurisdictions greatest change has occurred provides helpful guidance for iterative conservation planning. The most ecologically disturbing trend in the UMGLJV region was dramatic expansion of developed land (i.e., areas with constructed materials and 20% to 100% im-

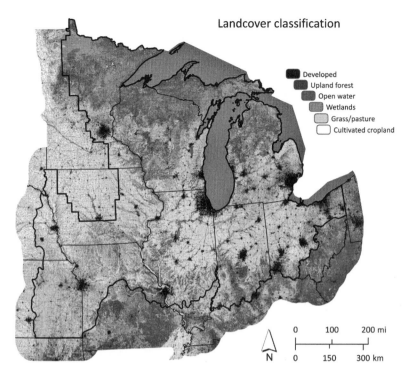

Landcover classification

Developed
Upland forest
Open water
Wetlands
Grass/pasture
Cultivated cropland

Figure 16.5. Landscape composition by primary cover types within the Upper Mississippi / Great Lakes Joint Venture boundary (heavy black line) and surrounding areas. The wetlands category includes emergent herbaceous and woody wetlands combined. Adapted from Soulliere et al. (2017); based on 2011 data from the National Land Cover Database.

0   100   200 mi

0   150   300 km

N

pervious surfaces; Table 16.3), a cover type considered unfriendly to priority bird species. Not surprising, when human population trends were compared across BCRs, population growth was highest where developed land coverage was also greatest and increasing (Table 16.2, Fig. 16.5).

In addition to anthropogenic landscape change, most natural plant communities are also in an endless state of fluctuation. For example, Great Lakes coastal wetlands undergo substantial change related to water levels. Seasonal and longer-term fluctuating water levels within Great Lakes emergent wetlands result in dynamic bird habitats. Water levels influence plant communities (Albert 2003) through lateral displacement (lakeward and landward shifts in wetland cover types) and horizontal zonation (varied plant species composition, stem densities, and height of adjacent stands). Following a long period of relatively high water from the 1970s through the 1990s, Great Lakes water levels were below average from 2000 to 2014, resulting in expansion of the coastal emergent marshes lakeward. Water levels

increased from 2013 to 2019, inundating reestablished emergent-plant areas and increasing habitat value for dabbling ducks and secretive marsh birds (Table 16.1) where these areas retained high native plant diversity (Monfils et al. 2014, 2015). Unfortunately, large areas of the transition zone (i.e., *temporarily flooded* wetland regime in NWI) exposed during the low-water period were colonized by invasive plants in portions of the UMGLJV region. An extended high-water period eventually suppresses emergent plant colonies, resulting in dabbling duck habitat transitioning back to diving duck habitat (Table 16.1).

CONSERVATION ESTATE

Conservation lands are areas in public ownership, land trust and conservancy lands, or private lands in long-term conservation easement, and the bird habitat they comprise is generally considered protected from development. Several sources of spatial data are available to help JVs identify conservation lands: Protected Areas Database, Conservation and Recreation

Table 16.2. Landscape and Social Characteristics Important to Wetland Bird Conservation Planning in Bird Conservation Regions Located in the Upper Mississippi / Great Lakes Joint Venture Region

Landscape Characteristics	BCR 22[a]	BCR 23	BCR (JV Region Only) 12	BCR (JV Region Only) 13	BCR (JV Region Only) 24	Total
Total area (ha)	51,762,267	25,827,603	20,583,051	2,174,150	3,547,207	103,894,278
Primary cover types[a]						
Herbaceous wetland (total)	433,352	1,188,564	656,863	13,534	20,164	2,312,477
Emergent (interior/inland)	383,596	1,115,819	634,493	12,704	18,631	2,165,243
Great Lakes coastal emergent[b]	248	18,497	12,643	48	0	31,437
Aquatic bed (submergent plant dominated)	49,508	54,248	9,727	782	1,533	115,798
Scrub/shrub wetland (woody deciduous <6 m tall)	52,788	474,069	1,051,908	22,493	7,073	1,608,330
Forested wetland (woody deciduous ≥6 m tall)	734,890	1,303,592	1,173,330	68,689	113,219	3,393,721
Total wetland (herbaceous and woody)	1,221,030	2,966,225	2,882,101	104,716	140,456	7,314,529
Upland grassland/herbaceous, hay, and pasture	12,242,050	3,748,567	1,259,628	333,491	512,927	18,096,663
Grassland/herbaceous (only)	4,064,150	520,347	573,110	45,298	77,936	5,280,841
Upland forest (all types)	6,077,593	5,219,278	10,095,671	668,725	1,671,927	23,733,194
Deciduous and mixed upland forest (only)	5,955,359	4,741,484	7,112,091	655,872	1,616,179	20,080,985
Scrub/shrub upland (all types)	79,087	180,833	1,478,420	474	4,466	1,743,280
Cultivated cropland	26,668,500	8,931,760	627,028	575,551	1,206,240	38,009,079
Open water (inland and coastal)[c]	1,118,258	1,366,554	2,321,634	96,670	143,441	5,046,557
Inland (lakes and rivers)	1,022,319	977,348	1,409,770	49,834	141,908	3,601,179
Coastal (≤1 km from Great Lakes shore)	46,431	334,957	902,138	46,055	0	1,329,581
0.5–5 m deep	19,819	109,949	337,024	20,601	0	487,392
≤0.5 m deep	1,869	39,431	29,393	2,006	0	72,699
Related landscape measures						
Hydric soils[d]	963,104	587,075	27,191	296,863	32,532	1,906,766
Prospective wetland (wet cropland)	783,065	276,383	8,342	19,252	24,310	1,111,352
Number of inland lakes (≥0.5 ha)[e]	22,689	20,657	23,552	1,703	2,728	71,329
Inland lake coverage (ha)[e]	170,200	441,109	497,900	25,476	16,313	1,150,998
Perennial river length (km)[e]	141,639	72,201	63,191	9,042	13,052	299,125
Social measures						
Number of people (residents), 2010[f]	31,743,779	20,560,074	1,875,258	4,584,002	1,482,640	60,245,753
Human density (people per hectare)[f]	0.613	0.796	0.091	2.108	0.418	0.585
Number of people (residents), 2000[f]	28,937,401	19,262,360	1,851,778	4,655,292	1,404,258	56,111,089
Population growth (%; 2000 to 2010)[f]	9.7	6.7	1.3	−1.5	5.6	7.4

Note: BCR, Bird Conservation Region; JV, joint venture. Area estimates are in hectares. Estimates for BCRs 22 and 23 encompass the entire BCR, including portions outside the joint venture boundary (i.e., 8% of BCR 22, 9% of BCR 23). Estimates for BCRs 12, 13, and 24 apply only to those areas within the joint venture boundary.

[a]Area of wetland and open-water cover types based on most recent National Wetland Inventory; other cover type measures are from the 2011 National Land Cover Database. Spatial data in metric units (1 ha = 2.5 acres).

[b]Great Lakes coastal emergent includes wetland ≤1 km from the Great Lakes coastline (generally, the recognized high-water mark); this coastal wetland area is dynamic, with emergent plant coverage changing in response to Great Lakes water levels.

[c]Open water includes all inland lakes and rivers with unconsolidated bottom, plus open aquatic bed wetlands, plus Great Lakes coastal waters (open coastal waters ≤1 km from shore); bathymetric data used to estimate water depth.

[d]Area with soils somewhat poorly drained and poorly drained to very poorly drained based on the Natural Resource Conservation Service Soil Survey Geographic Database. Prospective wetland was the intersection of hydric soils and cultivated cropland.

[e]Number and area (hectares) of inland lakes (includes ponds, reservoirs) and river length (kilometers) calculated using National Hydrologic Data Plus, Version 2.

[f]Number of residents, human population density, and population growth based on data from US Census Bureau (2010).

Table 16.3. Total Area and Recent Net Change in Primary Cover Types for Bird Conservation Regions Located within the Upper Mississippi/Great Lakes Joint Venture Region Based on Comparison of 2001 and 2011 National Landcover Data

	BCR		BCR (JV Region Only)			
	22	23	12	13	24	Total
Total area (ha)	51,762,267	25,827,603	20,583,051	2,174,150	3,547,207	103,894,278
Open water	+31,182	+27,434	−135	−246	+2,460	+60,696
Developed	+189,510	+152,779	+39,654	+15,827	+17,174	+414,944
Upland forest	−56,668	−71,646	−351,844[a]	−15,451	−12,008	−507,617
Shrub/scrub	−2,514	+10,096	+239,548[a]	+2,575	+377	+250,082
Grassland/herbaceous and hay/pasture	−131,752	−31,046	+70,928	+280	−4,711	−96,300
Grassland/herbaceous (only)	−33,345	+2,204	+71,243[a]	+1,398	+1,830	+43,331
Cultivated cropland	−85,740	−93,776	+2,570	−5,052	−6,414	−188,412
Wetland	+48,007	−7,485	−8,631	+155	+1,156	+33,201
Herbaceous wetland (only)	+51,147	+3,011	+58,452	+1,518	+719	+114,848
Woody wetland (only)	−3,140	−10,497	−67,084	−1,363	+437	−81,647

Note: BCR, Bird Conservation Region; JV, joint venture. Area measured in hectares; 1 ha = 2.5 acres.

[a]Apparent loss of upland forest and gain in shrub/scrub and grassland/herbaceous in BCR 12 reflects an active timber industry in this heavily forested landscape, with successional setback due to logging rather than permanent cover type change (Soulliere at al. 2017).

Lands Database, and National Conservation Easement Database. Having an image of the current conservation estate reveals the abundance and configuration of protected lands, an important complement to functional differences of landscapes when comparing management options across the JV region.

Armed with this information, partners can compare current ownership patterns with priority-bird conservation areas and develop a strategy for land acquisition and conservation easements. For example, parcels adjacent to existing smaller conservation lands (e.g., <5,000 ha) may be weighted for higher protection priority to expand the distribution of habitat complexes depending on focal species habitat requirements and related JV priorities, including social objectives (e.g., hunting, bird watching). Conversely, large (e.g., >10,000 ha) complexes of public land may be considered adequate to meet area bird habitat and social needs, allowing refocus of acquisition resources to other strategic locations. Bird habitats, particularly coastal areas proximate to people (areas ≤50 km from population centers), are considered to have the highest likelihood of public use by potential wetland bird supporters (Devers et al. 2017).

When an image of conservation estate was developed for the UMGLJV region (Fig. 16.6), the partnership was informed that most protected land was in the north part of the region (BCR 12), an area of lower relative importance to waterfowl and distant from human population centers. There were concentrations of protected land found in central and southern areas of the region, but much of this property was private land under perpetual and 30-year conservation easement through the Wetlands Reserve Program. Whereas private land is extremely important to birds, its social capital, such as public use for recreation, is typically limited. The distribution of Conservation Reserve Program lands was not determined by the JV during assessment of conservation estate, as these easement contracts are temporary (typically 10–15 years), and Conservation Reserve Program lands often revert back to agriculture (Morefield et al. 2016).

Targeting Conservation for Birds and People

The future of bird habitat conservation depends on political and financial societal support. This point

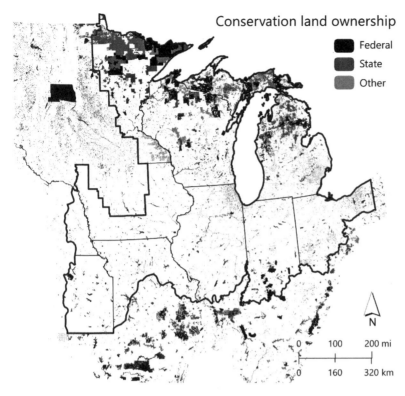

Conservation land ownership

■ Federal
■ State
■ Other

Figure 16.6. Location of federal, state, or other conservation lands in the Upper Mississippi/Great Lakes Joint Venture region. The other ownership category includes private land with perpetual and long-term conservation easements (e.g., Wetlands Reserve Program), conservancy land, and county, township, and city-owned land. Some large blocks encompass areas of acquisition interest rather than full ownership.

0   100   200 mi
0   160   320 km
N

was expressly emphasized in the latest versions of the North American Waterfowl Management Plan (2012, 2018), where all plan goals included a social component. A primary concern identified by NAWMP authors, and a growing concern among others in the bird conservation community, is the declining public understanding and support for nature, conservation, and other environmental interests. The NAWMP identifies a need for the conservation community to become more relevant to society or risk losing program support. Joint ventures are beginning to address this issue, and some have collaborated with Landscape Conservation Cooperatives, NABCI, and university partners to develop plans with biological and social objectives (Dayer et al. 2019).

Identifying conservation actions that balance mixed objectives for bird populations (biological) with the needs of current and potential conservation supporters (social) represents a key future challenge

to JVs. The task, however, may be simplified if we view bird habitat (i.e., quantity, quality, location) as the primary means to also achieve social objectives. For example, the 2012 NAWMP includes goals for "abundant and resilient waterfowl populations" and "growing numbers of waterfowl hunters, other conservationists, and citizens who enjoy and support conservation" (North American Waterfowl Management Plan 2012:2). Value models and decision support tools (DSTs) can be used to target habitat conservation to benefit birds and people. These tools, however, require reliable, scientific information, and recognition that social objectives and information required will vary depending on location (Krainyk et al. 2019). The UMGLJV recently developed an initial DST to help achieve NAWMP goals at the regional scale (Soulliere and Al-Saffar 2017). The process included development of a decision matrix with relevant objectives, spatially explicit model-based maps related to each objective, weighting of objective parameters, and amalgamation of spatial data to produce a single output map (the mixed-model DST).

The UMGLJV intention was to develop a DST process that was well documented, understandable, repeatable, and easily adjusted in response to new knowledge (Soulliere and Al-Saffar 2017). First, a decision matrix was built with components specific to the region (Table 16.4) and envisioned to help target conservation resources in a more transparent way. In addition to waterfowl population demography, the prioritization system accounted for waterfowl habitat features related to social values, thus addressing NAWMP goals. The process also allows future addition or deletion of alternate objectives and related spatial data, depending on the decision context. Conservation issues, objectives, and measurable attributes were identified and weighted by perceived importance (Table 16.4). Weights represented the relative value that decision makers placed on different objectives; weights used in the analysis were obtained from an UMGLJV management board exercise (Soulliere and Al-Saffar 2017).

The UMGLJV spatial analysis included converting numerical data into digital spatial data useful for modeling and map creation. A suite of geospatial tools and a kernel density approach was used to develop six spatial models, one for each of six relevant objectives (Table 16.4). Model development for some objectives was relatively complicated, depending on the type and size of available data and the number and complexity of required geospatial tools. Results were depicted in six different maps, and then spatial data were combined with the negotiated objective weights (Table 16.4) applied to each map. The aggregated output map (Fig. 16.7) was included in the 2017 UMGLJV waterfowl habitat strategy (Soulliere et al. 2017) to target conservation actions for waterfowl and people. Actual amounts and types of habitat delivery (e.g., restoration of emergent wetland and associated grassland) required to meet population objectives were estimated at JV regional and subregional scales (state × BCR polygons), and the DST

will guide conservation placement within these subregions (Soulliere et al. 2017). Similar spatial data and analyses are available to other JVs (Krainyk et al. 2019) that are interested in this landscape prioritization approach, and additional decision analysis may be employed at smaller scales to evaluate trade-offs among sites, habitat approaches, and return on investment (Wainger and Mazzotta 2011, Richardson et al. 2015).

Summary

Bird habitat management is implemented at local scales, but effective conservation of migratory birds—habitat restoration, and retention that results in sustaining target populations—requires an understanding and integration of population-level and landscape-scale priorities. Moreover, social considerations have become increasingly accepted as the key to growing support for bird conservation and an integral part of joint venture implementation plan development. For example, bird habitat restoration in locations providing society ecological goods and services (e.g., water filtration, floodwater storage, open space for outdoor recreation) is relevant beyond traditional stakeholders, potentially growing the base of future conservation supporters. These aspects of contemporary bird habitat planning are challenging JVs, and the need to be strategic when dedicating limited resources has never been greater. Fortunately, the science and technical tools available to planners continue to advance. The scientific process for assessing, planning, implementing, and evaluating is logically organized in SHC, a planning and implementation framework embraced by JVs.

Joint venture implementation plans typically step down population goals and conservation priorities from continental plans to JV regions and smaller scales, with BCRs providing an ecological planning framework for the bird conservation community. Joint venture technical committees provide scientific guidance to JV management boards, which oversee habitat implementation. Science guidance

Table 16.4. Upper Mississippi/Great Lakes Joint Venture Conservation Planning Issues, Objectives, Objective Weights, Spatial Data, and Measurable Attributes Used to Prioritize Landscapes for Waterfowl Habitat Retention and Restoration at the Joint Venture Regional Scale

Conservation Issue (NAWMP/JV Goal)	Objective	Weight	Spatial Data (Model-Based Maps)	Measurable Attribute
Populations and habitats				
Breeding habitat limitation (improve BPOP trend with 80th% LTA target).	Maximize focal species recruitment through conservation of high-quality breeding habitats.	0.3	Density and distribution of key breeding duck habitats (focal species combined and weighted for BPOP composition).	Breeding populations (BPOP survey data) and harvest data (fall age ratios) for demographic trends.
Nonbreeding habitat limitation (retain or increase BPOPs, considering full life cycle habitat needs).	Maximize focal species survival and body condition with habitat focus on cross-seasonal effect and spring fitness.	0.3	Duck harvest relative to wetland coverage, reflecting locations with wetland area limitations (data for county-level harvest, normalized by county wetland percentage coverage).[a]	Science-based estimate of nonbreeding habitat carrying capacity, body condition analysis, and tracking/marking data to determine winter and spring habitat use and survival trends.
Conservation supporters and social values				
Resource users: waterfowl hunters.	Maximize hunter retention and recruitment (NAWMP/JV target).	0.1	Harvest distribution reflects hunting community (county-level duck and goose harvest or hunter days, normalized by county size).	Active hunters and trends, hunter days and harvest (USFWS data), and/or model-based prediction of new user days (net) and trends.
Resource users: birders and other outdoor recreation.	Maximize waterfowl viewer/recreationist retention and recruitment (JV target).	0.1	Human population distribution and distance relative to potential habitat areas.	Active birders (USFWS data) and/or model-based prediction of new outdoor recreation days (net) and trends.
Gulf Hypoxia, water quality, and flood abatement associated with Mississippi River and its major tributaries.	Minimize nutrient and sediment runoff detrimental to system ecology (i.e., Gulf Hypoxia focus) and reduce flood damage.	0.1	Mississippi River subbasins (8-digit HU) with impairment due to altered hydrology and nutrified runoff (cultivated cropland and development coverage).	Water quality/nutrient monitoring at key river stations and flood insurance claims (number, cost, area).
Great Lakes coastal wetland function and lake/tributary water quality.	Maximize health, function, and biological diversity of Great Lakes coastal zones (coastal wetland and lake focus).	0.1	Coastal subbasins (8-digit HU) with impairment due to nutrient/sediment runoff, altered hydrology, and wetland loss (cultivated cropland and development coverage).	Great Lakes Restoration Initiative coastal wetland integrity measures and nearshore measures of water quality and biodiversity.
Weighting total		1.0		

Note: BPOP, breeding population; HU, hydrologic unit; JV, joint venture; LTA, long-term average; NAWMP, North American Waterfowl Management Plan; USFWS, US Fish and Wildlife Service. Objectives may be adjusted (added or deleted) and weighted differently depending on stakeholder input and scale of analysis (e.g., JV region, Bird Conservation Region, state, local). Once agreed upon, weights were applied to attribute spatial data to generate decision support maps used to focus resources for more effective JV and NAWMP goal achievement. Measurable attributes served as monitoring components for strategic habitat conservation. Weighted spatial data were combined to identify priority landscapes for conservation; this information can be filtered with hydric soils and cover type data (i.e., STATSGO, poorly drained; NLCD, cropland) to target habitat restoration in areas with historic wetland coverage.

[a]An alternative spatial analysis could compare estimated energy supply in available wetlands to energy demand based on estimated duck-use days (e.g., energy demand − energy supply = energy/habitat deficit at target areas).

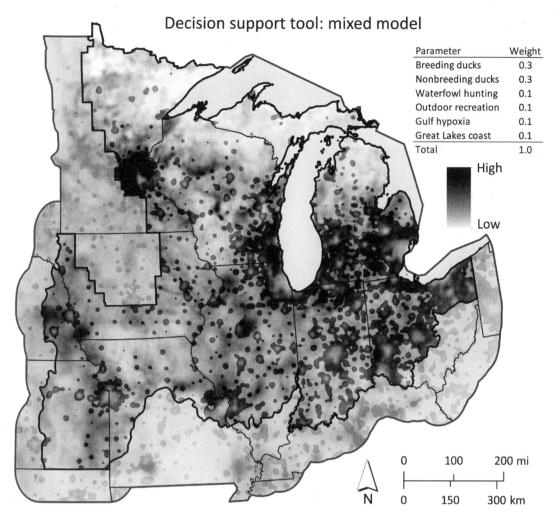

Decision support tool: mixed model

Parameter	Weight
Breeding ducks	0.3
Nonbreeding ducks	0.3
Waterfowl hunting	0.1
Outdoor recreation	0.1
Gulf hypoxia	0.1
Great Lakes coast	0.1
Total	1.0

High

Low

Figure 16.7. Decision support tool (DST) indicating higher- versus lower-priority areas to target waterfowl habitat conservation in the Upper Mississippi / Great Lakes Joint Venture region. The DST was established by integrating biological and social objectives using model-based maps weighted by regional waterfowl stakeholders. Biological objectives included focus on breeding and nonbreeding waterfowl habitats, whereas social objec- tives were related to providing locations for hunting and other outdoor recreation plus addressing gulf hypoxia and degraded coastal wetlands by identification of impaired watersheds. State and Bird Conservation Region (BCR) boundaries were used to designate state × BCR polygons with specific waterfowl habitat retention and restoration objectives downscaled from regional objectives. Repro- duced from Soulliere et al. (2017).

comes with estimating what, where, when, and how much habitat is needed to increase and sustain popu- lations of priority bird species to achieve established objectives. Although in early stages, JVs are also be- ginning to integrate human dimensions aspects into landscape designs for birds. At least one continental bird plan, the NAWMP, has included people (poten- tial supporters) in each of its strategic goals.

Population estimates and objectives are con- tinually being refined for birds, and with new in- formation resulting from evaluation, JV planning documents are updated regularly. They contain science-based recommendations that seek to increase landscape carrying capacity through more effectively targeting bird habitat retention and restoration. Joint venture regional population objectives are linked to continental conservation plans, and population objectives are used to generate habitat objectives for focal species. Each JV focal species represents a cover-type configuration, or habitat association, and JV partners assume that habitat treatments designed for focal species will accommodate populations of other birds within designated guilds. Biological models have been used to incorporate the best available information with expert-based knowledge and clearly identified assumptions. Joint venture planning models, however, have increased in complexity with integration of social considerations.

Regional bird population and habitat trends, in concert with focal species abundance estimates and an assessment of factors thought to limit their population growth, provide the JV biological planning foundation. Steps for contemporary JV planning efforts include characterizing and assessing key bird populations, landscapes, habitats vital to those populations, and the distribution and abundances of people who can theoretically benefit from strategically placed birds habitats. Biological models are used to predict species response to habitat actions, spatially identify conservation opportunities, and develop a landscape design with capacity expected to sustain current target bird populations and eliminate population deficits. Much of the technical information, including biological models and decision support maps, is developed for the focal species representing habitat guilds. Joint venture plans also include direction regarding monitoring and research needs, increasing management efficacy, adaptive management, and program coordination by the partnership.

Joint ventures, through their science staff and technical committees, establish explicit regional objectives for bird habitat conservation and identify and use available survey data and new technological tools to most efficiently work toward objectives. Lack of population and ecological information for some species of conservation concern has been a challenge for JVs. But they continue to use a scientific process for population and habitat objective setting coupled with explicitly stated assumptions and research needs to focus evaluation and to improve conservation effectiveness over time. Like the NAWMP, JV implementation plans are living documents, evolving with new knowledge regarding bird populations, landscapes and spatial data, social and scale considerations, and key relationships that influence birds and people.

LITERATURE CITED

Albert, D. A. 2003. Between land and lake: Michigan's Great Lakes coastal wetlands. Michigan Natural Features Inventory, Michigan State University Extension Bulletin E-2902, East Lansing, Michigan, USA.

Bartuszevige, A. M., K. Taylor, A. Daniels, and M. F. Carter. 2016. Landscape design: integrating ecological, social, and economic considerations into conservation planning. Wildlife Society Bulletin 40:411–422.

Brown, S., C. Hickey, B. Harrington, and R. Gill, editors. 2001. United States Shorebird Conservation Plan. Manomet Center for Conservation Sciences, Manomet, Massachusetts, USA.

Campellone, R. M., K. M. Chouinard, N. A. Fisichelli, J. A. Gallo, J. R. Lujan, R. J. McCormick, T. A. Miewald, B. A. Murray, D. J. Pierce, and D. R. Shively. 2018. The iCASS Platform: nine principles for landscape conservation design. Landscape and Urban Planning 176:64–74.

Commission for Environmental Cooperation. 1998. A proposed framework for delineating ecologically-based planning, implementation, and evaluation units for cooperative bird conservation in the U.S. Commission for Environmental Cooperation. Montreal, Quebec, Canada.

Dayer, A., A. Gramza, and J. Barnes. 2019. Incorporating human dimensions into joint venture implementation plans. March 2019 Bulletin, North American Bird Conservation Initiative, Washington, DC, USA. http://nabci-us.org/wp-content/uploads/2019/05/Incorporating-Human-Dimensions-into-Joint-Venture-Implementation-Plans.pdf. Accessed 26 Aug 2019.

Devers, P. K., A. J. Roberts, S. Knoche, P. I. Padding, and R. Raftovich. 2017. Incorporating human dimension objectives into waterfowl habitat planning and delivery. Wildlife Society Bulletin 41:405–415.

Federal Geographic Data Committee. 2013. Classification of wetlands and deep-water habitats of the United States. FGDC-STD-004-2013. Second Edition. Wetlands

Subcommittee, Federal Geographic Data Committee and US Fish and Wildlife Service, Washington, DC, USA.

Homer, C. G., J. A. Dewitz, L. Yang, S. Jin, P. Danielson, G. Xian, J. Coulston, N. D. Herold, J. D. Wickham, and K. Megown. 2015. Completion of the 2011 National Land Cover Database for the conterminous United States—representing a decade of land cover change information. Photogrammetric Engineering and Remote Sensing 81:345–354.

Krainyk, A., J. E. Lyons, M. G. Brasher, D. D. Humburg, G. J. Soulliere, J. M. Coluccy, M. J. Petrie, D. W. Howerter, S. M. Slattery, M. B. Rice, et al. 2019. Spatial integration of biological and social objectives to identify priority landscapes for waterfowl habitat conservation. US Geological Survey Open-File Report 2019–1029, Reston, Virginia, USA.

Kushlan, J. A., M. J. Steinkamp, K. C. Parsons, J. Capp, M. A. Cruz, M. Erwin, S. Hatch, S. Kress, R. Milko, S. Miller, et al. 2002. Waterbird conservation for the Americas: the North American Waterbird Conservation Plan. Version 1. Waterbird Conservation for the Americas, Washington, DC, USA.

Monfils, M. J., P. W. Brown, D. B. Hayes, and G. J. Soulliere. 2015. Post-breeding and early migrant bird use and wetland conditions of diked and undiked coastal wetlands in Michigan. Waterbirds 38:373–386.

Monfils, M. J., P. W. Brown, D. B. Hayes, G. J. Soulliere, and E. N. Kafcas. 2014. Breeding bird use and wetland characteristics of diked and undiked coastal marsh in Michigan. Journal of Wildlife Management 78:79–92.

Morefield, P. E., S. D. LeDuc, C. M. Clark, and R. Iovanna. 2016. Grasslands, wetlands, and agriculture: the fate of land expiring from the Conservation Reserve Program in the Midwestern United States. Environmental Research Letters 11:094005.

National Ecological Assessment Team. 2006. Strategic habitat conservation. Final report of the National Ecological Assessment Team. US Geological Survey and US Department of Interior, Fish and Wildlife Service, Washington, DC, USA.

North American Bird Conservation Initiative. 2000a. North American Bird Conservation Initiative: bringing it all together. http://www.nabci.net/wp-content/uploads/NABCI-Bringing-it-all-together.pdf. Accessed 26 Aug 2019.

North American Bird Conservation Initiative. 2000b. North American bird conservation initiative: bird conservation region descriptions. Map and supplement. US Fish and Wildlife Service, Division of Bird Habitat Conservation. Arlington, Virginia, USA.

North American Bird Conservation Initiative. 2017. 2017–2018 Work Plan for the U.S. NABCI Committee, within the framework of the 2017–2021 strategic plan. http://nabci-us.org/committee/governance. Accessed 26 Aug 2019.

North American Waterfowl Management Plan. 1986. North American Waterfowl Management Plan. US Department of Interior, Fish and Wildlife Service, Washington, DC, USA, and Environment Canada, Canadian Wildlife Service, Ottawa, Ontario, Canada.

North American Waterfowl Management Plan. 1994. Update to the North American Waterfowl Management Plan: expanding the commitment. US Department of Interior, Fish and Wildlife Service, Washington, DC, USA, and Environment Canada, Canadian Wildlife Service, Ottawa, Ontario, Canada.

North American Waterfowl Management Plan. 2004. North American Waterfowl Management Plan: strengthening the biological foundation (implementation framework). US Department of Interior, Fish and Wildlife Service, Washington, DC, USA, and Environment Canada, Canadian Wildlife Service, Ottawa, Ontario, Canada.

North American Waterfowl Management Plan. 2012. North American Waterfowl Management Plan 2012: people conserving waterfowl and wetlands. US Department of Interior, Fish and Wildlife Service, Washington, DC, USA; Environment Canada, Canadian Wildlife Service, Ottawa, Ontario, Canada; and Ministry of Environment and Natural Resources, Mexico City, Mexico.

North American Waterfowl Management Plan. 2018. North American Waterfowl Management Plan (NAWMP) update: connecting people, waterfowl, and wetlands. US Department of Interior, Fish and Wildlife Service, Washington, DC, USA; Environment Canada, Canadian Wildlife Service, Ottawa, Ontario, Canada; and Ministry of Environment and Natural Resources, Mexico City, Mexico.

Petrie, M. J., M. G. Brasher, G. J. Soulliere, J. M. Tirpak, D. B. Pool, and R. R. Reker. 2011. Guidelines for establishing joint venture waterfowl population abundance objectives. North American Waterfowl Management Plan Science Support Team Technical Report No. 2011-1, US Department of Interior, Fish and Wildlife Service, Washington, DC, USA.

Rich, T. D., C. J. Beardmore, H. Berlanga, P. J. Blancher, M. S. W. Bradstreet, G. S. Butcher, D. W. Demarest, E. H. Dunn, W. C. Hunter, E. E. Inigo-Elias, et al. 2004. Partners in flight, North American Landbird Conservation Plan. Cornell Lab of Ornithology, Ithaca, New York, USA.

Richardson, L., J. Loomis, T. Kroeger, and F. Casey. 2015. The role of benefit transfer in ecosystem service valuation. Ecological Economics 115:51–58.

Schermerhorn, J. R., J. G. Hunt, and R. N. Osborn. 1991.

Managing organizational behavior. John Wiley and Sons, New York, New York, USA.

Soulliere, G. J., and M. A. Al-Saffar. 2017. Targeting conservation for waterfowl and people in the Upper Mississippi River and Great Lakes Joint Venture Region. Upper Mississippi River and Great Lakes Region Joint Venture Technical Report No. 2017-1, Bloomington, Minnesota, USA.

Soulliere, G. J., M. A. Al-Saffar, J. M. Coluccy, R. J. Gates, H. M. Hagy, J. W. Simpson, J. N. Straub, R. L. Pierce, M. W. Eichholz, and D. R. Luukkonen. 2017. Upper Mississippi River and Great Lakes Region Joint Venture Waterfowl Habitat Conservation Strategy—2017 Revision. US Fish and Wildlife Service, Bloomington, Minnesota, USA.

Soulliere, G. J., M. A. Al-Saffar, R. L. Pierce, M. J. Monfils, L. R. Wires, B. W. Loges, B. T. Shirkey, N. S. Miller, R. D. Schultheis, F. A. Nelson, et al. 2018. Upper Mississippi River and Great Lakes Region Joint Venture Waterbird Habitat Conservation Strategy—2018 Revision. US Fish and Wildlife Service, Bloomington, Minnesota, USA.

US Fish and Wildlife Service. 2005. Service manual chapter 721 FW 6 Joint Ventures (supersedes 2002 Director's Order No. 146). US Department of Interior, Washington, DC, USA.

US Census Bureau. 2010. Population change for counties in the United States and for municipios in Puerto Rico: 2000 to 2010. www.census.gov/2010census. Accessed 26 Aug 2019.

US Fish and Wildlife Service. 2016. National wetlands inventory. US Department of the Interior, Washington, DC, USA.

Wainger, L. A., and M. Mazzotta. 2011. Realizing the potential of ecosystem services: a framework for relating ecological changes to economic benefits. Environmental Management 48:710–733.

— 17 — Translating Landscape Ecology to Management

Cynthia A. Jacobson,
Amanda L. Sesser,
Elsa M. Haubold,
Kevin M. Johnson,
Kimberly A. Lisgo,
Betsy E. Neely,
Fiona K. A. Schmiegelow,
Stephen C. Torbit, and
Greg Wathen

A Landscape Conservation Cooperatives Approach

Introduction

The scale and pace of conservation challenges caused by human population increase and related consumption of natural resources are unprecedented, resulting in profound effects on social-ecological systems (Folke et al. 2005). As insurmountable as these issues seem, it is important to look to the dynamics within the conservation institution (CI; Jacobson and Decker 2006) and the role of its organizational actors as key in leading or resisting the transformation needed to help ensure functional and resilient systems into the future. Jacobson and Decker (2006:532) defined the CI as "the people, processes, and rules as well as the norms, values and behaviors" associated with wildlife conservation. The authors applied institutional and organizational theory to understand behavior observed among many state fish and wildlife agencies and traditional conservation nongovernmental organizations in response to pressure for change (Jacobson et al. 2007). While the need for transformation of the CI to address contemporary challenges such as the effects of climate change, invasive species, and habitat loss has been identified for more than a decade, precisely how that transformation occurs has only just started to take shape (Nie 2004, Jacobson and Decker 2006, Decker et al. 2016).

One area where there has been agreement is the need for broadening the way the CI approaches conservation (Jacobson et al. 2010, Decker et al. 2016). Institutional reform literature stresses that change occurs through broadening (or narrowing) in three primary areas: goals, boundaries, and activities (Scott 2001). Broadening may include expanding organizational visions, goals, and missions to embrace a more holistic versus narrow perspective on conservation, as some state wildlife agencies have done to help ensure alternative funding for their agencies (Jacobson et al. 2010); working across organizational or jurisdictional boundaries to identify and achieve common desired outcomes; seeking input from a greater diversity of stakeholders; and integrating multiple scientific disciplines into conservation decision making. Pressure for reform comes from endogenous (e.g., normative and cultural shifts) and exogenous factors (e.g., changing environmental circumstances, human demographics; Jacobson and Decker 2006) and is often resisted in well-established institutions (Dorado 2005). Scholars (Decker et al. 2016, Kretser et al. 2016) contend that transformation will involve innovative, multidisciplinary thinking and action across jurisdictions and organizations.

To address contemporary challenges (e.g., effects of a rapidly changing climate, increasing pressure to develop land for residential and commercial purposes) and associated uncertainty, the CI will need to better integrate social, ecological, physical, and economic principles to ensure desirable outcomes (i.e., landscapes capable of sustaining fish and wildlife that people value). Resistance within the CI to such integration is apparent owing to real and perceived barriers to change (Jacobson 2016). Even so, some progress has been made in this area, including diverse organizations working together to identify common goals and objectives, and to combine resources to achieve shared desired outcomes (Bixler et al. 2016a). Moreover, advances in geospatial and computational technology and capacity now provide broader and more accessible tools to integrate landscape ecology into natural resource decision making. Wiens (2007:479) suggests that as "the emphasis of conservation has shifted from protecting species to include entire ecological systems or 'functional landscapes,' the need for closer linkages between conservation and landscape ecology has become obvious."

Contemporary thought in landscape ecology stresses that often it is best to start natural resources management discussions at ecologically meaningful scales and to think of humans as a part of versus apart from the natural world (Chapin et al. 2010). Landscape-scale conservation through partnership for specific guilds and habitats is an established model (e.g., migratory bird joint ventures [JVs], national fish habitat partnerships [NFHPs]). Acknowledging the importance of these partnerships but recognizing that a broader complementary approach was needed to address all natural and cultural resources, the US Department of Interior expanded the JV and NFHP models to create the Landscape Conservation Cooperatives Network* (Fig. 17.1; Millard et al. 2012).

The purpose of this chapter is to introduce a holistic, collaborative approach to applying landscape

ecology principles to address contemporary conservation and management challenges across geopolitical boundaries. Using an adaptive co-governance approach (Jacobson and Robertson 2012), the Landscape Conservation Cooperatives (LCCs) identify priority conservation targets such as species, habitats, and systems; establish goals and objectives for those targets; and develop strategies and provide or leverage funding to facilitate their landscapes' ability to sustain conservation targets. Essentially, LCCs are a forum for applying landscape ecology principles to conservation efforts, starting with nesting place-based management decisions in the context of broader landscapes (Wiens 2007). Although at different stages in their planning processes, most LCCs have adopted a collaborative and adaptive landscape conservation design (LCD) approach that has different manifestations according to the issues facing the landscape. This approach to conservation is not without challenges, as illustrated by case studies from three distinct LCCs. We conclude with considerations and lessons learned from landscape-scale initiatives in their efforts to affect management decisions to help conserve species and habitats for the benefit of current and future generations.

Landscape Conservation Cooperatives

Beginning in 2009, a network of 22 LCCs was established as self-directed landscape-scale partnerships transcending geopolitical boundaries within the North American continent and beyond, crossing state boundaries, and extending into Canada, the Caribbean islands, Mexico, and as far west as the Pacific islands (Fig. 17.1). Each LCC has a steering committee composed of a diversity of governmental and nongovernmental partners who have a stake in conservation decisions for that landscape. As self-

---

* When this chapter was originally drafted, LCCs were an intact landscape conservation delivery program. Beginning in 2017, the US Department of the Interior shifted priorities, and the department no longer supports staffing or provides

financial resources to many LCCs. Several LCCs supported by private partnerships continue to function, and the material in this chapter remains a good schematic for other landscape conservation delivery programs.

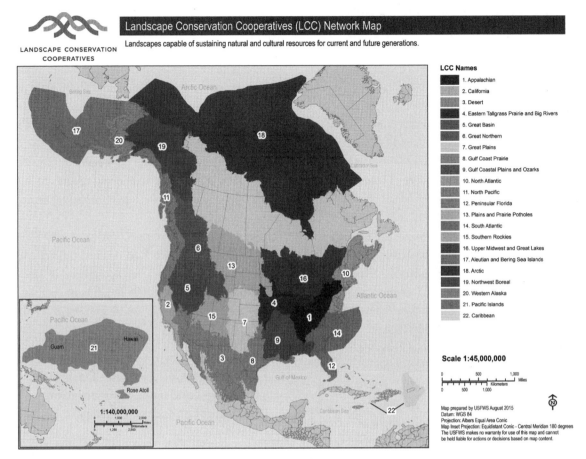

Figure 17.1. Landscape Conservation Cooperatives Network map.   Reproduced from Landscape Conservation Cooperative Network (2015).

directed partnerships, the members of each steering committee establish the agenda for the LCC according to the conservation priorities, objectives, and needs for its landscape. Landscape Conservation Cooperatives bridge multiple organizations and partnerships, jurisdictions, conservation issues, and stakeholders (Jacobson and Robertson 2012). Jacobson and Haubold (2014) described their potential to assist states, for example, in meeting public trust obligations in managing wildlife for the public now and in the future (Smith 2011). Further, the reach of LCCs depends on their ability to work closely with other conservation partners such as JVs and NFHPs. Researchers reported that each partnership has a unique, complementary role and emphasized

the importance of close collaboration between the partnerships (National Academies of Sciences, Engineering, and Medicine 2016). Although an LCC's area of emphasis may be somewhat different, each is focused on achieving the same overarching vision of maintaining "landscapes capable of sustaining natural and cultural resources for current and future generations" (Landscape Conservation Cooperative Network 2014:3).

In addition to adaptive co-governance and planning, LCCs address landscape-scale stressors including habitat fragmentation, climate change, and invasive species. For example, the Pacific Islands Climate Change Cooperative conducted an evaluation of the effects of increasing ocean temperatures and acidification on corals. Working with the National Oceanic and Atmospheric Administration, the cooperative developed an interactive tool for the public and governmental agencies to use to evaluate the effects of climate change on coral reefs (National Oceanic and Atmospheric Administration 2019). The North Pacific LCC helped tribal communities identify and address subsistence issues related to a changing climate. Projects addressed vulnerability and adaptation strategies for Pacific lamprey (*Entosphenus tridentatus*), Eulachon (*Thaleichthys pacificus*), salmon, and berry plants. Risk models and analysis of climate change projections for these projects are aiding in climate change adaptation and prioritization for these important tribal resources. Similarly, the North Atlantic LCC provided tools and information to facilitate community resilience related to hurricane. After Hurricanes Irene and Sandy washed out roads, it was apparent a regional planning approach was needed to increase road resilience in anticipation of future flooding. The LCC helped partners develop roads and improve fish passage by providing regional maps and condition assessments of road-stream crossings, fish passage, future flood predictions, risk assessments, and prioritization tools. This information is helping states and communities prepare for future intense storms and improve conditions for aquatic organisms.

Each of the 22 LCCs has made conservation inroads that were unlikely without the existence of its unique partnership. For example, the Desert LCC provided an inventory and database of aquatic species, something that no other entity was able to fund alone but was an essential tool for managing these species by state fish and wildlife agencies. The network collectively has great potential to conserve landscapes in ways never realized before. The ability and reach of the LCC Network surpass geographical boundaries, allowing collective governance at ecologically relevant scales. Overall, the network formally represents >280 conservation organizations and agencies, demonstrating the reach and ability

of this super-partnership to influence conservation outcomes. Further, the boundaries of LCCs are functionally fluid, with LCCs coalescing when needed to address challenges at scales different than what is represented by the LCC-specific boundary. For example, the Eastern Tall Grass Prairie and Big Rivers LCC have brought together six additional LCCs spanning the Mississippi River drainage basins to work collectively to consider development of wildlife corridors in an effort to address effects of Gulf hypoxia on aquatic systems that support priority species of fish and wildlife.

Regardless of some of these early contributions of LCCs, the LCC Network recognized the need to adopt a shared framework for application of landscape ecology principles. The framework that was adopted and modified was landscape design (Haila 2007).

Landscape Conservation Design: Bridging Landscape Conservation and Landscape Ecology

Landscape conservation design (LCD; Table 17.1) is a collaborative planning process that leads to identification of conservation action needed to achieve landscapes that support the sustainability of natural and cultural resources. Landscape conservation design is founded on a vision for a desired future condition and related goals and objectives that are shared by those involved. It is a method by which each partner can see its unique role in working collectively toward achieving that shared vision. The LCD framework lays out how to apply landscape ecology principles and methods to realize actual on-the-ground conservation across multiple scales. Required landscape ecology techniques include identifying and mapping priority areas for conservation, identifying potential future range shifts, and ensuring that management activities allow for ecological and evolutionary processes. Landscape conservation design links these scientific products to specific

Table 17.1. Common Elements Included in Landscape Conservation Design and a Short Description of Each Element

LCD Element	Description
A. Visioning	The typical first step in any planning process is to identify and bring the partners to the table. LCD is typically problem oriented: the scope and problem statements are defined by the conservation challenges, but by definition, LCD includes multiple stakeholders working across jurisdictional or geopolitical boundaries. The vision drives the entire process. The partners should articulate the desired future of this landscape. This vision is based on the multiple values articulated by those involved. Inclusion of diverse partners is critical, as is having partner engagement prior to problem definition. Diverse perspectives contribute innovative solutions, and the problem itself is shaped by who is contributing.
B. Landscape-scale goals, objectives, and priorities	What are the geospatial landscape-scale goals to help the partnership achieve its vision? Measurable, attainable objectives guide the partnership, or individual partners, in carrying out work to achieve the shared goals. What are the values (shared or unique) within the partnership? Once a suite of objectives that represent the multiple values across the landscape is identified, those objectives can be translated into priorities. Priorities can be selected on the basis of urgency, the need for collaboration to be successful, ecological or cultural relevance, or other criteria. It's important to be inclusive of multiple values, even if seemingly contradictory.
C. Landscape assessment and scenarios	Once a desired future is articulated and a path toward reaching it is set, it is important to bring together the best available science to assess the plausibility of reaching the vision. This is the nexus with landscape ecology. Geospatial and temporal impact assessments, climate envelope modeling, population growth, and other data sets are compiled (or created) to describe current and projected future landscape condition. Scenario planning may be useful to determine the feasibility of the vision, goals, and objectives, all of which may be revised as a result. Climate change should be explicitly considered.
D. Landscape-scale conservation action	Each partner has a unique contribution to identifying and implementing actions that will result in achieving the landscape-scale goals and objectives. Landowners have the ability to take action on their land; research institutions and nonprofits can support research and monitoring, guide education and outreach, or provide other support to landowners or the public. Each partner organization supports actions that fit within its own mission and the larger landscape context. The success of the LCD partnership depends on implementation—without it, there will be no landscape conservation.
E. Monitoring and evaluation	Landscape-scale conservation is complex. Many partners with different roles and capacities will take different approaches. This diversity is a natural experiment to test actions versus outcomes. Moreover, the partnership needs to assess whether the agreed-upon actions are having their desired effect in conserving resources or directing sustainable development at the landscape scale. Objectives may need to be revisited based on results. Indicators or metrics for success must be periodically evaluated. Often the partners each have a role in collecting and sharing data.

Note: LCD, landscape conservation design. LCD is not a linear process; thus any element may serve as a starting point. It is similar to other types of conservation planning but is distinguished by the landscape-scale focus, the number and diversity of partners included, and the holistic inclusion of multiple sectors and values in addition to conservation.

on-the-ground management activities and to measurable objectives. To implement the types of landscape ecology principles and products discussed in previous chapters of this book, it is imperative to use a social process that brings together multiple landowners and researchers in a manner that builds trust and encourages coordinated action. Landscape conservation design encompasses planning, implementation, monitoring, and learning components of adaptive landscape management (Hobbs 2007, Perera et al. 2007; Table 17.1). The LCC Network has adopted a coarse framework for the characteristics of LCD (Landscape Conservation Cooperative Network 2016); we treat LCD generally because it could apply to multiple large-scale conservation efforts.

Landscape-scale goals and objectives can be stepped down to inform local plans. For example, an individual park or refuge can contribute to overarching landscape objectives and thus can direct a portion of its workflow to meet those objectives. Management and land use plans at the single management-unit scale can specifically describe how an individual partner organization will contribute to the larger conservation vision. The LCDs also can be scaled up. Just as local priorities feed into landscape priorities, landscape priorities can inform larger conservation efforts at continental scales. Landscape ecology expertise is needed to help land and resource managers understand and work across multiple scales. Landscape conservation design is not the purpose of landscape conservation; it is the process and products that inform how landscape conservation takes shape (Landscape Conservation Cooperative Network 2016, Campellone et al. 2018).

Application of LCD varies by situation and context. Conservation challenges will help identify which stakeholders should be involved, geographic scope, and resources or capacities needed. There are, however, some common elements of the LCD process that can act as a guide. Many existing planning frameworks are appropriate to guide LCD, including Open Standards for Conservation Practice (Conservation Measures Partnership 2019), Climate Smart Conservation (Stein et al. 2014), and other adaptive planning processes (Groves and Game 2015). Regardless of planning approach, the foundation for every LCD is an articulated shared vision of the future (labeled A in Table 17.1). The partners within the landscape work collectively to define desired landscape conditions for the future (e.g., 50 or 100 years forward). Vision statements describe the desired future conditions and are based on social values (Manfredo et al. 2004). Healthy landscapes provide biodiversity, wildlife, ecosystem services, human well-being, sustainable development, access to natural resources, and economic growth, which ideally encompass the entire range of public values. The articulated desired future condition will guide the entire LCD process. Also, it will serve to ensure that the products are forward looking, rather than attempting to maintain status quo or manage to a historical state. Additionally, it is important to assess whether the desired future condition is attainable given climate change, development pressures, or geopolitical realities.

Landscape conservation design works to identify select species and habitats as part of a management strategy that is integrated among other values on the landscape. Species-specific management, however, is warranted in some cases, including species used for subsistence, enhancing desired species or controlling undesired species, stopping the spread of disease vectors or invasive species, and for the conservation of threatened or endangered species (Zavaleta and Pasari 2007). Using species as proxies to characterize health of a broader landscape or to monitor biological responses to change agents is also commonly used in wildlife management but can be controversial (Caro 2010). Regardless, it is important to identify the purpose for species or population management in the landscape context. Once objectives have been set, LCD uses geospatial and analytical techniques to spatially display priority habitats, corridors, breeding grounds, dynamic processes such as disturbance, or other important ecological factors for priority species assessment and conservation, including management actions such as habitat restoration or acquisition. The LCD can direct management to or development away from these areas while identifying those areas suitable for development and other uses. In this way, multiple objectives in addition to fish and wildlife management can be reached across a landscape. Once priority actions have been implemented, conservation effectiveness must be monitored to evaluate whether the objectives have been met as part of the adaptive management cycle (labeled E in Table 17.1). It is recommended to be inclusive of multiple sectors and diverse stakeholder groups from the beginning of the process (labeled A in Table 17.1). Multiple objectives must be considered equally, and conflicts must be addressed when they arise.

The challenges associated with multiple objectives are frequently related to scale. The geographic context of species or population-based policy and management decisions can range from local jurisdictions to state or continent-wide scales. Increasingly, wildlife professionals recognize the need to coordinate management of transboundary fish and wildlife. For example, in the United States, state wildlife action plans identify species of greatest conservation need, priority habitats, and management at statewide scales (Association of Fish and Wildlife Agencies 2019). Likewise, a network of 22 LCCs works toward continent-scale objectives in addition to each LCC's own self-directed priorities (Landscape Conservation Cooperative Network 2014). When working at larger geographic scales, temporal scale also becomes an issue. Many processes need to be considered when managing landscapes such as disturbance regimes, gene flow, and evolution. It is thus necessary, albeit challenging, to take the long view when designing landscapes and setting measurable objectives. The LCC partners are the ones who implement the landscape conservation goals and objectives at local or regional scales (e.g., within a park or state boundary, each partner contributes to the large landscape goals and objectives by stepping them down into local strategies and actions). Local and regional priorities also feed into landscape conservation priorities. The mismatch between the scale of landscape planning and management implementation is one of the greatest challenges to landscape conservation (Cash and Moser 2000). It is important that LCD and other activities at the landscape or ecosystem scale explicitly acknowledge scale and the importance of conservation at each level of organization.

Scale issues are interrelated and perhaps best illustrated through the organizational culture and structure of the LCC network itself. Each of the 22 LCCs was established to be a self-directed partnership recognizing that while landscape conservation occurs at large scales, those who have public trust responsibility for the local resources and who have interest in specific landscapes and seascapes are best positioned to establish goals and implement actions that will lead to sustainability locally. The LCCs are working within their established geographies to effect conservation on their landscape. Despite the established large geographies of LCCs that fre-

quently transcend state and international borders, however, there are often shared needs extending beyond a single LCC. There are already numerous examples of LCCs working across boundaries, such as the Southeast Conservation Adaptation Strategy (SECAS), which created a conservation blueprint across 6 LCCs and 15 states. Convening a steering committee of approximately 30 individuals from multiple jurisdictions, each with their own mandates and priorities, to create a governance structure and functional LCC partnership with shared vision and goals is challenging. More complicated still is bringing together those steering committees from multiple LCCs to develop a shared vision and goals across an even larger landscape. The challenges are magnified and involve negotiations and the willingness of the partners to compromise some of what they believe is important for their individual LCC to have even greater conservation effect than they would singularly. These are not technical challenges for which a known solution is available but rather are adaptive challenges. The partners must be able to exert adaptive leadership principles such as articulating a shared vision, regulating distress from perceived losses by partners, and ensuring the partners are working together to address the issues (Heifetz and Laurie 2001, Manolis et al. 2009, Haubold 2012). Challenges of working across jurisdictions and geopolitical boundaries within and across any landscape-scale partnership are not insurmountable but must be overcome to ensure effective conservation for wildlife and their habitats in the future.

Landscape-Scale Conservation: Case Studies from Three Landscape Conservation Cooperatives

Landscape Conservation Cooperative partnerships are having an effect on the conservation of wildlife. The following three case studies provide examples of how working collaboratively across large landscapes has resulted in improved management actions. The first, the Northwest Boreal LCC, demonstrates the benefits of working on ecosystems spanning international boundaries. The Northwest Boreal LCC steering committee, composed of an equal number of partners from Canada and Alaska, USA, has begun determining important habitats they should proactively work to conserve before the habitat is compromised permanently. Northwest Boreal LCC partners from multiple federal agencies, state and provinces, tribes and First Nations, and nongovernmental organizations are using LCD (highlighting LCD elements A-C in Table 17.1) to plan their desired landscape of the future. The second case study, from the Southern Rockies LCC, provides an example of how convening various partners including nongovernmental organizations, state, private lands, and academic stakeholders has facilitated adaptation efforts throughout the Gunnison River Basin, Colorado, USA, to enhance resilience of habitat for Gunnison sage-grouse (*Centrocercus minimus*) and other species that depend on the sagebrush system. The case highlights LCD elements D-E (Table 17.1) that focus on management actions and monitoring. Finally, SECAS showcases LCD elements applied across multiple LCCs; as conservation needs span geopolitical boundaries, they also can encompass multiple LCCs. The SECAS represents the effort of 6 LCCs to jointly produce an LCD that identifies important areas for wildlife and habitat conservation for the entire southeastern United States.

Using the Conservation Matrix Model to Inform Proactive Management of the Northwest Boreal Region

The Northwest Boreal LCC spans >133,546,262 ha, encompassing large portions of Alaska and northern regions of Canada (Fig. 17.2A). The LCC area is distinguished by relatively intact ecological conditions across its full extent, supporting the shared vision among partner organizations of "a dynamic landscape that maintains functioning, resilient boreal ecosystems and associated cultural resources" (Northwest Boreal Landscape Conservation Coop-

erative 2019). These conditions afford unparalleled opportunities to incorporate principles of landscape ecology in management, with active engagement of partner organizations key to forging the landscape and administrative linkages necessary to support planning and implementation over these vast landscapes.

For the Northwest Boreal LCD, an innovative approach known as the Conservation Matrix Model (CMM) is being employed in collaboration with the Boreal Ecosystems Analysis for Conservation Networks Project (2019). Whereas conservation science developed largely in highly altered systems, with a reactive focus on imperiled species and protection of remnant habitats, the CMM is a broad conceptual framework developed to support conservation of natural patterns of species distribution and abundance and the processes that support them, across large dynamic landscapes, in relatively intact systems (Schmiegelow et al. 2014). Consistent with a landscape perspective, the framework recognizes the role that all landscape elements play in maintenance of ecological integrity and combines principles of landscape ecology with the strengths of systematic conservation planning and the structured process of adaptive resource management to support integrated approaches over large regions by emphasizing shared stewardship responsibilities. The model places strong emphasis on the critical role the landscape matrix plays in supporting populations of species, regulating the movement of organisms, buffering sensitive areas, and maintaining the integrity of aquatic systems (Lindenmayer and Franklin 2002). The LCD goals addressed by the CMM include maintaining ecological flows, providing habitat for wildlife, planning for climate change, and identifying measurable objectives and ecologically sustainable activities to achieve shared goals.

Foundational to the CMM is recognition of the uncertainty and risk associated with management decisions and, consistent with tenets of adaptive management, treating associated activities as incremental experiments to be learned from and adapted

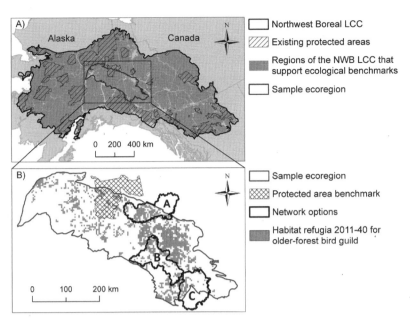

Figure 17.2. Components of the landscape conservation design process in the Northwest Boreal Landscape Conservation Cooperative (NWB LCC). (*A*) The geographic extent of the NWB LCC is outlined in black; the partnership includes the boreal ecoregion of Alaska, USA, and Yukon Territory, southwest Northwest Territories, and northern British Columbia, Canada. (*B*) A sample ecoregion used for identifying a network of ecological benchmarks, encompassing existing protected areas, active development sites, and projected habitat refugia in a matrix of relatively intact lands. Land managers use this information to prioritize conservation or development across the landscape while leveraging existing protected areas.

over time. In this context, ecological benchmarks are required to improve understanding of system dynamics and to act as reference areas or controls for distinguishing change due to human activities from underlying environmental dynamics. If carefully designed, such areas can also serve as cornerstones for conservation. Efforts to apply the CMM in the Northwest Boreal LCC have focused on the identification of ecological benchmarks in consideration of the time-limited opportunities that various planning processes underway by partner organizations present.

To meet their intended functions, system-level ecological benchmarks should be relatively devoid of human activities, representative of environmental variation, and sufficiently large to sustain key ecological processes. The primary criteria are thus physical intactness of landscapes, a proxy for the integrity of terrestrial systems; hydrological connectivity of intact water catchments, a measure of the integrity of aquatic systems (Anderson 2009); representation of environmental variation (landcover, productivity, climatic conditions) as coarse-filter surrogates for biodiversity; and adequate size for maintenance of representative landcover types over time in the face of an active natural disturbance regime (fire), as a measure of the resilience of the system (Leroux et al. 2007a, 2007b). Where available, species data—such as the location of habitat, minimum viable population estimates, and home range sizes—can also inform the size and location of ecological benchmarks, and implementation of appropriate management activities in other portions of the landscape.

To support the assessment of ecological benchmark potential across the Northwest Boreal LCC, transboundary data on landscape condition, surficial hydrology, fire regimes, and environmental variation were assembled and analyzed. Spatially dynamic landscape models were used to simulate fire and vegetation succession for the region, and streamflow networks supported analysis of hydrological connectivity. Data for a core set of focal species identified by the Northwest Boreal LCC steering committee— including caribou (*Rangifer tarandus*), moose (*Alces alces*), mountain sheep (*Ovis canadensis*), beaver (*Castor canadensis*), old-growth forest birds, waterfowl, and fish—also were evaluated. A review of management plans across jurisdictions revealed few specific population or distribution targets for these species, so in most cases the ecological benchmark selection process attempted to maximize the amount of species habitat maintained over time given an active fire regime. Existing protected areas may serve as ecological benchmarks if they meet the specified criteria, although most were not designed with this goal in mind.

Overall, the system-level ecological benchmark potential of the Northwest Boreal LCC is high, with most areas supporting the identification of numerous options on the basis of size, intactness, and hydrology at the ecoregion scale (Fig. 17.2A). Ecoregions defined by vegetation, soil, and climate are appropriate stratification units for assessing and representing environmental variation across the landscape. Often, more than one ecological benchmark or a network of benchmarks is required to represent an ecoregion. Before identifying new ecological benchmark options, existing protected areas were evaluated for their benchmark potential. In the transboundary ecoregion (Fig. 17.2B), there is one existing protected area with the potential to serve as an ecological benchmark based on size and intactness criteria; however, it does not adequately represent the environmental variation of the ecoregion; thus a complementary network of benchmarks is needed. Note that protected areas that do not meet the criteria for system-level benchmarks may still serve a benchmark function, depending on the attributes of management interest. Three ecological benchmarks (A-C) from >1,000 potential benchmarks designed for the ecoregion provide examples

(Fig. 17.2B). When selecting from options for completing an ecological benchmark network to complement existing protected areas, several additional benchmark characteristics can be considered, such as shape, proximity to existing protected areas, and the ability to represent the needs of focal species. In this example, spatial output from a predictive model of projected climate refugia for older forest birds (Stralberg et al. 2015) is shown as an underlay to the options. Given this consideration, ecological benchmark A would be the best addition to the network, as it captures the greatest amount of habitat refugia for this focal species group (Fig. 17.2B).

This case study illustrates elements A-C of the LCD process (Table 17.1), and the associated work has served as an important catalyst for galvanizing efforts to establish goals, assemble data, consider options, and improve understanding of the opportunities and constraints facing the region. Principles of landscape ecology that address issues of pattern (representation of environmental variation), process (natural disturbance and hydrologic connectivity), and scale (extent relative to spatial and temporal dynamics) are central to the establishment of criteria for ecological benchmarks and in evaluating options. While these criteria are intended to act as surrogates for biodiversity, inclusion of species' needs complemented the analyses and provided a focus for enhanced interagency and interjurisdictional collaboration. At the time of writing, this work is one of the best examples in the region that shows the application of landscape ecology to on-the-ground management decisions. The Bureau of Land Management and US Fish and Wildlife Service, in cooperation with partners, are actively using the products from this work to inform regional planning in Alaska that simultaneously considers protected areas, managed lands, and their relationship in a broader landscape context. Once the agencies receive potential networks of ecological benchmarks, planners then bring in social, economic, cultural, and other considerations to choose the best network configurations to meet the needs and mandates for that agency. In this way the ecological benchmarks become social-ecological benchmarks, and they capture the LCC vision of maintaining "landscapes capable of sustaining natural and cultural resources for current and future generations" (Landscape Conservation Cooperative Network 2014:3).

Gunnison Basin Climate Vulnerability Assessment and Resiliency Strategies for Gunnison Sage-Grouse

The geography of the Southern Rockies LCC encompasses portions of Arizona, Colorado, New Mexico, and Utah, USA, and smaller portions of Idaho, Nevada, and Wyoming, USA. The area is geographically complex, including 4,267-m mountain peaks, the Grand Canyon, and cold desert basins. The Southern Rockies LCC steering committee is composed of federal, state, tribal, and nongovernmental organizations involved in the conservation of key natural resources such as water, fish, wildlife, and plants. The partnership is working to identify where and how to achieve conservation with a landscape-scale impact.

Understanding climate vulnerability of habitats in the West was one of the early priorities for the Southern Rockies LCC. The steering committee encouraged vulnerability assessments as part of its integrated LCD that would assess climate change effects and lead to and implement adaptation strategies to address vulnerabilities and demonstrate practical approaches for adaptation at a local scale. The Southern Rockies LCC prioritized a comprehensive landscape-scale climate change vulnerability assessment of the Upper Gunnison River Basin in Colorado (Neely et al. 2011) in cooperation with the Gunnison Climate Working Group and other nongovernmental organizations. The team used methods developed by the Manomet Center for Conservation Sciences and Massachusetts Division of Fisheries and Wildlife (2010) and the NatureServe Climate Change Vulnerability Index (Young et al. 2011).

The Upper Gunnison River Basin encompasses ~1,011,714 ha of public and private lands with eleva-tions ranging from 2,286 m to >4,267 m. The basin supports a diversity of plant communities, ranging from sagebrush shrublands to alpine tundra, therefore providing habitat for a variety of wildlife and plant species. The Gunnison sage-grouse, a federally threatened species inhabiting the sagebrush ecosystem, is considered a distinct species of sage-grouse and is characterized by a small and declining population. The sagebrush ecosystem upon which the Gunnison sage-grouse depends is declining because of fragmentation and other anthropogenic developments. Sagebrush is highly vulnerable to invasive plant species that can increase its vulnerability to catastrophic wildfire. These factors suggest that sagebrush and the numerous wildlife species that depend on the sagebrush system during their annual cycle are also highly vulnerable to climate change. Gunnison sage-grouse are particularly vulnerable to climate change because of their dependence on wetlands, wet meadows, seeps and springs, and low vegetation riparian areas for rearing their young chicks. These mesic areas are important for brood rearing because they have a greater diversity and abundance of plants and insects than nearby uplands. These habitats are predicted to become less suitable for sage-grouse brood rearing as a result of increased drought and intense runoff events.

Building on the vulnerability assessment and strategies developed at a climate adaptation workshop for natural resource managers, the Gunnison Climate Working Group successfully implemented a collaborative conservation strategy to enhance the resilience of wet meadows and riparian areas to help the Gunnison sage-grouse and other wildlife species adapt to a changing climate. In one of the earliest efforts of its kind, the Southern Rockies LCC and the Gunnison Climate Working Group worked to deliver the identified conservation activities and monitor their success or failure. The group has designed and built a variety of simple restoration structures (e.g., one-rock dams; Zeedyk and Clothier 2014) to slow water during spring runoff, increase sediment deposition and water retention, and reduce erosion in areas prone to instream erosion. By the end of 2015, the group constructed 750 structures, restoring ~51 ha along 37 km of stream, working with volunteers, youth field crew, partners, and ranchers. Restoration activities occurred on public and private lands within multiple watersheds across the Gunnison Basin.

Long-term ecological monitoring of vegetation response to the restoration treatments is an integral component of this climate adaptation project (Nature Conservancy and Gunnison Climate Working Group 2015). Monitoring effectiveness of priority actions identified by the LCD is an important step of the process, one that is often overlooked in natural resource management. Between 2012 and 2017, the average cover of sedges, rushes, willows, and wetland forbs increased significantly, and bare ground and upland species decreased in treated areas. Before treatment at one site, for example, bare ground and litter occupied 41% in 2012. After treatment, bare ground and litter occupied 12% in 2017 (Fig. 17.3). This effort provides some of the foundational work in designing a restoration project to support adaptation of the Gunnison sage-grouse and other species that rely on the sagebrush system.

This case study exemplifies that LCD does not need to be a linear process, and any element (Table 17.1) may serve as a starting point. In this case, stakeholders formulated a quasi-vision (i.e., they recognized it was important to collaborate on making the landscape resilient to climate change and identified the landscape of interest and important entities within the landscape). They did not articulate a desired future for the landscape, nor did they articulate their landscape-level goals, objectives, and priorities. Just knowing climate change was a common thread, they placed an initial emphasis on the third LCD element (landscape assessment and scenarios). The results of the initial assessment (Gunnison Basin Climate Vulnerability Assessment; Neely et al. 2019) are what helped stakeholders ultimately identify their specific targets, goals, and objectives, and reach collaborative conservation actions on the ground.

Southeast Conservation Adaptation Strategy

The SECAS initiative was launched in 2011 to provide a shared, long-term vision of lands and waters needed to sustain fish and wildlife populations and improve human quality of life in the southeastern United States (G. Wathen et al., Tennessee Wildlife Resources Agency, unpublished manuscript). The southeastern United States is undergoing unprecedented landscape change, influenced by several factors (e.g., urbanization and population growth, climate change, energy development). These global changes are accelerating and are expected to continue throughout the remainder of this century. The effects of these changes on fish and wildlife are uncertain at best but will require a strategic, systems-scale response if society hopes to achieve landscape sustainability in the Southeast.

SECAS is a regional response to today's and tomorrow's most pressing conservation challenges, employing the collective efforts of six LCCs (Appalachian, Caribbean, Gulf Coastal Plains and Ozarks, Gulf Coast Prairie, Peninsular Florida, and South Atlantic), two US Department of Interior Climate Adaptation Science Centers (Southeast and South Central), and their partners. The LCCs provide the forums where SECAS partners can articulate and shape the future conservation landscape of the Southeast (Fig. 17.4; labeled A and B in Table 17.1). The Climate Adaptation Science Centers provide critical scientific research and products that are incorporated into the decision-making frameworks for SECAS (labeled C in Table 17.1). Important sponsors of the SECAS initiative are the fish and wildlife agencies of 15 states, through the regional Southeastern Association of Fish and Wildlife Agencies, and their federal counterparts, through the Southeast Natural Resources Leadership Group. The agencies and organizations of the state, federal, and nongovernmental CI provide the critical backbone of SECAS and will be called upon for implementation of the SECAS vision (labeled D in Table 17.1). SECAS, however, also envisions a much broader reach into other societal

Figure 17.3. The Upper Gunnison Basin habitat restoration project for Gunnison sage-grouse (*Centrocercus minimus*) in Colorado, USA. Before treatment (*top*), bare ground and litter occupied 41% in 2012. After treatment (*bottom*), bare ground and litter occupied 12% in 2017. Photos by R. Rondeau, Colorado Natural Heritage Program.

sectors to facilitate a comprehensive outcome of landscape sustainability. Nonconservation sectors such as urban planning, agricultural commodities, health care, transportation, and energy development are important components of society that shape our landscapes in ways that are not always conducive to fish and wildlife conservation goals.

An important goal of SECAS is to develop a conservation blueprint of a future conservation landscape as well as a new way of working that reflects a comprehensive and collaborative framework for achieving landscapes able to sustain the resources that the public needs and desires. This conservation blueprint represents a regional plan that transcends individual LCC boundaries. SECAS provides the regional framework that facilitates the alignment of

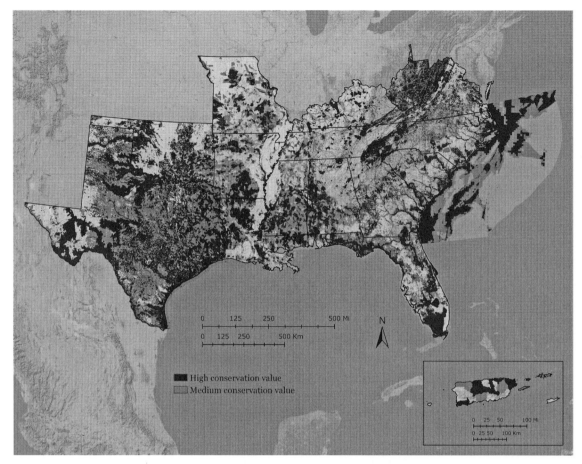

Figure 17.4. Blueprint for the Southeast Conservation Adaptation Strategy (Version 3.0) in the United States. High conservation-value areas are most important for ecosystem health, function, and connectivity. Medium conservation-value areas might require more restoration but are important for buffering high-value areas and maintaining connectivity. The Southeast Conservation Blueprint will continue to be refined through the collaborative partnership. Reproduced from Southeast Conservation Adaptation Strategy (2019).

conservation planning across LCC partnerships, so that the LCC Network can realize its strategic goal of an ecologically connected network of landscapes and seascapes (Landscape Conservation Cooperative Network 2014). The six LCCs that are working toward the larger vision of SECAS are composed of a diverse set of partners and landscapes and are addressing challenges to maintaining sustainable landscapes. This diversity can easily lead toward divergent conservation outcomes and approaches that are incompatible. Thus an initiative such as SECAS is needed to foster an alignment of conservation ob-

jectives that can be rolled up into a regional conservation goal.

SECAS builds on other efforts in the southeastern United States that attempt to articulate a future conservation vision for this vital region (labeled A in Table 17.1). Notably, the Southern Forest Futures Project (Wear and Greis 2012) provides a regional outlook for the future of southeastern forests, concluding that four interacting primary factors (population growth, climate change, timber markets, invasive species) will have dramatic effects on the future of forests in the South and on the fish and wildlife that depend on those forested landscapes. Similarly, the Southeast Regional Assessment Project was funded by the Southeast Climate Adaptation Science Center to provide critical data sets such as downscaled climate projections; regional terrestrial, aquatic, and coastal assessments; and other scientific information necessary for conservation planners to assess the effects of future changes on natural systems in the Southeast (Dalton and Jones 2010).

In 2006, anticipating a future in which the human population could double by 2060, Florida, USA, developed a first-of-its-kind planning tool called the Critical Lands and Waters Identification Project, a geographic information system database of statewide conservation priorities for a broad range of natural resources, including biodiversity, landscape function, surface water, groundwater, and marine resources (Oetting et al. 2016). The project, now on version 4.0, has become a primary conservation planning tool for the Peninsular Florida LCC and an important foundation for Florida's contribution to SECAS. The Peninsular Florida LCC is using the Critical Lands and Waters Identification Project to aid in identifying priority resources in the development of their blueprint and is working with multiple entities to develop additional geographic information system data sets that will be incorporated into the Florida LCD. These scenario planning analyses include multiple factors such as sea level rise, the amount and type of population growth, and financial resources for conservation. Consequently, future scenario planning efforts that incorporate climate change indicators such as sea level are important considerations for conservation planners in Florida.

Already, SECAS is providing on-the-ground benefits to wildlife managers and other partners throughout the southeastern region. The conservation blueprint is being used by the South Atlantic LCC to secure funding toward a directed conservation portfolio of projects ranging from coastal wetlands protection, wildland fire resilient landscapes, aquatic

connectivity, and National Wildlife Refuge planning. The LCC has also initiated an innovative collaboration with the American Planning Association to improve its Urban Open Space Indicator so that planners in urban environments can more effectively employ the tool in their community plans. On the Gulf Coast, a suite of projects will provide guidance on future conservation priorities that incorporate information on species and ecosystem vulnerabilities, the likely migration corridors of coastal wetlands as sea level rises, and resilient communities. These projects are being integrated into a Gulf Coast adaptation strategy that will become an important component of the larger SECAS vision.

What will the landscape of the Southeast look like in 2060? This is a fundamental question for SECAS, which uses all of the LCD elements (Table 17.1) as it seeks to establish a vision, goals, and objectives for the landscape of the future, incorporating the changes that are projected to occur over the next several decades and shaping a LCD that is capable of sustaining fish and wildlife. A functional landscape in 2060, one that sustains not only fish and wildlife but also provides for the ecosystem goods and services needed by human populations, will require a transformation in the way that the CI currently works. Resource agencies will need to transform their way of working by considering future changes on the front end; developing landscape-scale solutions that are built on a solid scientific foundation; investing time and resources into a portfolio of conservation actions and monitoring that promotes resiliency and can mitigate the most adverse effects of these changes; and entering into a collaborative enterprise with other landscape-shaping sectors to promote a more sustainable future. Conservation actions include the suite of species and land protection policies and tools that the CI has relied upon for the past 40-plus years, but they will need to include transformative decision-making processes that incorporate governance principles representative of contemporary societal expectations (Decker et al.

2016). Collaborative dialogue and governance structures designed to promote sustainability across multiple sectors of society are needed to ensure that conservation decision making includes a broad public discourse and broad public support. SECAS intends to show what the future conservation landscape of the Southeast should look like, it but will also explore the decision frameworks under which conservation decisions are made. This exploration will use recommendations from the National Research Council (2013:5) to employ a decision framework that "reflects relevant connections, identifies those in charge and those affected, and surveys what can be done to integrate the needs and responsibilities of all." SECAS will use the decision framework to make recommendations on improving governance structures that address key influences of ecosystem change in the Southeast, and to suggest governance structures to address future changes such as climate change-related stresses.

Case Study Summary

Collectively, these case studies demonstrate the benefit of using a collaborative landscape conservation approach that combines landscape ecology and social science principles to define and achieve desired social-ecological outcomes. The LCCs need to continue to evolve to address challenges to ensure achievement of outcomes desired by the partners. Those challenges include uncertainty associated with system stressors; the need for integration of human dimensions data and public engagement into the process from identification of conservation targets and desired future condition to implementation of actions and even monitoring; the need to clearly connect landscape-scale thinking and decisions to on-the-ground actions; and finding a balance between the outcomes desired by individual LCCs as self-directed partnerships and the recognition that desired ecological outcomes often require blurring or expansion of these already large geographic boundaries. These challenges are not unique

to LCCs and may apply to any large-landscape conservation effort, but the LCC model illustrates contemporary application of the principles of landscape ecology and provides an opportunity to evaluate and adapt future efforts. Working across scales, from local to continental, is the source of many of these challenges, but it is also where the potential lies in landscape-scale collaborative conservation.

Lessons Learned and Considerations for Landscape Conservation Cooperatives
Moving from Ecological to Social-Ecological Considerations

Although at different stages in their development, each of the cases had considerable and early participation of key conservation organizations, emphasizing the importance of working across bureaucratic and jurisdictional lines. Inherent in the LCC model is engagement of organizational stakeholders as steering committee or working group members or as collaborating entities (e.g., regional associations of fish and wildlife agencies, JVs). Jacobson and Robertson (2012) describe LCCs as bridging entities that convene diverse organizations in an adaptive co-governance process to work toward identifying and achieving shared visions for the future of their respective landscapes. In the case of the Gunnison Basin, partner involvement was demonstrable from selection of conservation targets by the steering committee all the way to site-specific implementation of climate adaptation activities. Because the restoration activities occurred mostly on public land, broad public buy-in may not have been as necessary as a scenario involving the cooperation of private landowners, such as in the Southeast blueprint. Successful implementation of LCDs involves full integration of social and ecological considerations into the process from visioning to monitoring and in some cases implementation and evaluation (e.g., citizen science, monitoring networks). By engaging stakeholders early and often, the Gunnison Climate Working Group effort demonstrated peripheral benefits (e.g., youth participation) that may help ensure support for other management actions recommended as part of a larger LCD for the sage-steppe geography.

Jacobson and Robertson (2012:342) note that "To fulfill their full potential as bridging entities, LCCs must institutionalize the use of social science information, along with ecological and other scientifically collected input, as a critical part of an adaptive co-governance framework to ensure that functional and resilient social-ecological systems persist into the future." Incorporation of social science data such as population growth and economic trends as part of the baseline information used to develop vulnerability assessments and identify likely future scenarios should be customary to initiating any LCD. To inform development of the conservation blueprint, the South Atlantic LCC is working with the American Planning Association Green Communities Research Center to examine how to incorporate green infrastructure and urban-scale definitions as part of the process to identify areas of shared conservation interest across the rural-urban continuum (South Atlantic Landscape Conservation Cooperative 2019). Further, the Gulf Coastal Plains and Ozarks LCC is assessing how to engage private landowners and provide incentives to sustain ecosystem services (e.g., clean water, biodiversity, wildlife habitat, recreation, aesthetics) in the southeastern United States through a focused case study in three major habitat types (bottomland hardwoods, open pine stands, and grasslands).

The Northwest Boreal LCC, conversely, is a largely intact geography, with relatively few people compared to the southeastern United States; however, climate change occurring at rates two to three times the global average amplifies the pressure for development potentially resulting in loss of terrestrial and aquatic connectivity and other effects (Wolken et al. 2011). Forecasting future scenarios of land use and emigration rates will be important in proactively identifying a desired future condition that ensures benefits for humans (e.g., economic

development, ecosystem services) and nonhuman species (e.g., unobstructed or spoiled habitat). The Northwest Boreal LCC is incorporating social demographic information into the LCD for this geography to facilitate its ability to proactively plan for anticipated changes on the landscape. For example, the partnership is planning for multi-jurisdictional landscape connectivity now, while the landscape is still intact, in addition to the work described in the case study above. Further, the LCC has funded work to evaluate the social network that exists among its partnership and beyond to facilitate maximum interaction and integration among collaborating entities. Through social network analysis, the LCC will identify communication flows and barriers and be able to target efforts toward strengthening collaboration (Bixler et al. 2016a). Moreover, similar to the work in the Crown of the Continent region (Bixler et al. 2016b), the Northwest Boreal LCC will use these social processes to set more comprehensive landscape conservation goals and objectives, identify priority conservation actions, and evaluate performance.

Although LCCs recognize the importance of public support for implementation of their landscape planning and design efforts, the social versus biological dimensions tend to be underemphasized (Scott and Tear 2007). Like other large partnership efforts, LCCs struggle with questions such as, Who are the stakeholders, and how and when should they become involved? Engaging diverse stakeholders in landscape-scale conservation will require creativity and new models of communication among the public, research communities, and managers. Typical science communication in journal publications and technical reports is not adequate to reach and inform decision makers. The decision makers need to be part of formulating research questions (based on management goals and objectives) and involved in every stage of the scientific process. This iterative model takes longer and is more complicated but ultimately should help bridge the gap between science and management. In the 21st century, scientists have a seat at the decision-making table and assist in finding complex, value-based solutions to conservation challenges (Chapin et al. 2010).

The work of the LCCs described here highlights that the limiting factors for implementing landscape-scale conservation often occur in the social realm, rather than stemming from lack of scientific data or technological capacity. The more that landscape ecologists can integrate with land and resource managers in the decision-making process, the more the scientific products will be applicable, and the more that experts in ecological pattern and process will be active in informing decisions.

Summary

Collaborative landscape conservation efforts—diverse partnerships working together at ecologically meaningful scales across jurisdictions to achieve collectively identified conservation outcomes—are important to address contemporary environmental problems. In a review of LCCs, the National Academies of Sciences, Engineering, and Medicine (2016:4) concluded that the "nation needs to take a landscape approach to conservation and that the U.S. Department of the Interior is justified in addressing this need with the Landscape Conservation Cooperatives." Since their beginning in 2010, LCCs individually and collectively have made progress in informing management decisions. While LCCs face challenges, the potential benefit of the LCC vision is undeniable. If considerations such as engaging the public and decision makers in the LCD process early and making appropriate linkages connecting landscape design to site-specific decisions are addressed, the LCC model in its current form or otherwise will continue to thrive. Critical to achieving desired conservation outcomes will be institutionalized consideration of the human dimensions of landscapes identified for conservation. Application of the principles of landscape ecology combined with social science inquiry, public participation, and evaluation of human demographic information—all of which may help support the sociopolitical feasibility of conservation outcomes—will likely play instrumental roles throughout the adaptive co-governance process, from visioning to management implementation, evaluation, and refinement (Doyle-Capitman et al. 2018). Ultimately, garnering social support for broadscale conservation will not be a simple undertaking. For example, landscape-scale goals may be too abstract for many stakeholders (e.g., ensuring that wildlife has habitat and connectivity among habitat areas sufficient to adapt to changing climatic conditions; Lee 2011). Support of management actions recommended in a LCD will therefore likely rely on promoting place-based connections (e.g., to local parks, reserves, or waterways) and increasing the relevance of conservation issues (e.g., demand for wildlife-related recreation, water quality) to local stakeholders with the power to support or impede conservation.

Further, conservation success considering challenges such as land use and climate change will depend on individual members of the CI to view their roles and efforts within an overarching landscape-scale context to facilitate adaptation to changing social-ecological conditions. Wiens (2007) notes that most conservation efforts focus on some scales (e.g., management units, species management) while ignoring others. By starting first at the larger, broader scale and considering connections among smaller scales on landscapes, landscape ecology principles help wildlife managers better address contemporary challenges. Adaptive co-governance partnerships such as LCCs are needed to engage stakeholders in setting achievable goals to ensure functional and resilient landscapes that sustain valued wildlife and other resources now and into the future. Transformation of the CI is needed to break out of jurisdictional and organizational silos and maximize conservation influences. Quite simply, if those of us who value fish, wildlife, and habitat do not maximize our collaborative efforts and engage the public in achieving functional and resilient landscapes, we will continue to lose conservation ground, both literally and figuratively.

LITERATURE CITED

Anderson, L. G. 2009. Quantitative methods for identifying ecological benchmarks in Canada's boreal forest. Thesis, University of Alberta, Edmonton, Alberta, Canada.

Association of Fish and Wildlife Agencies. 2019. State wildlife action plans—blueprints for conserving our nation's fish and wildlife. https://www.fishwildlife.org/afwa-informs/state-wildlife-action-plans. Accessed 15 Aug 2019.

Bixler, R. P., S. Johnson, K. Emerson, T. Nabatchi, M. Reuling, C. Curtin, M. Romolini, and J. M. Grove. 2016b. Networks and landscapes: a framework for setting goals and evaluating performance at the large landscape scale. Frontiers in Ecology and Environment 14:145–153.

Bixler, R. P., D. M. Wald, L. A. Ogden, K. M. Leong, E. W. Johnson, and M. Romolini. 2016a. Network governance for large-scale natural resource extraction and the challenge of capture. Frontiers in Ecology and Environment 14:165–171.

Boreal Ecosystems Analysis for Conservation Networks Project. 2019. The Canadian BEACONs Project: Boreal Ecosystem Analysis for Conservation Networks. http://www.beaconsproject.ca/. Accessed 15 Aug 2019.

Campellone, R., T. Chouinard, N. Fisichelli, J. Gallo, J. Lujan, R. McCormick, T. Miewald, B. Murry, J. Pierce, and D. Shively. 2018. The iCASS Platform: nine principles for landscape conservation design. Landscape and Urban Planning 176:64–74.

Caro, T. 2010. Conservation by proxy: indicator, umbrella, keystone, flagship and other surrogate species. Island Press, Washington, DC, USA.

Cash, D. W., and S. C. Moser. 2000. Linking global and local scales: designing dynamic assessment and management processes. Global Environmental Change 10:109–120.

Chapin, F. S., III, S. R. Carpenter, G. P. Kofinas, C. Folke, N. Abel, W. C. Clark, P. Olsson, D. M. S. Smith, B. Walker, O. R. Young, et al. 2010. Ecosystem stewardship: sustainability strategies for a rapidly changing planet. Trends in Ecology and Evolution 25:241–249.

Conservation Measures Partnership. 2019. The open standards for the practice of conservation. http://cmp-openstandards.org/. Accessed 15 Aug 2019.

Dalton, M. S., and S. A. Jones, compilers. 2010. Southeast Regional Assessment Project for the National Climate Change and Wildlife Science Center. US Geological Survey Open-File Report 2010–1213, Washington, DC, USA.

Decker, D. J., C. Smith, A. Forstchen, D. Hare, E. Pomeranz, C. Doyle-Chapman, K. Schuler, and J. Organ. 2016. Governance principles for wildlife conservation in the 21st century. Conservation Letters 9:290–295.

Dorado, S. 2005. Institutional entrepreneurship, partaking and convening. Organization Studies 26:385–414.

Doyle-Capitman, C. E., D. J. Decker, and C. A. Jacobson. 2018. Toward a model for local stakeholder participation in landscape-level wildlife conservation. Human Dimensions of Wildlife 23:375–390.

Folke, C., T. Hahn, P. Olsson, and J. Norberg. 2005. Adaptive governance of social-ecological systems. Annual Review of Environment and Resources 30:441–473.

Groves, C., and E. Game. 2015. Conservation planning: informed decisions for a healthier planet. First edition. Macmillan Learning, New York, New York, USA.

Haila, Y. 2007. Enacting landscape design: from specific cases to general principles. Pages 22–34 in D. B. Lindenmayer and R. Hobbs, editors. Managing and designing landscapes for conservation: moving from perspectives to principles. Blackwell, Malden, Massachusetts, USA.

Haubold, E. M. 2012. Using adaptive leadership principles in collaborative conservation with stakeholders to tackle a wicked problem: imperiled species management in Florida. Human Dimensions of Wildlife 17:344–356.

Heifetz, R. A., and D. L. Laurie. 2001. The work of leadership. Harvard Business Review 79:131–140.

Hobbs, R. J. 2007. Goal, targets and priorities for landscape-scale restoration. Pages 511–526 in D. B. Lindenmayer and R. Hobbs, editors. Managing and designing landscapes for conservation: moving from perspectives to principles. Blackwell, Malden, Massachusetts, USA.

Jacobson, C. A. 2016. Adoption and diffusion of wildlife governance principles: challenges and suggestions for moving forward. Transactions of the 81st North American Wildlife and Natural Resources Conference. 81:178–184.

Jacobson, C. A., and D. J. Decker. 2006. Ensuring the future of state wildlife management: understanding challenges for institutional change. Wildlife Society Bulletin 34:531–536.

Jacobson, C. A., D. J. Decker, and L. Carpenter. 2007. Securing alternative funding for wildlife management: insights from agency leaders. Journal of Wildlife Management 71:2106–2113.

Jacobson, C. A., and E. M. Haubold. 2014. Landscape conservation cooperatives: building a network to help fulfill public trust obligations. Human Dimensions of Wildlife 19:427–436.

Jacobson, C. A., J. F. Organ, and D. J. Decker. 2010. Fish and wildlife conservation and management in the 21st century: understanding challenges for institutional transformation. Transactions of the 75th North American Wildlife and Natural Resources Conference 75:107–114.

Jacobson, C. A., and A. L. Robertson. 2012. Landscape conservation cooperatives: bridging entities to facilitate adaptive co-governance of social-ecological systems. Human Dimensions of Wildlife 17:333–343.

Kretser, H. E., M. V. Schiavone, D. Hare, and C. A. Smith. 2016. Applying wildlife governance principles: opportunities and limitations. Transactions of the 81st North American Wildlife and Natural Resources Conference 81:152–160.

Landscape Conservation Cooperative Network. 2014. LCC network strategic plan. https://lccnetwork.org/resource /landscape-conservation-cooperative-network-strategic -plan. Accessed 15 Aug 2019.

Landscape Conservation Cooperative Network. 2015. 2015 LCC Network areas. https://www.sciencebase.gov /catalog/item/55b943ade4b09a3b01b65d78. Accessed 15 Aug 2019.

Landscape Conservation Cooperative Network. 2016. Characteristics of landscape conservation design. https://lccnetwork.org/resource/lcc-network-landscape -conservation-design-characteristics. Accessed 15 Aug 2019.

Lee, C. W. 2011. The politics of localness: scale-bridging ties and legitimacy in regional resource management partnerships. Society and Natural Resources 24(5):439–454.

Leroux, S. J., F. K. A. Schmiegelow, S. G. Cumming, R. B. Lessard, and J. Nagy. 2007b. Accounting for system dynamics in reserve design. Ecological Applications 17:1954–1966.

Leroux, S. J., F. K. A. Schmiegelow, R. B. Lessard, and S. G. Cumming. 2007a. Minimum dynamic reserves: a conceptual framework for reserve size. Biological Conservation 138:464–473.

Lindenmayer, D. B., and J. F. Franklin. 2002. Conserving forest biodiversity: a comprehensive multi-scaled approach. Island Press, Washington, DC, USA.

Manfredo, M. J., T. L. Teel, and A. D. Bright. 2004. Applications of the concept of values and attitudes in human dimensions of natural resources research. Pages 271–282 in M. J. Manfredo, J. J. Vaske, B. L. Bruyere, et al., editors. Society and natural resources: a summary of knowledge. Modern Litho, Jefferson, Missouri, USA.

Manolis, J. C., K. M. Chan, M. E. Finkelstein, S. Stephens, C. R. Nelson, J. B. Grant, and M. P. Dombeck. 2009. Leadership: a new frontier in conservation science. Conservation Biology 23:879–886.

Manomet Center for Conservation Sciences and Massachusetts Division of Fisheries and Wildlife. 2010. Climate change and Massachusetts fish and wildlife. Volumes 1–3. Manomet Center for Conservation Sciences and Massachusetts Division of Fisheries and Wildlife, West-borough, Massachusetts, USA. https://www.cakex.org /documents/climate-change-and-massachusetts-fish -and-wildlife-volume-1-introduction-and-background.

Millard, M. J., C. A. Czarnecki, J. M. Morton, L. A. Brandt, J. S. Briggs, F. S. Shipley, and J. Taylor. 2012. A national geographic framework for guiding conservation on a landscape scale. Journal of Fish and Wildlife Management 3:175–183.

National Academies of Sciences, Engineering, and Medicine. 2016. A review of the Landscape Conservation Cooperatives. National Academies Press, Washington, DC, USA.

National Oceanic and Atmospheric Administration. 2019. http://coralreefwatch.noaa.gov/climate/projections /piccc_oa_and_bleaching/index.php. Accessed 15 Aug 2019.

National Research Council. 2013. Sustainability for the nation: resource connection and governance linkages. National Academies Press, Washington, DC, USA.

Nature Conservancy and Gunnison Climate Working Group. 2015. Annual report to the Terra Foundation. Restoration/resilience building of riparian and wet meadow habitats: Upper Gunnison Basin, Colorado. https://tnc.box.com/s/qppy9r2ju6tcf2yjf4o3zfgcoxp 57ur8. Accessed 15 Aug 2019.

Neely, B., R. Rondeau, J. Sanderson, C. Ague, B. Kuhn, J. Siemers, L. Grunau, J. Robertson, P. McCarthy, J. Barsugli, et al., editors. 2011. Gunnison Basin: climate change vulnerability assessment for the Gunnison climate working group by The Nature Conservancy, Colorado Natural Heritage Program, Western Water Assessment, University of Colorado, Boulder, and University of Alaska, Fairbanks. Project of the Southwest Climate Change Initiative. https://www.conservation-gateway.org/ConservationByGeography/NorthAmerica /UnitedStates/Colorado/science/Pages/gunnison-basin -climate-ch.aspx. Accessed 15 Aug 2019.

Nie, M. 2004. State wildlife policy and management: the scope and bias of political conflict. Public Administration Review 64:221–233.

Northwest Boreal Landscape Conservation Cooperative. 2019. Vision and mission. https://Northwest Boreallcc .org/?page_id=82. Accessed 15 Aug 2019.

Oetting, J., T. Hoctor, and M. Volk. 2016. Critical lands and waters identification project (CLIP): Version 4.0 technical report. http://fnai.org/pdf/CLIP_v4_technical_report .pdf. Accessed 15 Aug 2019.

Perera, A. H., L. J. Buse, and T. R. Crow. 2007. Knowledge transfer in forest landscape ecology: a primer. Pages 1–18 in A. H. Perera, L. J. Buse, and T. R. Crow, editors. Forest landscape ecology: transferring knowledge to practice. Springer, New York, New York, USA.

Schmiegelow, F. K. A., S. G. Cumming, K. A. Lisgo, S. J. Leroux, and M. A. Krawchuk. 2014. Catalyzing large landscape conservation in Canada's boreal systems: the BEACONs Project experience. Pages 97–122 *in* J. N. Levitt, editor. Conservation catalysts: the academy as nature's agent. Lincoln Institute of Land Policy, Cambridge, Massachusetts, USA.

Scott, M. J., and T. H. Tear. 2007. What are we conserving? establishing multiscale conservation goals and objectives in the face of global threats. Pages 494–510 *in* D. B. Lindenmayer and R. Hobbs, editors. Managing and designing landscapes for conservation: moving from perspectives to principles. Blackwell, Malden, Massachusetts, USA.

Scott, W. R. 2001. Institutions and organizations. Second edition. Sage, Thousand Oaks, California, USA.

Smith, C. A. 2011. The role of state wildlife professionals under the public trust doctrine. Journal of Wildlife Management 75:1539–1543.

South Atlantic Landscape Conservation Cooperative. 2019. Improving the natural and built environment connection in the Blueprint. http://www.southatlanticlcc.org/projects/. Accessed 15 Aug 2019.

Southeast Conservation Adaptation Strategy. 2019. The Southeast Conservation Blueprint. http://secassoutheast.org/blueprint. Accessed 15 Aug 2019.

Stein, B. A., P. Glick, N. Edelson, and A. Staudt, editors. 2014. Climate-smart conservation: putting adaptation principles into practice. National Wildlife Federation, Washington, DC, USA.

Stralberg, D., E. M. Bayne, S. G. Cumming, P. Solymos, S. J. Song, and F. K. A. Schmiegelow. 2015. Consideration of future boreal forest bird communities considering lags in vegetation response to climate change: a modified refugia approach. Diversity and Distributions 21(9):1112–1128.

Wear, D. N., and J. G. Greis. 2012. The Southern Forest Futures Project: summary report. General Technical Report SRSGTR168. USDA Forest Service, Southern Research Station. Asheville, North Carolina, USA.

Wiens, J. A. 2007. Does conservation need landscape ecology? a perspective from both sides of the divide. Pages 479–493 *in* D. B. Lindenmayer and R. Hobbs, editors. Managing and designing landscapes for conservation: moving from perspectives to principles. Blackwell, Malden, Massachusetts, USA.

Wolken, J. M., T. N. Hollingsworth, T. S. Rupp, F. S. Chapin III, S. F. Trainor, T. M. Barret, P. F. Sullivan, A. D. McGuire, E. S. Euskirchen, P. E. Hennon, et al. 2011. Evidence and implications of recent and projected climate change in Alaska's forest ecosystems. Ecosphere 2(11):1–35.

Young, B., E. Byers, K. Gravuer, K. Hall, G. Hammerson, and A. Redder. 2011. Guidelines for using the NatureServe Climate Change Vulnerability Index, Release 2.1. NatureServe, Arlington, Virginia, USA.

Zaveleta, A., and J. R. Pasari. 2007. Managing landscapes for vulnerable, invasive and disease species. Pages 311–329 *in* D. B. Lindenmayer and R. Hobbs, editors. Managing and designing landscapes for conservation: moving from perspectives to principles. Blackwell, Malden, Massachusetts, USA.

Zeedyk, B., and V. Clothier. 2014. Let the water do the work: induced meandering, an evolving method for restoring incised channels. Second edition. Chelsea Green, Hartford, Vermont, USA.

18 — Mapping Priority Areas for Species Conservation

Casey A. Lott,
Jeffery L. Larkin,
Darin J. McNeil,
Cameron J. Fiss, and
Bridgett E. Costanzo

Introduction

Throughout the Appalachian Mountains, USA, a reduction in the frequency and severity of disturbance, coupled with heavy browsing pressure from white-tailed deer (*Odocoileus virginianus*), has resulted in a shortage of young forest stands that are important to wildlife (Askins 2001, Dey 2014, King and Schlossberg 2014). Most acutely, steep population declines (Sauer et al. 2014) and range contractions (Rosenberg et al. 2016) for the golden-winged warbler (*Vermivora chrysoptera*) resulted in a petition to list the species under the US Endangered Species Act (US Fish and Wildlife Service 2011). Currently, young forest patches must be deliberately created using silvicultural or other management practices to supplement natural disturbance enough to meet nesting habitat needs for some species of early-successional wildlife (DeGraaf and Yamasaki 2003, Brose et al. 2008). Many different state and federal agencies, and several private conservation groups, are currently working to increase the amount of young forest to benefit declining wildlife populations (Roth et al. 2012).

Over the past century, active forest management has been focused primarily on public and industry lands where extensive forest tracts occur (Mac-Cleery 2011). In the Appalachian Mountains, however, a high proportion of forested lands is privately owned. For example, in Pennsylvania, USA, 68% of the state's 6.84 million ha of forestland is privately owned by 736,000 individual, nonindustrial, private landowners (Metcalf et al. 2012). Conservation action on private forests is essential to improve regional habitat conditions for early-successional birds (Thogmartin and Rohweder 2009, Stauffer et al. 2017) and has the added benefit of reestablishing more diverse and valuable timber stands for private landowners.

Through the US Farm Bill, a number of technical and financial assistance programs have become available to private forest owners working with the US Department of Agriculture. Farm Bill conservation funding is significant ($29 billion [USA] in 2014), and these funds have the potential to influence habitat management and restoration on private forests at large spatial extents (Ciuzio et al. 2013). The US Department of Agriculture Natural Resources Conservation Service has recognized this opportunity through the implementation of Working Lands for Wildlife (WLFW) partnerships that focus on restoring ecosystem health through habitat management for target species in decline (e.g., golden-winged warbler). The WLFW program provides leverage for collaboration in the implementation of land management practices that support the economic viability of forests, ranches, and farms while also conserving at-risk wildlife populations (US Natural Resources Conservation Service 2015, 2016a). Examples already exist (greater sage-grouse [*Centrocercus urophasianus*], New England cottontail [*Sylvilagus transitionalis*]) where the WLFW model has been instrumental in keeping species off of the list of threatened and endangered wildlife, thereby reducing regulatory burden on private landowners (US Fish and Wildlife Service 2015).

Priority Area Mapping

Species distribution models and potential habitat maps are commonly promoted as essential tools for conservation planning (Guisan et al. 2013). In theory, if resource managers have maps showing the environmental attributes, they can manipulate management to improve species-specific habitat conditions, and they have the information they need to successfully meet conservation objectives for target species. In practice, however, implementation of any type of active management to benefit a target species is subject to multiple constraints related to land ownership, other sensitive species needs, conflicting stakeholder perspectives, budgetary realities, remote area access issues, and many other factors. It is only when potential habitat maps are intersected with maps of these constraints that locations providing real-world opportunity for objective-oriented management begin to crystallize.

For this reason, WLFW has developed spatially explicit Priority Areas for Conservation (PAC) maps for each of its target species. Methods for PAC creation have varied across species with different life histories and interactions with management (US Natural Resources Conservation Service 2015, 2016a, 2016b). Common objectives have been to focus private lands habitat conservation efforts supported by WLFW in areas that will have the greatest benefit to target species, and to illustrate how private

and public land conservation efforts may need to be combined to reach regional or range-wide targets for priority species.

Our objective in this chapter is to describe methods of PAC development for golden-winged warblers in the Appalachian Mountains. For golden-winged warblers, as with other species, the Natural Resources Conservation Service requested that PAC development follow the best available science, supplemented by additional analyses if necessary. Several best management practice documents are already available to support local, site-specific nesting habitat creation efforts for golden-winged warbler, mostly at a scale of one to several breeding territories (Bakermans et al. 2011, Roth et al. 2012, Golden-Winged Warbler Working Group 2013). Therefore we focused golden-winged warbler PAC development on identifying areas where landscape context is most appropriate for the creation of golden-winged warbler nesting habitat and young forest creation.

We followed a series of 7 steps to develop spatially explicit PAC for golden-winged warblers. Each of these steps is commonly taken in the development of priority area maps. First, we conducted a review of peer-reviewed articles and documents on best management practices to identify specific environmental characteristics that relate to golden-winged warbler occurrence during the breeding season. Second, we identified and evaluated a range of geographic information system (GIS) landcover products across our region of interest that could be used to map these characteristics. Third, we consolidated golden-winged warbler presence-absence data sources and filtered these to remove low-quality records relative to our mapping objectives. Fourth, we constructed statistical models of relationships between the occurrence of golden-winged warbler and landcover characteristics (identified during step 1) by extracting environmental data from GIS layers (step 2) at high-quality bird presence-absence locations (step 3). Fifth, we prepared a set of PAC maps for golden-winged warblers using statistical models from step 4 that were designed to either support targeted conservation near recent breeding territories or identify larger areas with suitable landscape context that currently lack golden-winged warbler populations but might support them if nesting habitat were available. Sixth, we removed locations within priority areas with local constraints that would preclude habitat management (e.g., open spaces near development, wetlands with sensitive species). Finally, we intersected final PAC maps with protected area boundaries and illustrated how management authority within PAC is spread across public and private ownership domains.

The last three steps do not always receive the emphasis they deserve for maps to work as effective tools for setting real-world conservation targets, informing effective management implementation plans that include multiple partners, and evaluating progress toward the attainment of local to ecoregional conservation targets. Next, we discuss how we dealt with challenging elements of spatial extent and landscape context in PAC delineations related to the patchy distribution of golden-winged warbler and spatial bias inherent to our final data set of filtered observations.

Spatial Extent and Landscape Context

We initially considered mapping potential golden-winged warbler habitat relationships across the entire Appalachian Mountains Level 3 ecoregion (Omernik and Griffith 2014; Fig. 18.1A). But much of this large region may be unavailable to golden-winged warblers, at least in the short term, owing to the absence of recent breeding populations (Roth et al. 2012, Rohrbaugh et al. 2016). Hijmans (2012) suggested that studies of species-habitat relationships should incorporate distances among presence samples to inform the selection of the spatial extents from which absence samples are selected. Following this suggestion, we chose our modeling area extent based on an analysis of nearest neighbor distances among 1,337 unique warbler presence locations (Fig. 18.1B). All

golden-winged warbler observations were <30 km from another golden-winged warbler observation; 97.5% were <2.5 km, and 90% were <4.3 km. We conservatively buffered all golden-winged warbler presence samples by 30 km (the max distance between any golden-winged warbler observation and its closest neighboring observation) and conducted all habitat analyses within this buffer (Fig. 18.1A). We consider areas outside of this buffer, many of which have been heavily sampled by eBird users and have not yielded golden-winged warbler observations, as unlikely to be heavily used, at least in the next few decades, owing to the absence of nearby source populations. We recognize that other studies of golden-winged warbler habitat relationships could come to different conclusions if they included large numbers of samples from outside of this buffered area (Crawford et al. 2016). The criteria listed above resulted in a 13.5-million-ha analysis extent representing all locations within 30 km of a golden-winged warbler detection during the breeding season that we compiled between 2005 and 2017 (Fig. 18.1B).

Golden-winged warblers usually nest within landscapes dominated by deciduous forest (Confer et al. 2011). While regular local disturbances that create early-successional nesting habitat patches are necessary, extensive forest loss and fragmentation at landscape scales may have detrimental effects on golden-winged warbler populations (Confer et al. 2011). Our focus for golden-winged warbler PAC development was to map landscape-scale patterns in forest cover and human-dominated cover types within our analysis extent, and to identify areas with suitable landscape context where the site-focused creation of young forest nesting patches could attract golden-winged warblers.

Literature Review and Landcover Associations

During our literature review we found that different studies used different combinations of landcover

A

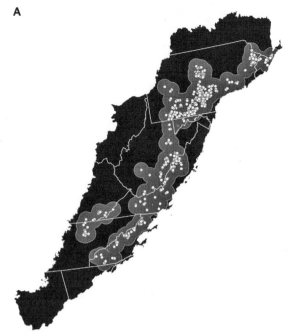

B

Figure 18.1. (*A*) Level 3 Appalachian Mountains ecoregion of the eastern United States (large, dark shape) and our analysis extent (smaller, lighter shape) formed by a dissolved 30-km buffer around 1,337 unique locations where golden-winged warblers were observed between 2005 and 2017. (*B*) All 11,492 golden-winged warbler absence records within the Appalachian Mountains ecoregion are from field surveys or eBird observations between 2005 and 2017. A subset of these absence records (5,961) is within the 30-km buffer surrounding golden-winged warbler locations.

classes in analyses of golden-winged warbler habitat. We also found that percentage forest cover was measured across a wide range of spatial scales, from 500 m to 24 km (Desrochers et al. 2010). Consequently, studies varied in their landscape recommendations for managed golden-winged warbler nesting habitat. For example, Bakermans et al. (2011) recommended landscape-scale forest cover of >70% within 800 m of nesting habitat but did not specify which combination of landcover classes comprised their definition of forest. Roth et al. (2012) and the Golden-Winged Warbler Working Group (2013) made general recommendations that nesting habitat creation occurs within landscapes of >50% and >60% forest cover, respectively, at the 2.5-km scale. Percentage landcover threshold recommendations also varied in thematic focus and spatial scale across different focal areas, with separate percentage forest cover recommendations for deciduous forest (positively associated with golden-winged warbler nesting) and evergreen forest (a negative association; Golden-Winged Warbler Working Group 2013). Other investigations of bird-habitat relationships in eastern forests have pooled all four of the National Land Cover Database (NLCD; Wickham et al. 2014) forest classes (deciduous forest, mixed forest, coniferous forest, forested wetlands) into a single forest category for analysis (Shake et al. 2012, Frazier and Kedron 2017). The decision about which mapped forest types are included or excluded in map-based analyses of percentage forest cover can make a difference of hundreds of thousands of hectares that meet a specific percentage cover threshold in priority area maps covering a large landscape. Therefore we conducted percentage cover analyses with different combinations of forest type classifications and chose the combination that best predicted golden-winged warbler occurrence.

In contrast to GIS-based analyses that rely on broad classifications of forest cover, field-based research reported that many Appalachian forest bird species, including golden-winged warbler, can be associated with specific deciduous forest types (e.g., maple; Rodewald and Abrams 2002, Roth and Lutz 2004). Each of the forest cover threshold recommendations for golden-winged warblers relates to the NLCD, which lacks the thematic resolution to parse these relationships. We identified two GIS products that map tree basal area by species across our entire study extent (Wilson et al. 2012, Ellenwood et al. 2015). We believe this would be the preferable thematic resolution to investigate the effects of forest composition on golden-winged warbler occurrence. Only one of these products has undergone peer review, and accuracy assessment has occurred only at coarse spatial grains (Wilson et al. 2012). We decided to use the NLCD to map landscape-scale forest cover for PAC development, despite its lower thematic resolution, because it has undergone systematic design-based accuracy assessment, where high classification accuracies have been reported for different forest types (Wickham et al. 2013). We recognize that this decision may result in errors of commission in the delineation of priority areas if golden-winged warblers strongly avoid different forest types that are lumped together as deciduous under the more general classification scheme of the NLCD.

Warbler Occurrence Data Sources

We first compiled golden-winged warbler occurrence data from a variety of sources associated with field research studies (see the acknowledgments in this volume). We restricted this data set to observations from 2005 to 2017, bracketing the 2011 NLCD map that we used for cover type analyses by 6 years on each side. Some golden-winged warbler occur-

rence data sources reflected narrowly focused field research studies with relatively small spatial extents (Bakermans et al. 2015, Leuenberger et al. 2017), while others represented coordinated efforts among multiple partners to collect golden-winged warbler occurrence data from multiple different nesting habitats across broad extents (Vallender et al. 2009, Aldinger et al. 2015, McNeil et al. 2017). All occurrence data from field research studies included geographic coordinates representing sampling locations. Golden-winged warbler occurrence points in these data sets had consistently high spatial and temporal precision relative to our analysis objective of developing criteria for landscape context surrounding nesting habitat.

In addition to data from field studies, we acquired breeding season presence-absence records for golden-winged warbler from eBird, an extensive citizen science data set that is frequently used in species distribution modeling (Sullivan et al. 2014). While eBird survey locations are reported as precise x and y coordinates in downloadable data sets, observations tend to reflect nonstationary birding sessions where observers may cover locations that are distant from this recorded point location. In practice, there can be considerable differences between the exact point locations recorded in the database and the actual locations of bird observations (Munson et al. 2015). Thus x and y coordinates associated with eBird observations are often not precisely related to the exact habitat where the bird was observed. Because positional accuracy can affect inference in habitat modeling (Naimi et al. 2014, Munson et al. 2015), we discarded all eBird observations with an effort distance value >0 prior to PAC delineation to record habitat features (albeit at landscape scales) associated only with specific breeding territories. We used eBird presence-absence records only from a narrow date window of 1–30 June. This date range was selected to remove observations of late spring migrants and post-fledging broods where observations may not accurately reflect the location of breeding territories, as adults and broods can move considerable distances from nest sites and defended song territories shortly after fledging (King et al. 2006, Vitz and Rodewald 2006, Streby et al. 2016).

Filtering Bird Occurrence Data

Many eBird locations had multiple observations (in some cases hundreds) from the same sampling point both within and among years. Similarly, some of our field research data sets had >1 observation from the same location. To minimize spatial sampling bias in analyses of habitat associations, we combined field research and eBird data sets and used only a single observation and exact location (for both presence and absence records) in subsequent analyses. For locations with multiple observations within 100 m of each other, we retained only a single location, assuming that observations this close together were likely to relate to the same breeding habitat patch. This approach resulted in a final filtered data set of 7,298 unique locations: 1,337 with golden-winged warbler presence records and 5,961 reflecting golden-winged warbler absence.

Our final set of observations still exhibited spatial clustering, where some areas had higher densities of observations than others (Fig. 18.1B). Filtered observations were neither spatially random, systematic, nor balanced across our study area. This is often the case with citizen science and collated research data sets that do not employ design-based sampling at regional scales (Aiello-Lammens et al. 2015). Therefore, as a precursor to statistical modeling, we employed several approaches to select individual points from the full pool of filtered locations to minimize different aspects of spatial bias inherent to these data sets (Holland et al. 2004, Robinson et al. 2018). Specifically, we used a tool developed by Holland et al. (2004) to generate 50 unique sets of randomly selected and spatially independent points for presence and absence samples. This resampling approach repetitively selects subsets of points from the total pool of samples, with replacement, by randomly selecting a first point and then randomly select-

ing additional points with the constraint that each sample in the final set of points has to be spatially independent (e.g., nonoverlapping distance buffers) at the given scale of analysis. This approach helps to minimize spatial autocorrelation and avoid the generation of spurious results based on overfitting a model to the specific properties of any one group of samples (Verbyla and Litvaitis 1989, Wenger and Olden 2012). The selection of an equal number of presence-absence observations at each spatial scale is important because we analyzed habitat relationships using logistic regression, where results can be sensitive to uneven sample sizes among classes (Keating and Cherry 2004). We also used model selection to identify characteristic scales of habitat relationships (Jackson and Fahrig 2012) and direct comparison of Akaike information criterion, corrected for small sample sizes (AICc) scores among models requiring that models have similar sample sizes (Anderson 2008).

Modeling Cover Type Relationships across Spatial Scales

We used logistic regression and model selection to evaluate the effect of different combinations of land-cover classes, measured at different landscape scales, on golden-winged warbler occurrence. On the basis our literature review we selected three spatial scales for analyses of percent cover: 1 km, 2.5 km, and 5 km. These numbers reflect the radii of circles around any single sampling point where habitat characteristics are measured and summarized. The 1-km and 2.5-km scales have been explored in other golden-winged warbler studies (Roth et al. 2012, Golden-Winged Warbler Working Group 2013, Crawford et al. 2016, Wood et al. 2016). We added the 5-km scale to this analysis because landscape-scale forest cover associations for golden-winged warblers have been documented at scales as large as 10,000 m in the Great Lakes (Thogmartin 2010), and the abundance of some forest bird species increase up to this scale (Schlossberg and King 2007). We elected not to

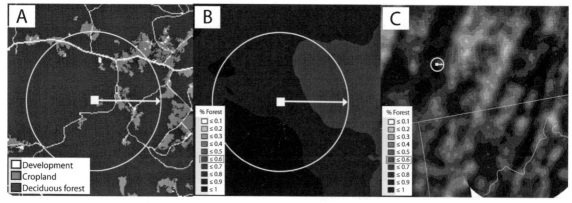

Figure 18.2. (*A*) A focal cell (the white square) on top of the National Land Cover Database at a randomly selected location within the Appalachian Mountains, USA, 2005-2017, where roads and other developed areas are white, cropland areas are light gray, and forested areas are dark gray. The white line with an arrow at the end, in all three panels, is a 2.5-km radius indicative of the 2.5-km landscape scale. This radius creates a circle that has an area of 1,973 ha. For percentage cover analyses at landscape scales, the percentage of any given landcover type (e.g., developed, cropland, deciduous forest) is calculated across this entire area, and values (%) are assigned back to the focal cell. In a moving window analysis, each of the 30-m² pixels of the landcover data set is treated as a focal cell. This creates a map (*B*) that shows deciduous forest (%) at the 2.5-km scale for each 30-m² pixel. Percentage forest cover values are higher in the western part (*B*) and lower in the eastern part, as one might expect from patchy patterns of landcover (*A*). (*C*) Percentage forest cover zoomed out to show percentage deciduous forest cover values for an area of ~60 km². The same focal cell (*A, B*) is shown for comparison (*C*). For habitat modeling, percentage cover values for different combinations of landcover classes were extracted from landscape context maps for every golden-winged warbler presence or absence point at each of our three spatial scales of analysis: 1 km, 2.5 km, and 5 km.

investigate scales >5 km in our study region because buffers of this size around sampling points overlapped such that locations were no longer spatially independent. At each spatial scale we performed moving window analyses (McGarigal and Marks 1995) using the 2011 NLCD to create landscape context maps by calculating the percentage of each cover type at each scale for every 30-m² NLCD raster cell within our golden-winged warbler PAC study extent (Fig. 18.2). These analyses resulted in maps of percentage forest cover, and percentage anthropogenic cover types, for buffered areas surrounding bird presence-absence records of 314 ha (e.g., a circle with a 1-km radius), 1,973 ha (a circle with a 2.5-km radius), and 7,864 ha (a circle within a 5-km radius). We extracted percentage cover values from landscape context maps for each resampled location in our golden-winged warbler presence-absence data set and used these values in subsequent logistic re-

gression analyses of the effects of different types of landscape composition on golden-winged warbler occurrence.

Given inconsistency in prior analyses of landcover effects on golden-winged warbler, we took a robust approach to clarifying which single or combined cover type groups were most strongly associated with golden-winged warbler occurrence. For each of the 10 individual NLCD landcover classes and each of the 7 combined landcover types (Table 18.1), we constructed logistic regression models of the effect of cover type on golden-winged warbler occurrence. Following the resampling protocols described above, we generated 50 replicate logistic regression models for each of these 17 landcover groups, where each model had an equal number of spatially independent presence-absence records and where the exact composition of samples used for each model varied across all 50 replicates. We constructed separate models for each of the 17 landcover type or groups of types (Table 18.1) at each of our 3 spatial scales. As a first assessment of model adequacy, we viewed each landcover x-scale combination where intercept-only models performed consistently better than models with explanatory variables as evidence for no relationship between explanatory variables and bird occurrence. For the remaining landcover x-scale combinations, we calculated the percentage of models where 95% confidence intervals for the effect of explanatory variables on golden-winged warbler occurrence did not overlap zero for ≥60% of our 50 replicates, indicating either positive or negative effects of a landcover x-scale combination on bird occurrence. From this remaining pool of landcover–golden-winged warbler occurrence relationships, we identified the spatial scale with the lowest AICc value, averaged across all 50 replicates, for each landcover group.

We found a positive relationship between the proportion of deciduous forest cover and golden-winged warbler occurrence at relatively large landscape scales (2.5 km and 5 km), with the 2.5-km scale having the most support. We did not find support for a relationship between golden-winged warbler occurrence and any other NLCD forest class (forested wetland, mixed forest, or evergreen forest) at any of the three scales we investigated. While others have recommended landscape context thresholds for evergreen forest cover for golden-winged warblers (Roth et al. 2012, Golden-Winged Warbler Working Group 2013), we chose not to use the proportion of evergreen forest cover in predictive models, or as exclusion criteria in final PAC, based on the evi-

Table 18.1. Single and Combined National Land Cover Database Cover Type Classes Used in Logistic Regression Analyses Comparing Landcover at 1-, 2.5-, and 5-km Scales and Golden-Winged Warbler Occurrence in the Appalachian Mountains, USA, 2005–2017

NLCD Class Name	NLCD Class Number(s)	Golden-Winged Warbler Priority Areas for Conservation Study Extent (%)	Full Appalachian Area (%)
Deciduous forest	41	63.4	57.5
Forested wetland	90	0.9	0.9
Deciduous forest and forested wetland	41 and 90	64.3	58.3
Mixed forest	42	3.6	4.8
Evergreen forest	43	4.1	5.3
Mixed forest and evergreen	42 and 43	7.7	10.2
All forest types	41, 42, 43, 90	72.0	68.5
Open space	20	6.0	5.6
Light development	21	2.0	1.9
Open space and light development	20 and 21	8.0	7.5
Moderate development	22	0.8	0.8
Heavy development	23	0.2	0.2
Moderate and heavy development	22 and 23	1.1	1.0
All development	21–24	9.1	8.5
Cropland	82	3.3	3.9
Hay pasture	81	11.5	13.1
All cultivated	81 and 82	14.7	17.0

Source: Wickham et al. (2014).

Note: NLCD, National Land Cover Database. Columns show the area of each NLCD cover type (or cover type combination) within our golden-winged warbler Priority Areas for Conservation study extent (where analysis points were restricted to locations within 30 km of recent golden-winged warbler observations) and for the entire Appalachian ecoregion.

dence presented here. As with previous studies, we found support for negative landscape-scale effects of development on golden-winged warbler occurrence that were strongest at the 2.5-km scale. Effects were slightly stronger for the moderate and heavy development category than for the open space and light development category, but confidence intervals overlapped, and the moderate and heavy development classes combined comprised only 1.1% of the analysis extent. Therefore we modeled the negative effect of percentage development cover on golden-winged warbler occurrence with all four development classes pooled at the 2.5-km scale.

Rohrbaugh et al. (2016) proposed a maximum percentage cover criterion of 25% for cultivated land, with cropland and hay and pasture classes combined. When we evaluated these cover types separately, slopes differed for their effect on golden-winged warbler occurrence. We found no effect for percentage hay and pasture on golden-winged warbler occurrence and a negative effect for percentage cropland cover on golden-winged warbler occurrence that was strongest at the 1-km scale. On the basis of these results, we used quantitative relationships describing the positive effect of percent deciduous forest on golden-winged warbler occurrence at 2.5 km, negative effects of development at 2.5 km, and negative effects of percentage cropland at the 1-km scale to develop PAC. In addition to percentage cover relationships, we also investigated whether proximity to other golden-winged warbler observations affected the probability of occurrence. We used 50 randomly selected sample sets of presence-absence observations for this analysis and found that golden-winged warbler occurrence probability was highest when other golden-winged warblers were nearby. Next, we compared models that included all three cover types (at their characteristic scale) with models that included all three cover types and distance to the nearest golden-winged warbler observation. Models including the distance variable had the most support; however, golden-winged warbler presence-absence data are not equally available for

all areas across our analysis extent (Fig. 18.1). Therefore we constructed two final statistical models to generate PAC maps: one that included landcover variables and a distance variable (which managers may want to use in areas with high-density data on golden-winged warbler presence or absence) and another based on landcover variables only, which may be more appropriate for use in areas where golden-winged warbler presence-absence data are limited. While only the best-performing maps that included landcover and distance to other golden-winged warbler observation variables are shown in this chapter, both map types have been provided to the Natural Resources Conservation Service and are available from C. A. Lott.

Translating Model Outputs into Actionable Maps

Statistical techniques that use environmental data to predict binary outcomes like presence or absence typically start by predicting occurrence probability on a continuous scale from 0 to 1. Model coefficients are then used to create spatially explicit maps where occurrence probabilities are presented for all locations within a region (Fig. 18.3). When binary maps are desired (e.g., spatially explicit PAC with discrete PAC and non-PAC boundaries), continuous probability maps can be converted into binary maps by selecting a threshold value for occurrence probability, above which all cells are classified as priority areas and below which all cells are classified as non-priority areas. Approaches for selecting occurrence probability thresholds to create binary maps have been debated for decades, and a range of procedures have been proposed (Fielding and Bell 1997, Allouche et al. 2006). Comparative studies report that the same statistical model can produce strikingly different maps depending on which method is used for threshold selection (Freeman and Moisen 2008). As the occurrence probability threshold will directly influence the spatial characteristics of the resulting map (e.g., how many hectares are predicted to be

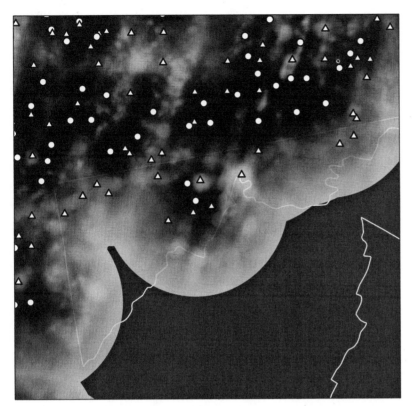

Figure 18.3. Example of a continuous predicted probability of occurrence map for golden-winged warblers showing western Maryland, southern Pennsylvania, northwestern Virginia, and northeastern West Virginia, USA, 2005-2017. State boundaries are white. Shades from white to black represent predicted probabilities of occurrence from 0% to 100% on a continuous scale. The contrasting edge between dark and light background is the boundary of the analysis extent, formed by a 30-km buffer around 1,337 unique locations where golden-winged warblers were observed between 2005 and 2017. Golden-winged warbler occurrence is not predicted outside this boundary. Circles indicate golden-winged warbler presence observations, and triangles are recorded absences. This map is based on the four-variable model, which includes percentage deciduous forest, development, croplands, and distance to the nearest golden-winged warbler location as predictors. The strong effect of proximity to other golden-winged warbler observations influences the bull's-eye pattern.

suitable and where they occur) and the ratio of errors of commission (false positives) to errors of omission (false negatives), it may be useful to make the decision about which threshold to use based on a priori mapping objectives (Freeman and Moisen 2008).

In practical conservation applications, errors of commission and errors of omission have different implications. Conservation practitioners often know in advance that they would like to avoid one type of error or the other. Yet the most commonly used metrics for selecting occurrence probability thresholds to create binary maps (kappa or the true skill statistic) are designed to balance or minimize overall error rates. If one type of error is more important than another for a specific conservation application, this may not be the most informative approach. Rather, the deliberate selection of an objective-influenced probability threshold can help conservation practitioners choose, strategically, which type of errors they would most like to minimize with their model predictions (Freeman and Moisen 2008). To illustrate this strategic approach to priority map generation, we used two different a priori occurrence probability threshold criteria (Fig. 18.4) to create alternative golden-winged warbler PAC maps based on the common conservation planning objectives of prioritizing locations for the development of a protected area network, selecting among areas with suitable landscape context for habitat restoration actions to increase nesting habitat availability across a broad landscape.

For a reserve network we want to minimize false positives (errors of commission) where our map predicts a species will be present when it really is not. If we acquire a property based on our priority area map and our target species is not there, we will have wasted funds for land acquisition. For the WLFW objective of increasing the availability of nesting habitat across a broad landscape, we aim to minimize false negatives (errors of omission). If our model predicts the species will be absent when it really is present, we could be missing opportunities to create nesting habitat in areas with suitable landscape context. The broader the net our model can cast to find these opportunities, the more likely we are to find landowners who are willing to support habitat restoration and increase habitat availability in different locations.

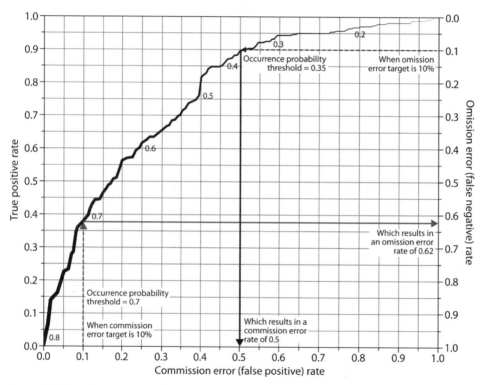

Figure 18.4. Receiver operating characteristic curve for the cover type plus distance to nearest golden-winged warbler model. The dark black line, increasing in thickness from top right to bottom left, displays occurrence probability values predicted by the model (labeled with gray numbers along the line). Dotted lines with arrows illustrate how desired commission error (gray line from x-axis) or omission error (black line from right y-axis) rates can be used to identify occurrence probability thresholds (intersections between arrows and occurrence probability values) for the creation of priority maps that minimize one type of error or the other. Solid lines extend from these intersections to perpendicular axes to show omission (horizontal grey line) or commission (vertical black line) error rates resulting from the selection of a priori occurrence probability thresholds for this model. Curves reflect error rates associated with internal model validation, not true accuracy assessment with independent data.

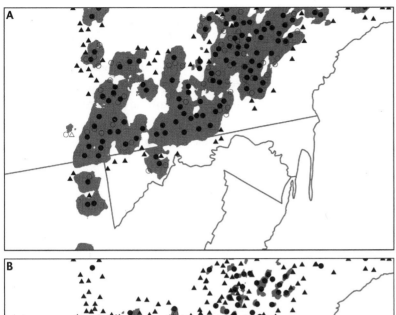

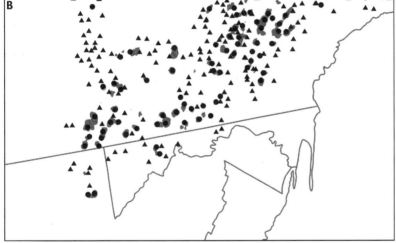

Figure 18.5. A randomly selected location covering parts of Maryland, Pennsylvania, and West Virginia, USA, showing golden-winged warbler Priority Areas for Conservation (PAC) landscape context maps based on models with two different a priori occurrence probability threshold values: (*A*) an omission error rate of 10% (designed to find areas with suitable landscape context for potential habitat restoration) and (*B*) a commission error rate of 10% (designed to help select areas to include in a protected area network). Black circles indicate true positives, open circles indicate false positives, and black triangles indicate true negatives.

Receiver operating characteristic (ROC) curves illustrate true positive and true negative error rates across the full range of predicted occurrence probabilities from 0 to 1 (Fawcett 2006, Freeman and Moisen 2008). We used ROC curves specific to each of our alternative predictive models (e.g., cover type only and cover type plus distance to nearest golden-winged warbler) to select occurrence probability thresholds that minimize commission errors (for our reserve network application) or omission errors (for our habitat restoration application). For our reserve network application, to reduce errors of commission, we decided a priori to select an occurrence probability threshold resulting in a false positive error rate of 10%. For the habitat restoration application, to reduce errors of omission, we decided a priori to select an occurrence probability threshold resulting in a false negative error rate of 10% (Fig. 18.4). We then used these thresholds to create binary maps where only the cells with occurrence probabilities above these thresholds on our continuous probability map (Fig. 18.3) were mapped as priority areas (Fig. 18.5).

Practical Management Considerations

The use of a priori occurrence probability threshold criteria to create priority area maps meeting specific conservation objectives results in maps with priority areas of different sizes, in different locations, with different types of error rates. These a priori criteria allow us to use the same statistical model of bird occurrence and habitat relationships to create maps that effectively minimize either commission or omission error rates to a desired level. In doing so, we increase the opposite error rate (e.g., minimizing the commission error rate results in a higher omission error rate). If we had used occurrence threshold probability criteria like kappa or the true skill statistic that take an error balancing approach, neither of our maps would have effectively reduced the specific error rate that mattered most to our conservation objective.

After we used statistical models and occurrence probability thresholds to generate initial priority area maps based on landscape context, we took the next essential step of removing all locations where local landcover is not compatible with nesting habitat creation (e.g., suburban areas, open water). Without taking this step, we would overestimate our operational area for nesting habitat creation, because this action requires that we meet both landscape context criteria and have a local landcover type that is appropriate for habitat restoration. Priority Areas for Conservation based on landscape context are constrained by local landcover characteristics (Fig. 18.6). For the PAC map (Fig. 18.6), only 83% of all areas with suitable landscape context have local landcover types where the primary management approach will be forestry. An additional 11% of areas with suitable landscape context have cultivated landcover, which will only allow for nesting habitat restoration through planting or land abandonment. Finally, 6% of all areas with suitable landscape context for golden-winged warbler nesting is not practically available for restoration owing to incompatible local landcover types like open water or development. Final versions of PAC maps, with locally incompatible landcover types removed, have been shared with the Natural Resources Conservation Service and are available from C. A. Lott.

The Mosaic of Management Authority in Priority Areas

When it comes to the implementation of regional conservation programs, conservation actions are often required by many different resource management agencies to meet regional targets (e.g., golden-winged warbler nesting habitat). The distribution of hectares within PAC across our area of interest illustrates the high proportion of private lands with suitable landscape context for nesting habitat (Table 18.2), which suggests that federal assistance programs for private landowners like WLFW can play a major part in achieving regional conservation goals. Each state typically has a small group of federal or state partners that manage the remaining fraction of PAC (Fig. 18.7). These summaries suggest that multiagency conservation action may be necessary to reach statewide or regional conservation targets.

Spatially explicit PAC maps (Fig. 18.5), intersected with cadastral and protected area data, will provide more specific insights into opportunities for complementary conservation actions by partners with adjacent or proximate landholdings. These maps can provide a useful starting point and a common frame of reference for agency-specific planning, where operational constraints also can be mapped to narrow the opportunity space for conservation action and to minimize the negative effects of a golden-winged warbler nesting habitat creation program on other management agency objectives or sensitive resources. This final level of planning, where a larger set of suitable areas is reduced to specific sites where entity-specific conservation action may occur, is often most effective when left to individual management authorities (rather than prescribed by priority area maps). Ideally, priority area maps strike a balance between identifying areas where conservation action has high potential for success and not narrowing the landscape so much that land management agencies or private landowners lose the operational flexibility they need to meet their other management objectives (Hiers et al. 2016).

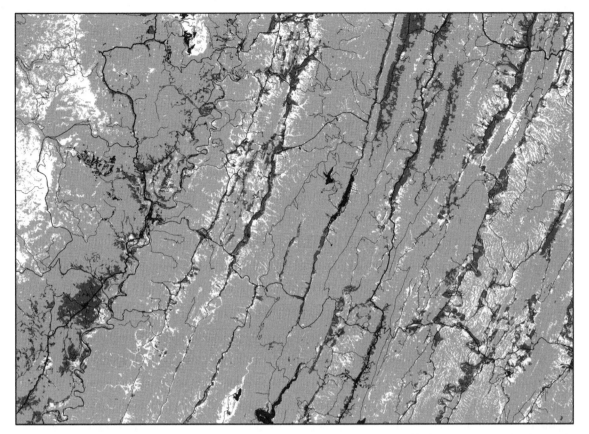

Figure 18.6. A randomly selected location covering parts of Maryland, Pennsylvania, and West Virginia, USA, 2005-2017, with a Priority Areas for Conservation (PAC) map that includes information on local land use. All black areas are outside of PAC because they do not have suitable landscape context for golden-winged warbler nesting. White areas have suitable landscape context but are not locally suitable for golden-winged warbler nesting habitat creation because they have landcover types where habitat management is impossible (e.g., open water, development). Dark gray areas have suitable landscape context with local landcover types (e.g., grassland, pasture, cropland) where methods like planting, grazing reduction, or farm abandonment would be necessary to convert existing lands to young forest nesting areas. The remaining light gray areas have suitable landscape context with local landcover (e.g., forest) that could support golden-winged warbler nesting habitat management using silvicultural techniques.

Future Improvements

The PAC developed here reflect published management recommendations (Bakermans et al. 2011, Roth et al. 2012, Golden-Winged Warbler Working Group 2013) and have room for improvement. Recent literature on golden-winged warbler breeding ecology raises several issues of landscape context for nesting habitat that we were not able to address owing to the inadequacy of existing spatial data. For example, researchers suggested that golden-winged warbler nesting habitat may need to be adjacent to more mature forest patches that are important after fledglings have left the nest, prior to their independence (Streby et al. 2015, Rohrbaugh et al. 2016). Others have suggested that there may be minimum patch size requirements for nesting habitat (Schlossberg and King 2007, Roth et al. 2012). Unfortunately, we were unable to explore either of these topics in a geographic information system because golden-winged warbler nesting habitat has not been mapped explicitly across the Appalachians. The absence of

Table 18.2. Distribution of Golden-Winged Warbler Priority Areas for Conservation Hectares in the Appalachian Mountains, USA, 2005–2017, by Management Authority and Land Type

Management Authority	Forested Hectares	Cumulative Forested Hectares (%)	Cultivated Hectares	Cumulative Cultivated Hectares (%)
Private	2,577,126	0.615	452,876	0.971
US Forest Service	825,165	0.812	3,819	0.980
State agency	336,893	0.892	1,717	0.983
State fish and wildlife	209,694	0.942	3,484	0.991
National Park Service	62,540	0.957	989	0.993
State parks	54,109	0.970	1,063	0.995
State departments of conservation	35,162	0.978	396	0.996
Joint	33,966	0.986	309	0.997
Nongovernmental organizations	21,123	0.991	507	0.998
Other	14,065	0.995	299	0.998
Unknown	13,379	0.998	299	0.999
US Army Corps of Engineers	4,296	0.999	164	0.999
County	3,194	1.000	22	0.999
City	693	1.000	1	0.999
Natural Resources Conservation Service	470	1.000	265	1.000
State land	107	1.000	35	1.000
Water district	2	1.000		1.000
Total hectares	4,191,984		466,245	

Note: Private lands comprise >2.5 million (61.5%) of all forested Priority Areas for Conservation (PAC) hectares and >450,000 (97.1%) of all cultivated hectares in PAC.

nesting habitat maps prevents a detailed evaluation of local characteristics such as patch size, configuration, adjacency, or proximity that may be important to predicting golden-winged warbler occupancy during the breeding season. As methods are developed to map golden-winged warbler nesting habitat at broad spatial extents, and to define different types of forest patches used during different parts of the breeding season, additional elements of landscape context could be used to refine golden-winged warbler PAC.

Cover type models likely have limited information content compared with models that also include vegetation structure, particularly for early-successional birds (Schlossberg and King 2009). Information on forest structure across large, forested ecoregions has been relatively coarse to date (Lemaître et al. 2012). Recent advances in lidar show promise for being able to map vertical vegetation structure with enough specificity to develop spatially explicit maps of nesting habitat for different types of forest structure that may be particularly important

for golden-winged warbler (Dickinson et al. 2014). The combination of higher-resolution data on vegetation composition, such as the tree species basal area maps of Wilson et al. (2012), combined with vegetation structure information, should provide the ability to map forest types with far greater and more biologically relevant thematic resolution than has been done to date. This increased capability will open the door to the analyses of patch size, isolation, and cover type as well as structure-adjacency elements of golden-winged warbler habitat use. The result should be predictive models of golden-winged warbler occurrence and revised PAC maps that can more specifically target areas for golden-winged warbler conservation based on habitat needs across the full breeding season (e.g., nesting, brood rearing, post-fledging, premigration dispersal).

Summary

We developed priority areas for golden-winged warbler conservation by conducting a literature review

Figure 18.7. Number of hectares by state, management authority, and land type (forest or cultivated) in the golden-winged warbler Priority Areas for Conservation designed to find potential nesting habitat restoration sites in areas with suitable landscape context in the Appalachian Mountains, USA, 2005–2017. This area summary uses the cover type and nearest golden-winged warbler distance model and an occurrence probability threshold designed to minimize errors of omission. Silviculture will be the primary technique for nesting habitat management on forested lands, whereas nesting habitat creation on cultivated lands would require planting, grazing management, or farm abandonment.

of scale-dependent landcover associations, consolidating presence-absence data from a large number of sources, and fitting a statistical model to provide spatially explicit predictions of probability of occurrence given landcover conditions and proximity to other golden-winged warblers. We then used cross-validation and receiver operating curve analysis to select separate occurrence probability thresholds to meet two different, yet common conservation objectives: (1) maximizing population growth by creating new nesting habitat in areas with suitable landscape context that are close to recent warbler nesting territories and (2) increasing the size of the breeding distribution by creating new nesting habitat in areas with suitable landscape context that occur within a minimum distance of recent warbler nesting territories.

The selection of different a priori probability of occurrence thresholds, based on different conservation strategies, had substantial effects on the size and spatial distribution of PAC, and how they intersected with different types of land ownership. Comparing how alternative approaches to delineating PAC result in different distributions of habitat management opportunity (e.g., the spatial distribution and total number of hectares and parcels where nesting habitat creation could benefit golden-winged warblers) may be helpful to inform complementary conservation actions by different land management authorities during the implementation phase of collaborative, multi-partner conservation initiatives.

LITERATURE CITED

Aiello-Lammens, M. E., R. A. Boria, A. Radosavljevic, B. Vilela, and R. P. Anderson. 2015. spThin: an R package for spatial thinning of species occurrence records for use in ecological niche models. Ecography 38:541–545.

Aldinger, K. R., T. M. Terhune II, P. B. Wood, D. A. Buehler, M. H. Bakermans, J. L. Confer, D. J. Flaspohler, J. L. Larkin, J. P. Loegering, K. L. Percy, et al. 2015. Variables associated with nest survival of golden-winged warblers (Vermivora chrysoptera) among vegetation communities commonly used for nesting. Avian Conservation and Ecology 10(1):6.

Allouche, O., A. Tsoar, and R. Kadmon. 2006. Assessing the accuracy of species distribution models: prevalence, kappa and the true skill statistic (TSS). Journal of Applied Ecology 43:1223–1232.

Anderson, D. R. 2008. Model based inference in the life sciences: a primer on evidence. Springer, New York, New York, USA.

Askins, R. A. 2001. Sustaining biological diversity in early successional communities: the challenge of managing unpopular habitats. Wildlife Society Bulletin 20:407–412.

Bakermans, M. H., J. L. Larkin, B. W. Smith, T. M. Fearer, and B. C. Jones. 2011. Golden-winged warbler habitat best management practices for forestlands in Maryland and Pennsylvania. American Bird Conservancy, The Plains, Virginia, USA.

Bakermans, M. H., B. W. Smith, B. C. Jones, and J. L. Larkin. 2015. Stand and within-stand factors influencing golden-winged warbler use of regenerating stands in the central Appalachian Mountains. Avian Conservation and Ecology 10(1):10.

Brose, P. H., T. K. Gottschalk, S. B. Horsley, P. D. Knopp, J. N. Kochenderfer, B. J. McGuinness, G. W. Miller, T. E. Ristau, S. H. Stoleson, and S. L. Stout. 2008. Prescribing regeneration treatments for mixed-oak forests in the mid-Atlantic region. US Forest Service General Technical Report NRS-33, Newtown Square, Pennsylvania, USA.

Ciuzio, E., W. L. Hohman, B. Martin, M. D. Smith, S. Stephens, A. M. Strong, and T. Vercauteren. 2013. Opportunities and challenges to implementing bird conservation on private lands. Wildlife Society Bulletin 37:267–277.

Confer, J. L., P. Hartman, and A. Roth. 2020. Golden-winged warbler (Vermivora chrysoptera), version 1.0. In A. F. Poole, editor. Birds of the world. Cornell Lab of Ornithology, Ithaca, New York, USA. https://doi.org/10.2173/bow.gowwar.01.

Crawford, D. L., R. W. Rohrbaugh, A. M. Roth, J. D. Lowe, S. B. Swarthout, and K. V. Rosenberg. 2016. Landscape-scale habitat and climate correlates of breeding golden-winged and blue-winged warblers. Pages 41–66 in H. M. Streby, D. E. Andersen, and D. A. Buehler, editors. Golden-winged warbler ecology, conservation, and habitat management. CRC Press, Boca Raton, Florida, USA.

DeGraaf, R. M., and M. Yamasaki. 2003. Options for managing early-successional forest and shrubland bird habitats in the northeastern United States. Forest Ecology and Management 185:179–191.

Desrochers, A., C. Renaud, W. M. Hochachka, and M. Cadman. 2010. Area-sensitivity by forest songbirds: theoretical and practical implications of scale-dependency. Ecography 33:921–931.

Dey, D. C. 2014. Sustaining oak forests in eastern North America: regeneration and recruitment, the pillars of sustainability. Forest Science 60:926–942.

Dickinson, Y., E. K. Zenner, and D. Miller. 2014. Examining the effect of diverse management strategies on landscape scale patterns of forest structure in Pennsylvania using novel remote sensing techniques. Canadian Journal of Forest Research 44:301–312.

Ellenwood, J. R., F. J. Krist Jr., and S. A. Romero. 2015. National individual tree species atlas. US Forest Service Publication FHTET-15-01, Fort Collins, Colorado, USA.

Fawcett, T. 2006. An introduction to ROC analysis. Pattern Recognition Letters 27:861–874.

Fielding, A. H., and J. F. Bell. 1997. A review of methods for the assessment of prediction errors in conservation presence/absence models. Environmental Conservation 24:38–49.

Frazier, A. E., and P. Kedron. 2017. Comparing forest fragmentation in eastern U.S. forests using patch-mosaic and gradient surface models. Ecological Informatics 41:108–115.

Freeman, E. A., and G. G. Moisen. 2008. A comparison of the performance of threshold criteria for binary classification in terms of predicted prevalence and kappa. Ecological Modelling 217:48–58.

Golden-Winged Warbler Working Group. 2013. Best management practices for golden-winged warbler habitat in the Appalachian Mountains: a guide for land managers and land owners. Cornell Lab of Ornithology, Ithaca, New York, USA.

Guisan, A., R. Tingley, J. B. Baumgartner, I. Naujokaitis-Lewis, P. R. Sutcliffe, A. I. Tulloch, T. J. Regan, L. Brotons, E. McDonald-Madden, C. Mantyka-Pringle, et al. 2013. Predicting species distributions for conservation decisions. Ecology Letters 16:1424–1435.

Hiers, J. K., S. T. Jackson, R. J. Hobbs, E. S. Bernhardt, and L. E. Valentine. 2016. The precision problem in conservation and restoration. Trends in Ecology and Evolution 31:820–830.

Hijmans, R. J. 2012. Cross-validation of species distribution models: removing spatial sorting bias and calibration with a null model. Ecology 93:679–688.

Holland, J. D., D. G. Bert, and L. Fahrig. 2004. Determining the spatial scale of species' response to habitat. BioScience 54:227–233.

Jackson, H. B., and L. Fahrig. 2012. What size is a biologically relevant landscape? Landscape Ecology 27:929–941.

Keating, K. A., and S. Cherry. 2004. Use and interpretation of logistic regression in habitat selection studies. Journal of Wildlife Management 68:774–789.

King, D. I., R. M. Degraaf, M. L. Smith, and J. P. Buonaccorsi. 2006. Habitat selection and habitat-specific survival of fledgling ovenbirds (Seiurus aurocapilla). Journal of Zoology 269:414–421.

King, D. I., and S. Schlossberg. 2014. Synthesis of the conservation value of the early-successional stage in forests of eastern North America. Forest Ecology and Management 324:186–195.

Lemaître, J., M. Darveau, Q. Zhao, and D. Fortin. 2012. Multiscale assessment of the influence of habitat structure and composition on bird assemblages in boreal forest. Biodiversity and Conservation 21:3355–3368.

Leuenberger, W., D. J. McNeil, J. Cohen, and J. L. Larkin. 2017. Characteristics of golden-winged warbler territories in plant communities associated with regenerating forest and abandoned agricultural fields. Journal of Field Ornithology 88:169–183.

MacCleery, D. W. 2011. American forests: a history of resiliency and recovery. Forest History Society, Durham, North Carolina, USA.

McGarigal, K., and B. J. Marks. 1995. FRAGSTATS: spatial pattern analysis program for quantifying landscape structure. US Forest Service General Technical Report PNW-GTR-351, Portland, Oregon, USA.

McNeil, D. J., K. R. Aldinger, M. H. Bakermans, J. A. Lehman, A. C. Tisdale, J. A. Jones, P. B. Wood, D. A. Buehler, C. G. Smalling, L. Siefferman, et al. 2017. An evaluation and comparison of conservation guidelines for an at-risk migratory songbird. Global Ecology and Conservation 9:90–103.

Metcalf, A. L., J. C. Finley, A. E. Luloff, and A. B. Muth. 2012. Pennsylvania's private forests: 2010 private landowner survey summary. Final report to the Pennsylvania Department of Conservation and Natural Resources Bureau of Forestry, Harrisburg, Pennsylvania, USA.

Munson, M. A., K. Webb, D. Sheldon, D. Fink, W. M. Hochachka, M. Illiff, M. Riedewald, D. Sorokina, B. L. Sullivan, C. Wood, et al. 2015. The eBird reference dataset, Version 5.0. Cornell Lab of Ornithology and National Audubon Society, Ithaca, New York, USA.

Naimi, B., N. A. S. Hamm, T. A. Groen, A. K. Skidmore, and A. G. Toxopeus. 2014. Where is positional uncertainty a problem for species distribution modelling? Ecography 37:191–203.

Omernik, J. M., and G. E. Griffith. 2014. Ecoregions of the conterminous United States: evolution of a hierarchical spatial framework. Environmental Management 54:1249–1266.

Robinson, O. J., V. Ruiz-Gutierrez, D. Fink, and R. Heikkinen. 2018. Correcting for bias in distribution modelling for rare species using citizen science data. Diversity and Distributions 24:460–472.

Rodewald, A. D., and M. D. Abrams. 2002. Floristics and avian community structure: implications for regional changes in eastern forest composition. Forest Science 48:267–272.

Rohrbaugh, R. W., D. A. Buehler, S. B. Swarthout, D. I. King, J. L. Larkin, K. V. Rosenberg, A. M. Roth, R. Vallender, and T. Will. 2016. Conservation perspectives: review of new science and primary threats to golden-winged warblers. Pages 207–215 in H. M. Streby, D. E. Andersen, and D. A. Beuhler, editors. Golden-winged warbler ecology, conservation, and habitat management. CRC Press, Boca Raton, Florida, USA.

Rosenberg, K. V., T. Will, D. A. Beuhler, S. B. Swarthout, W. E. Thogmartin, R. E. Bennett, and R. B. Chandler. 2016. Dynamic distributions and population declines of golden-winged warblers. Pages 3–28 in H. M. Streby, D. E. Andersen, and D. A. Beuhler, editors. Golden-winged warbler ecology, conservation, and habitat management. CRC Press, Boca Raton, Florida, USA.

Roth, A. M., and S. Lutz. 2004. Relationship between territorial male golden-winged warblers in managed aspen stands in northern Wisconsin, USA. Forest Science 50:153–161.

Roth, A. M., R. W. Rohrbaugh, T. Will, and D. A. Beuhler, editors. 2012. Golden-winged warbler status assessment and conservation plan. www.gwwa.org. Accessed 15 June 2019.

Sauer, J. R., J. E. Hines, J. E. Fallon, K. L. Pardieck, D. J. Ziolkowski Jr., and W. A. Link. 2014. The North American breeding bird survey, results and analysis 1966-2013, Version 01.30.2015. USGS Patuxent Wildlife Research Center, Laurel, Maryland, USA.

Schlossberg, S. R., and D. I. King. 2007. Ecology and management of scrub-shrub birds in New England: a comprehensive review. Report submitted to Natural Resources Conservation Service, Resource Inventory and Assessment Division, Beltsville, Maryland, USA.

Schlossberg, S., and D. I. King. 2009. Modeling animal habitats based on cover types: a critical review. Environmental Management 43:609–618.

Shake, C. S., C. E. Moorman, J. D. Riddle, and M. R. Burchell. 2012. Influence of patch size and shape on occupancy by shrubland birds. Condor 114:268–278.

Stauffer, G. E., D. A. W. Miller, A. M. Wilson, M. Brittingham, and D. W. Brauning. 2017. Stewardship responsibility of Pennsylvania public and private lands for songbird conservation. Biological Conservation 213:185–193.

Streby, H. M., S. M. Peterson, and D. E. Andersen. 2016. Survival and habitat use of fledgling golden-winged warblers in the western Great Lakes region. Pages 127–140 in H. M. Streby, D. E. Andersen, and D. A. Beuhler, editors. Golden-winged warbler ecology, conservation, and habitat management. CRC Press, Boca Raton, Florida, USA.

Streby, H. M., S. M. Peterson, G. R. Kramer, and D. E. Andersen. 2015. Post-independence fledgling ecology in a migratory songbird: implications for breeding-grounds conservation. Animal Conservation 18:228–235.

Sullivan, B. L., J. L. Aycrigg, J. H. Barry, R. E. Bonney, N. Bruns, C. B. Cooper, T. Damoulas, A. A. Dhondt, T. Dietterich, A. Farnsworth, et al. 2014. The eBird enterprise: an integrated approach to development and application of citizen science. Biological Conservation 169:31–40.

Thogmartin, W. E. 2010. Modeling and mapping golden-winged warbler abundance to improve regional conservation strategies. Avian Conservation and Ecology 5(2):12.

Thogmartin, W. E., and J. J. Rohweder. 2009. Conservation opportunity assessment for rare birds in the midwestern United States: a private lands imperative. Pages 419–425 in Proceedings of the Fourth International Partners in Flight Conference, 13–16 Feb 2008. Partners in Flight, McAllen, Texas, USA.

US Fish and Wildlife Service. 2011. 90-day finding on a petition to list the golden-winged warbler as endangered or threatened. Federal Register 76:31,920–31,926.

US Fish and Wildlife Service. 2015. 12-month finding on a petition to list greater sage-grouse (Centrocercus urophasianus) as an endangered or threatened species. Federal Register 80:59,857–59,942.

US Natural Resources Conservation Service. 2015. Lesser prairie chicken initiative FY16–18 conservation strategy. US Department of Agriculture, Washington, DC, USA.

US Natural Resources Conservation Service. 2016a. Healthy forests for golden-winged warbler. working lands for wildlife FY17–21 conservation strategy. US Department of Agriculture, Washington, DC, USA.

US Natural Resources Conservation Service. 2016b. Gopher tortoise working lands for wildlife FY17–18 conservation strategy. US Department of Agriculture, Washington, DC, USA.

Vallender, R., S. L. Van Wilgenburg, L. P. Bulluck, A. Roth, R. Canterbury, J. Larkin, R. M. Fowlds, and I. J. Lovette. 2009. Extensive rangewide mitochondrial introgression indicates substantial cryptic hybridization in the golden-winged warbler (Vermivora chrysoptera). Avian Conservation and Ecology 4(2):4.

Verbyla, D. L., and J. A. Litvaitis. 1989. Resampling methods for evaluating classification accuracy of wildlife habitat models. Environmental Management 13:783–787.

Vitz, A. C., and A. D. Rodewald. 2006. Can regenerating clearcuts benefit mature-forest songbirds? an examination of post-breeding ecology. Biological Conservation 127:477–486.

Wenger, S. J., and J. D. Olden. 2012. Assessing transferability of ecological models: an underappreciated aspect of statistical validation. Methods in Ecology and Evolution 3:260–267.

Wickham, J., C. Homer, J. Vogelmann, A. McKerrow, R. Mueller, N. Herold, and J. Coulston. 2014. The Multi-Resolution Land Characteristics (MRLC) Consortium—20 years of development and integration of USA National Land Cover Data. Remote Sensing 6:7424–7441.

Wickham, J. D., S. V. Stehman, L. Gass, J. Dewitz, J. A. Fry, and T. G. Wade. 2013. Accuracy assessment of NLCD 2006 land cover and impervious surface. Remote Sensing of Environment 130:294–304.

Wilson, B. T., A. J. Lister, and R. I. Riemann. 2012. A nearest-neighbor imputation approach to mapping tree species over large areas using forest inventory plots and moderate resolution raster data. Forest Ecology and Management 271:182–198.

Wood, E. M., S. E. Barker Swarthout, W. M. Hochachka, J. L. Larkin, R. W. Rohrbaugh, K. V. Rosenberg, and A. D. Rodewald. 2016. Intermediate habitat associations by hybrids may facilitate genetic introgression in a songbird. Journal of Avian Biology 47:508–520.

19 — Nongovernmental Organizations

JODI A. HILTY,
KARL A. DIDIER, AND
JON P. BECKMANN

Their Role in and Approach to Landscape Conservation

Introduction

In 2010, a young male wolverine (*Gulo gulo*) referenced as M556 journeyed ~800 km for more than
2 months from the Greater Yellowstone Ecosystem
(GYE) into Colorado, USA, becoming the first wolverine recorded in Colorado since 1919 (Inman et al.
2012). In 2016, M556 was killed in North Dakota,
USA. It was the first wolverine in North Dakota in
>150 years (Mazza 2016). The scale of movement
of this animal is symbolic of why conservationists
around the world have come to advocate for large-
landscape conservation to stem the loss of global
biodiversity. The early movements of M556 were recorded by a nongovernmental organization (NGO),
the Wildlife Conservation Society (WCS), as part of
a collaborative study including multiple federal and
state agency partnerships to establish and lead the
first intensive wolverine research effort in the GYE.
The movements of M556 and other wolverines in
the study and the natural low density of wolverines
in the northern Rockies of the United States captured the attention of many NGOs and other entities. These organizations recognized the need to
conserve wolverines and other animals at the massive landscape scale of the Rocky Mountains, not just
in the United States but also into Canada.

The wolverine story hints to how NGOs can
and have played a key role in shifting the conservation paradigm from protected area conservation
to large-landscape conservation or management of
connected networks of protected areas. Compared
to other sectors (e.g., government, academia), NGOs
can better use a number of tools (e.g., science, policy
advocacy) to influence landscape-scale conservation
of wildlife and habitats, although they are weak in
other arenas (e.g., inability to set policy). This chapter explores how the breadth of NGOs, including
their approaches, tools, and mechanisms, has helped
shift the conservation paradigm to large-landscape
conservation. We aim to understand one of the most
significant paradigm shifts in the field of conservation biology from the heavy focus on parks and protected areas in the 20th century to large-landscape
conservation in the 21st century. We provide case
studies to illustrate how NGOs are supporting this
paradigm shift through wildlife science and initiatives such as Yellowstone to Yukon (Y2Y) because
of their ability to work across boundaries and form
collaborative partnerships.

Making the Case for Large-Landscape Conservation

Landscape ecology is the study of the pattern and interaction between ecosystems, and how the interactions affect ecological processes, particularly the effects of spatial heterogeneity on these interactions (Clark 2010). During the 1980s, advances in technologies such as computing, remotely sensed satellite and aerial imagery, geographic information systems, global positioning systems, and spatial statistics led to the emergence of landscape ecology, a subdiscipline of ecology (Clark 2010). For natural science, they revolutionized ways that humans could perceive landscape-level patterns and effects. For conservation, they transformed the focus of conservation from single populations of an animal or individual protected areas to large-scale conservation.

Large-landscape conservation is now globally accepted as the key to conserving biological diversity and the ecological processes on which biodiversity and human civilization depend. We now know that few protected areas are large enough to sustain the biodiversity contained within them without connectivity to other natural areas (Hilty et al. 2019a). Protected areas that are completely isolated from other protected areas owing to surrounding human development and activities are likely to lose species and often suffer compromised ecological processes (Saintato 2015). Additionally, climate change science reviews repeatedly highlight the importance of expanding protected areas and connecting them as a key strategy for conservation of biodiversity (Heller and Zavaleta 2009). Connectivity between protected areas is increasingly being accepted as an important tool for large-landscape conservation. For example, the International Union for Conservation of Nature Connectivity Working Group drafted international standards for connectivity to supplement protected areas tools (Hilty et al. 2019b). Finally, with new technologies such as global positioning system collars, scientists are increasingly understanding that many different types of wildlife, in addition to the aforementioned wolverine, make enormous movements far bigger than any protected area. For example, scientists in Iran tracked a female cheetah (*Acinonyx jubatus venaticus*) that covered more than 3,000 km² in a 3-year period across two reserves (Farhadinia et al. 2013).

Implementation of large-landscape conservation has followed science more slowly, and conservationists have had an instrumental role in advancing progress. The below case studies reflect various efforts by the NGO community and offer insights into how conservationists are also continuing to influence this shift forward across governmental institutions that were not originally set up to operate at such scales.

Typology of Conservation Nongovernmental Organizations

Before moving to the case studies, it is important to have a better understanding of the breadth of NGOs that may engage in various aspects of large-landscape conservation. A wide range of conservation NGOs with differing philosophies, goals, and tools exist throughout the world, all working to advance biodiversity conservation. The mission of these organizations varies significantly, with some entities focused on wildlife species and others focused on ecosystems or even the intersection between human well-being and conservation. Some organizations are information producers using science, while others are information consumers that fall further out on the advocacy and policy spectrum (Table 19.1). Such a typology is necessarily oversimplified in that conservation NGOs may engage in a range of actions, and their activity and emphasis may change depending on the goals of a project.

Engagements of NGOs in any landscape also vary in time and scope of commitment. In some cases, organizations focus on a particular landscape for a number of years, often with a directed scope of activity, and then dissolve when the issues are resolved or resources are depleted. Other entities act as umbrella visions or consortiums of other organizations

Table 19.1. Simplified Typology of Conservation Nongovernmental Organizations

Type	Focus	Primary Method	Examples
Land trusts	Private land	Easements or acquisition	The Nature Conservancy, regional land trusts
Litigatory or advocacy groups	Policy and public land management	Litigation or lobbying, communications	Defenders of Wildlife, Center for Biological Diversity
Science-based groups	Information to guide conservation	(Collaborative) applied research	Wildlife Conservation Society, Conservation Biology Institute
Community-based groups	Engaging communities to conserve in their backyards	Training, information sharing, communications	Yellowstone to Yukon Conservation Initiative, local community-based organizations

Note: Some nongovernmental organizations employ multiple methods.

but have few, if any, paid staff. An example would be the Rio Negro Network in the central Amazon, South America, which is a consortium that advocates and collaborates for basin-scale management of the Rio Negro. Still other entities engage paid staff and become part of the fabric of the landscape. Some, such as Y2Y—which promotes a vision of an interconnected system of wild lands and waters stretching from Yellowstone, USA, to Yukon, Canada, harmonizing the needs of people with those of nature—maintain partnerships at the heart of what they do. Yellowstone to Yukon amplifies partner messaging and communications about their work to broader audiences, raises funds for local groups, and provides support for localized efforts such as through strategic support and government relations work, which advances shared priorities. Collectively, the wide range of NGO approaches offers broad ways in which they advance landscape conservation, whether through scientific enquiry, collaboration, advocacy, litigation, or even civil disobedience. Because NGOs tend to specialize in one or a few approaches, several NGOs will often work on different aspects of conservation on the same landscape whether or not they work in a formal partnership.

Nongovernmental organizations play a distinct role compared to academia and governmental agencies. Scientific inquiry and contributions to peer-reviewed literature are activities that any of the three entities (i.e., conservation NGOs, academia, governmental agencies) can undertake; however, their approaches may vary. Generally, the pressure for

academics is to conduct research with a strong theoretical basis and publish the results in peer-reviewed journals. Most academic field research lasts about three years (Blumstein 2012). Agency scientists tend to focus on resource management questions and are sometimes limited or obstructed by political ideologies (Jacobson et al. 2010). Agencies have the most flexibility in terms of maintaining long-term studies and monitoring. Conservation NGOs tend to focus on questions around biodiversity conservation and documenting status of species and systems as well as proposing tools and approaches to better conservation. Another difference between NGOs and governmental agencies or academia is that NGOs often engage same-sector partners to help define research needs and priorities, fund the research, and publicize results in a way that advances a campaign goal, such as influencing management approaches or advocating for changes by governmental agencies.

Whereas agencies often directly manage wildlife and lands and tend to be bureaucratic, slower, and deliberate in making change, conservation organizations often seek to influence wildlife and land management agencies and can be nimbler and more responsive, although NGOs range from very flexible to highly inflexible. One reason for this is their structure and function. Like a business, they tend to have a president who reports to a volunteer board, administrative support staff, and program staff, who may be scientists or play various conservation roles. If accepted within the culture of the organization, a president or a small leadership group can decide to quickly shift an organization's direction. Likewise, conservation NGOs and academics generally can speak out on issues much more freely than governmental agencies and lobby for changes within management and policy (Humphreys 1996). In the context of shifting from the management of individual protected areas and parcels to large-landscape conservation, conservation NGOs played a significant and early role in visioning and promoting transboundary large landscape visions. As discussed in more detail below, Y2Y was one of the first transboundary large-landscape visions, and it has inspired other similar efforts around the world, such as Alps to Atherton (now Great Eastern Ranges Corridor), Baja to Bering, and Two Countries One Forest in North America's northern Appalachian forest (M. Hebblewhite, University of Montana, unpublished data).

As an example of how conservation NGOs positively influence agencies to achieve a transboundary conservation outcome, consider the Path of the Pronghorn. Using data collected in the field, conservation scientists at WCS discovered the longest known and described (at the time) ungulate migration in North America: that of pronghorn (*Antilocapra americana*) between Grand Teton National Park and the Upper Green River Basin in Wyoming, USA (Berger and Cain 2014). This research and a clear vision that conservation of pronghorn in the park required protection of the migration pathway across multiple jurisdictions led to the creation of the first federally designated wildlife corridor, Path of the Pronghorn (Berger and Cain 2014). It also led to a signed memorandum of understanding by three federal agencies agreeing to protect the migration route (Bridger-Teton National Forest, US Fish and Wildlife Service, National Park Service). This work and discussions with a state governor also were the inspiration for the Western Governors' Association (WGA) resolution to identify and protect wildlife corridors across the western United States. This resolution led to the WGA's crucial habitat assessment

tool process, where all 17 western states developed maps and models detailing crucial habitat and areas of connectivity for priority species and a regional, landscape-scale model. This example is representative of a larger conservation community push to conserve wildlife corridors, based on strong science, as part of moving the conservation focus from individual protected areas to an interconnected network of protected areas at a landscape level.

Path of the Pronghorn also demonstrates some of the ways that NGOs use publicity to advance conservation initiatives. First, the name is catchy and has been widely adopted. Today it is written on roadside exhibits, and most organizations and agencies in the region refer to the migration by name. Such wide adoption of the name occurred because it appeared in peer-reviewed papers (e.g., Berger and Cain 2014) and newspapers (e.g., Nuwer 2012) and was heard at talks (e.g., at scientific conferences such as The Wildlife Society Annual Conference and Restoring the West Conference) and discussed in conversations. Success in protecting the Path of the Pronghorn or succeeding at other conservation priorities requires raising the profile of an issue, particularly to the key audiences. While generally it is the agencies who can act to implement the desired goal, they are more likely to act when they feel public pressure to do so (Humphreys 1996). It is not uncommon for NGOs to form partnerships with agencies for the purpose of advancing science or forwarding large-landscape conservation (Beckmann and Lackey 2018). Nongovernmental organizations also often excel at crafting messages to inspire public action and at garnering the publicity necessary to reach key audiences.

As alluded to above, a strength of conservation NGOs as related to large-landscape conservation is their ability to work beyond institutional, administrative, and geographic boundaries. Whereas agencies tend to focus the bulk of their energies on the lands that they manage or the species for which they are responsible, NGOs are less bound by geographical constraints. Nongovernmental organizations can move and invest resources more easily across geographic and thematic jurisdictions, working across public and private lands, across state and provincial boundaries, and even across international boundaries. Agencies traditionally have had a more difficult time working across multiple jurisdictions. Agencies do set up formal structures for transboundary cooperation to support transboundary work, such as the North American Free Trade Agreement's Commission for Environmental Cooperation or the US Fish and Wildlife Service's Landscape Conservation Cooperatives, but funding for these can wax and wane with politics.

Another strength of NGOs is their ability to bring a broad spectrum of entities together for longer-term applied collaborations. Nongovernmental organizations work across natural resource-type organizations and with private individuals, companies, landowner associations, scientists, local communities, indigenous entities, and government agencies. In fact, it is through pooling of expertise from such sectors that NGOs are often most effective at achieving management or policy change. Relatedly, where government agencies are weak, NGOs can help empower bottom-up leadership, such as support from local or indigenous communities for shared conservation priorities.

Limitations of Nongovernmental Organizations

While NGOs have many strengths, such as those identified above, there are also limitations, some of them significant. Nongovernmental organizations may in many cases have more flexible funding than academia and government agencies, but they generally have access to fewer and less fixed (i.e., soft money) income streams. The availability of financial resources can be fickle, both related to the overall global economy and to the changing interests of private funders. Many philanthropists lose interest in

a region, species, or project after a limited period of funding that is often far short of the time needed to achieve a successful outcome (i.e., funder fatigue). Additionally, some NGOs lack scientists on staff and do not have access to journals, limiting their conversancy with science. Some important conservation goals at the landscape scale are simply not attractive to funders, limiting the ability of NGOs to advance those goals. Because of the limitations of resources, the effectiveness of NGOs depends on one of three factors: (1) access to the right policy makers and managers who will carry forward the NGO's recommendations, (2) ability to gain a plurality of public voices that turns the tide on issues or win a legal case that supports or forces a decision maker to move in the direction of the NGO's goals, or (3) a compelling story or vision with obvious solutions that are easy to implement and seen as win-win.

Nongovernmental organizations, for the most part, do not directly create policy or manage lands (except for the relatively small area managed by land trusts) or wildlife, although some NGOs (e.g., WCS) do manage protected areas in developing countries that lack the resources to do so across the globe. In most cases, NGOs can only hope to persuade those who do hold the power to make decisions that enhance biodiversity conservation. In this way, NGOs are limited to helping identify the mechanisms of change and steering decision makers toward those mechanisms, but they cannot make the policy change. Any progress toward enhancing conservation also runs the risk of being undone with the change of a policy, elected officials, or management regime. This means that a significant role for many conservation NGOs is that of watchdogs holding governments to their policies and good management practices. In Australia, a past government administration announced a National Wildlife Corridors Plan, a great stride forward toward large-landscape conservation. Calls for weakening of the legislation soon thereafter, however, became an immediate conservation concern (Pulsford et al. 2013).

Case Studies

The following three case studies demonstrate the different ways in which an NGO can make positive contributions to landscape-scale management.

Case Study 1: Yellowstone to Yukon Conservation Initiative

The Y2Y represents one of the best-known and most advanced large-landscape conservation efforts in the world (Chester 2006). Stretching some 3,200 km in length, the Y2Y region is one of the last intact mountain ecosystems left on Earth. It is home to the full suite of wildlife that existed when European explorers first arrived, and it is the source of clean drinking water to at least 15 million North Americans. Natural processes such as flooding, fire, and wildlife migration still occur unimpeded across much of the region. To conserve this global gem, scientists and conservation came together to form a vision for connecting and protecting habitat up the mountainous spine stretching between Yellowstone National Park in Wyoming and the Arctic Circle in the Yukon so that people and nature can thrive (Chester 2006).

The need for such a vision to move to a large-landscape approach in the Y2Y region was grounded in new scientific research and technologies that demonstrated conservation would fail if we continued to focus on individual protected areas. Scientific studies showed that animals were moving across both countries and through many different land management jurisdictions. For example, the journey of a radio-collared wolf (*Canis lupus*) named Pluie was one of the many animals tracked by scientists that made immense journeys around the time of the creation of the Y2Y vision. Pluie traveled more than 100,000 km² and encountered >30 different jurisdictions, including two countries, two provinces, two states, and many different public and private lands (Dean 2006; Fig. 19.1). Since then, scientists have tracked many different animals over similar extensive distances, reconfirming that many ani-

mals need immense room to roam and are impaired by increasing human alteration of the environment (Mikle et al. 2019). Likewise, analyses suggest that most designated protected areas are high-elevation, steeper terrain with private lands dominating lower-elevation and higher-productivity areas that are also important to wildlife (Weber 2004). Finally, with climate change disproportionately influencing this and all mountain ecosystems, scientists have increasingly pointed to the need to expand protected areas to retain biodiversity and connect these areas, especially in topographically diverse areas such as mountains (Graumlich and Francis 2010).

The challenge in the Y2Y region is that some parts of the landscape are rapidly becoming disconnected and fragmented mainly because of sprawling human developments, increasing recreational pressure, roads, dams, and oil, gas, and mineral extraction (Proctor et al. 2018). At the time of the vision's inception in 1993, wolves were gone from Yellowstone National Park, the world's first national park, and grizzly bears (*Ursus arctos*) had almost disappeared from Yellowstone as well. The Trans-Canada Highway began to have significant effects on wildlife in Canada's first national park in Banff, both contributing to significant roadkill and also being a barrier to connectivity of more northern and more southern populations (Clevenger and Waltho 2005). Most of the major rivers in the Alberta, Canada, portion of Y2Y already had one or more significant dams. Even with a rebound in numbers of grizzly bears in the Yellowstone area in the past 20 years, bears in Yellowstone and probably other species are still genetically disconnected from the populations farther north in Montana and Idaho, USA, and the fast-occurring immigration of people into the region and resulting development sprawl threaten to sever the potential connection (Theobald 2001). Grizzly bears also are experiencing genetic separation across Highway 3 in southern British Columbia, Canada (Proctor et al. 2012; Fig. 19.2). Few protected lands exist in Alberta between Waterton Lakes and Banff National Parks to help ensure connectivity in that region, with wolverine's genetic disconnectivity already detectable (Stewart et al. 2016). Farther north, massive linear disturbance from fossil fuel extraction and a proposed hydroelectric dam in the Peace River Break of northern British Columbia threaten to sever the Y2Y region at its narrowest point (Apps 2013). Likewise, while some species have recovered, others such as caribou (*Rangifer tarandus*; Ray et al. 2014), westslope cutthroat trout (*Oncorhynchus clarki lewisi*; Shepard et al. 2005), and pika (*Ochotona princeps*; Erb et al. 2011) have declined. These and other effects are clearly threatening this globally significant mountain ecosystem.

Addressing this range of challenges requires a variety of tools and approaches. In some places, a priority is to create new legal protections, generally on public lands, to ensure enough secure core areas where vulnerable populations can reproduce and grow, or to create connectivity zones. In other areas the challenge is to work with expanding communities to ensure that development does not occur in key corridors or accidentally isolate core habitat areas. Educational efforts to inform communities about how to avoid human-wildlife conflicts are key in some areas. Highways experiencing increasing numbers of vehicles in the Y2Y region require infrastructure improvements to reduce wildlife-vehicle collisions. For these and other challenges, a conservation vision based on solid science is necessary, sociopolitical buy-in must be achieved through planned communications and pressure strategies, and enduring conservation solutions must be implemented through appropriate policy or management actions. Yellowstone to Yukon approaches this vision by working across a broad range of organizations, indigenous peoples, local communities, and other entities.

The Cabinet-Purcell Mountain Corridor Project is illustrative of the way in which Y2Y works to complement and magnify the work of many partners, including government agencies (Proctor et al. 2018). In 1990, about 10 grizzly bears were thought to occupy the Cabinet-Yaak region of northwest Montana. The US Fish and Wildlife Service embarked on an augmentation program to help recover the population by capturing bears from more populous areas and releasing them in the Cabinet Mountains Wilderness (Proctor et al. 2018). Understanding the significance of keeping grizzly populations in the United States genetically connected to their Canadian cousins, Y2Y launched in 2005 what became a 65-plus member transboundary collaboration of scientists, NGOs, tribes, academics, and agencies in the Cabinet-Purcell transboundary landscape (Proctor et al. 2018, Yellowstone to Yukon 2019). Its goal was to help ensure the long-term survival of grizzlies being reintroduced to the area by increasing available habitats and reducing sources of mortality. Priorities were selected following science recommendations, such as that of grizzly bear researchers identifying critical habitat and key wildlife corridors (Proctor et al. 2012). Project work conducted by Y2Y and its partners has included removing and recontouring closed forest roads to increase core secure habitat, acquiring private lands in wildlife corridors (Fig. 19.2), providing landowners with electric fencing and other tools to reduce wildlife conflicts, implementing wildlife warning systems on regional highways, and advocating for public lands management. More than 10 years later, the Cabinet-Yaak's grizzly population is thought to number as many as 50 individuals (Kendall et al. 2015), almost 1000 km of road were closed to vehicles, and 3 bottleneck corridors in the broader Cabinet-Purcell mountains are significantly better protected. The Yellowstone to Yukon Conservation Initiative was uniquely placed to work across an international boundary, promoting transboundary science and work with public and private partners on both sides of the border to advance conservation.

The approach in the Cabinet-Purcell mountains is indicative of other efforts in Y2Y that together add up to significant progress toward the Y2Y vision. In the first 20 years since the articulation of the Y2Y region, the amount of the region in strictly protected areas increased substantially, and many of the new designated protected areas can be directly linked

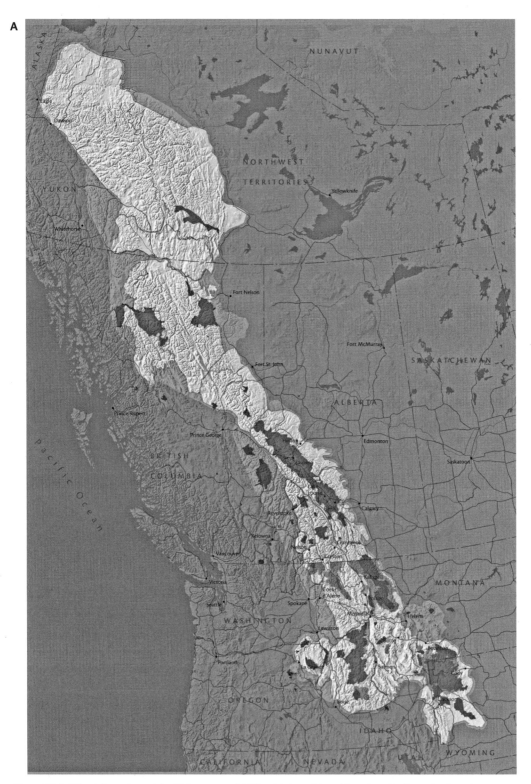

Figure 19.1. The delineation of the Yellowstone (USA) to Yukon (Canada) region. Darkest shading shows protected areas, and lighter shading shows other areas with improved conservation in 1993 (*A*) versus 2013 (*B*).

B

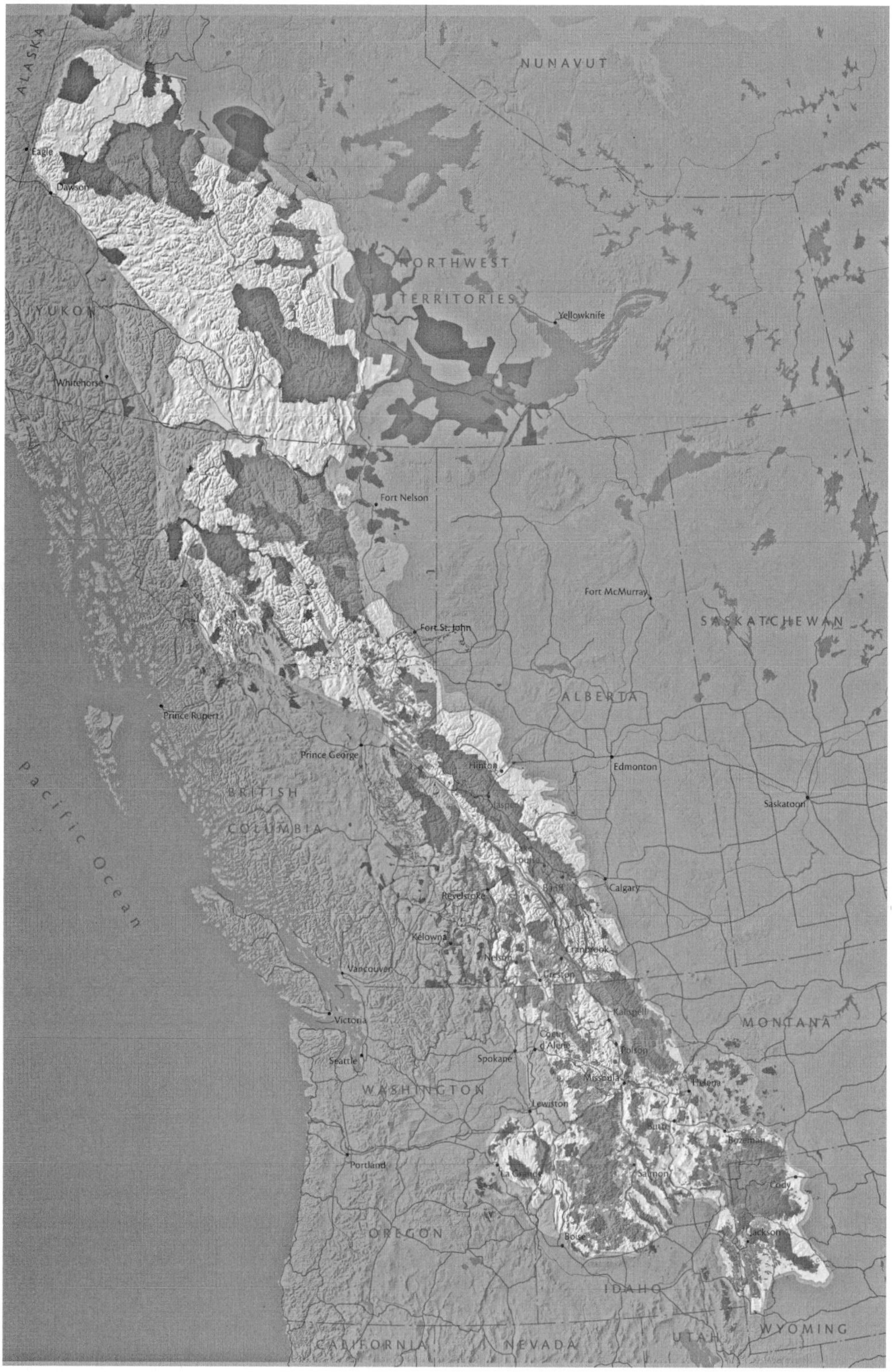
ALASKA
NUNAVUT
Eagle
Dawson
NORTHWEST
TERRITORIES
YUKON
Yellowknife
Whitehorse
Fort Nelson
Fort McMurray
SASKATCHEWAN
Fort St. John
ALBERTA
Prince Rupert
Prince George
Edmonton
Hinton
Saskatoon
Pacific Ocean
BRITISH
COLUMBIA
Calgary
Revelstoke
Kelowna
Nelson
Cranbrook
Vancouver
Creston
Victoria
Kalispell
MONTANA
Seattle
Spokane
Polson
Missoula
Helena
WASHINGTON
Lewiston
Butte
Bozeman
Portland
Salmon
Cody
OREGON
Boise
Jackson
IDAHO
WYOMING
CALIFORNIA
NEVADA
UTAH

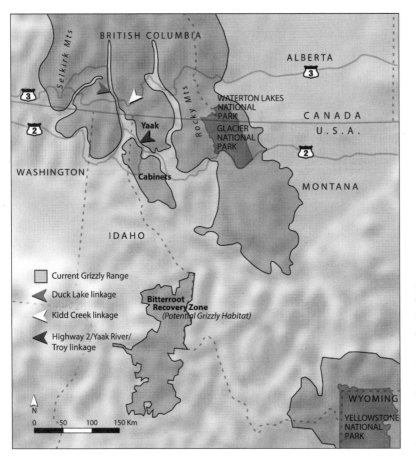

Figure 19.2. Three corridors in the Cabinet-Purcell Mountains, USA. Duck Lake is now 98% secure, Kidd Creek is now 50% secure, and Yaak will be 98% secure pending final 2020 purchases as a result of the Yellowstone to Yukon Conservation Initiative and partner nongovernmental organization collective acquisition efforts.

in part to the organization of Y2Y or the vision (M. Hebblewhite, University of Montana, unpublished data; Fig. 19.1). Other gains in conservation, while perhaps temporary or incomplete (e.g., closures to off-road vehicles, new wilderness study areas, conservation easements, interim protection of caribou habitat), occur across an additional 31% of the landscape. Another major effort that has already seen significant advances is the accelerating pace and extent of efforts in the region to mitigate the effects of highways on wildlife. There are 44 over- and underpasses for wildlife along an 80-km stretch of the Trans-Canada Highway through Banff National Park (Parks Canada 2014), and 40 structures span a 90-km length of road through the Flathead Indian Reservation in western Montana (Huijser et al. 2016). A number of other structures exist across the Y2Y region, and even more are being proposed (J. A. Hilty,

Yellowstone to Yukon, unpublished data). Increasing connectivity across roads may seem localized but is critical for connectivity at the landscape scale.

These represent just a few of the more significant conservation outcomes since the Y2Y vision was first articulated. While many of these efforts can be traced to the organization or the Y2Y vision, some happened independently. All of this progress was accomplished by natural resource entities working on a multitude of projects in an array of partnerships. While significant progress has been made since the inception of the Y2Y vision, promoting conservation at the landscape scale will require ongoing engagement by diverse entities, including NGOs. The Yellowstone to Yukon Conservation Initiative will continue to promote its large-landscape vision, develop partnerships in priority locations on the landscape, commission science to guide solutions to conservation challenges, and advocate for key policy and management changes. Yellowstone to Yukon will continue to work with and through a diversity of partners across the region. Promoting the Y2Y vision and inspiring others to act toward its implementation are two of Y2Y's most powerful tools. The vision will be achieved only when the myriad jurisdictions that make up the Y2Y region and their political leadership embrace their roles in contributing to the larger-landscape vision. As a promising sign, J. B. Jarvis, former director of the US National Park Service, recognized the importance of Y2Y in creating connectivity between parks (Shockman 2016). Even with much to be done to achieve the vision, there are few places in the world that remain so intact and with such local and global public support for conservation as the Y2Y region. Here, it is possible to achieve a vision of protected and connected habitat where people and nature can thrive today and into the future.

Case Study 2: The Amazon Waters Initiative and Integrated River Basin Management

The Amazon River Basin is, by most measures, the largest and most diverse watershed on the planet. The basin covers large portions of Brazil, Colombia, Ecuador, Peru, and smaller fractions of Guyana and Venezuela, draining ~680 million hectares and about 38% of the South American continent (Goulding et al. 2010).

The basin qualifies as a large landscape because many of its species, ecosystems, and natural resources cannot be effectively managed solely within existing, individual management units (e.g., individual protected areas, municipalities, public or private lands, states, or even countries). In addition, key principles of landscape management are needed to maintain biodiversity and ecosystem services, especially maintaining the heterogeneity and connectivity of the basin. Instead of using the term *landscape management*, WCS and many of its partners working in the Amazon have adopted the term integrated

river basin management (IRBM; Watson 2004, Barthem et al. 2017) as representative of how aquatic ecosystems and dynamics are central to the ecology, economy, and culture of the Amazon, although the concepts behind IRBM are similar to those of landscape ecology.

Landscape management or IRBM is required for the conservation of many Amazonian species, including a large group of food fishes, turtles such as the giant South American turtle (*Podocnemis expansa*), and wide-ranging mammals such as white-lipped peccaries (*Tayassu pecari*) and jaguars (*Panthera onca*). The need for IRBM and the role of NGOs in that management, however, is probably best demonstrated by the case of the dourada (*Brachyplatystoma rousseauxii*), a species of catfish that has the longest known reproductive migration of any freshwater fish (Batista and Alves-Gomes 2006, Barthem et al. 2017; Fig. 19.3). Dourada migrate from their spawning grounds in the Amazon headwaters nearly 3,900 km to the Amazon estuary on the Atlantic coast, where they feed as juveniles for 2 to 3 years (Fig. 19.4). Once they reach 60–80 cm in size, dourada return to the Amazon lowlands, feeding in the Amazon main stem and channels of its principal tributaries. At 3 to 4 years of age (90–110 cm) the species returns to the Amazon headwaters to spawn, completing a migration over its life cycle of >8,000 km (Batista et al. 2005, Barthem et al. 2017). During its migration along the main stem of the Amazon, a dourada individual would likely pass through parts of two countries (Brazil and Peru) and four provinces and states (Amapá, Amazonas, Loreto, Pará), and through or along the border of at least 14 protected areas and ~61 indigenous territories.

Several human activities threaten the health of dourada and other migratory species, and they make IRBM important. Dourada are exposed to fishing pressure along their entire migration. Mining, oil exploration, human colonization, and associated deforestation severely alter the ecological conditions of the headwater streams where dourada spawn (Asner et al. 2013), increasing sedimentation and

Figure 19.3. Dourada catfish (*Brachyplatystoma rousseauxii*) in a market in the Ucayali River Basin, Peru. Photo by M. Goulding.

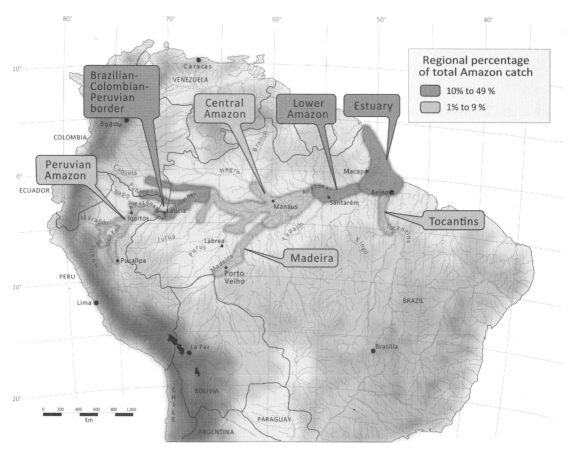

Figure 19.4. Fishing catch for the Dourada catfish in the Amazon, South America. The fish uses a larger portion of the Amazon Basin, but the regions shown represent the areas of greatest commercial harvest. Map courtesy of the Wildlife Conservation Society Amazon-Andes Program.

concentrations of heavy metals, altering water temperatures, and likely harming reproduction (Castello et al. 2013, Barthem et al. 2017). The accumulation of heavy metals (e.g., mercury) and other toxins, the former of which can bioaccumulate especially in predatory fish like dourada, results in unhealthy toxin levels in fish consumed by humans (Castello et al. 2013, Barthem et al. 2017). Deforestation along other parts of the basin also affects water quality, runoff, streamflow, and the timing of the annual flood dynamics, and it reduces important riparian and flooded forest habitats where many migratory fish feed (Coe et al. 2009, Castello et al. 2013).

Hydroelectric dams, of which there are >150 in the basin, probably best exemplify the need for landscape and IRBM management. Dams affect the ability of dourada and other fish to migrate longitudinally and latitudinally (Castello et al. 2013, Barthem et al. 2017). More than 200 more dams are planned, primarily for power generation, particularly in the Andes (Finer and Jenkins 2012, Castello et al. 2013, Lees et al. 2016). Although there are no published environmental impact statements examining existing dams in the Amazon (Castello et al. 2013), prior experience in other regions and predictive studies suggest that if they are not managed well, dams could have vast effects on the hydrology of the Amazon Basin, potentially reducing flow rates, sediment and nutrient discharge, and the annual variability in flooding regimes that characterize the system and influence its productivity, including fisheries production (Lees et al. 2016, Forsberg et al. 2017).

Finally, climate change is already altering drought and flooding dynamics (Coe et al. 2011, Sorribas et al. 2016) and will continue to increase annual variability and frequency of extreme events in terms of dry seasons, precipitation discharge, and inundation extent (Sorribas et al. 2016). These changes will increase the need to address the threats above and to maximize the probability that migratory and other species can adapt to unavoidable climate changes.

These numerous, diverse, and urgent threats call for landscape, basin-scale decision making and man-agement, involving multiple countries, public and private sectors, and stakeholders. Since 2014, WCS has been leading an initiative called Amazon Waters (Wildlife Conservation Society 2017) to create an integrated, basin-scale management system for the entire Amazon Basin, focusing on management of wetlands, fisheries, and rivers, and streams. To encourage use of IRBM concepts, the initiative has chosen to focus on migratory fisheries as a resource.

The first two years of the initiative focused primarily on data collection, construction of a model fisheries database and monitoring system, development of an IRBM case study in a Peruvian watershed, assessment of the potential impacts of climate change on discharge and inundation rates, development of basin-wide stakeholder networks, and initial engagement of stakeholders, government agencies, and political representatives in key subbasins. With these foundations in place, the key next steps to establish effective IRBM in the Amazon will be to develop IRBM systems for 2 to 3 major subbasins, develop and implement mechanisms to disseminate scientific information to stakeholders and decision makers, evaluate the role of existing protected areas in the management of migratory species, identify gaps, and facilitate discussions to improve public policies to support IRBM.

While nearly all sectors of society need to be involved in IRBM at the landscape scale, NGOs are playing several key roles. First, international organizations such as WCS, World Wildlife Fund, and The Nature Conservancy have a unique capacity to work across jurisdictions in a coordinated fashion, especially in different countries that contain parts of the Amazon Basin. Such coordination is possible because these NGOs have offices and staff operating in multiple countries that share the same programmatic objectives, communicate regularly, and share financial and project management systems.

Second, NGOs in the Amazon Basin are well positioned to engage local stakeholders and bring them into landscape and basin management processes. Effective engagement of local stakeholders, particularly at state and municipal scales, is important because it requires only one or a few local stakeholders acting solely for their own good (e.g., a municipality who wants to build a dam) to cause major problems for much of the landscape. Local communities and business associations, however, often have negative views of government agencies for various reasons, and government agencies often do not have sufficient resources to effectively engage and deliver services to communities in the vast rural areas of the Amazon. National NGOs (e.g., Fundação Vitória Amazônica and Instituto Socio Ambiental in Brazil and Sociedad Peruana de Derecho Ambiental and Instituto del Bien Comun in Peru) have long-standing positive relationships with local communities, business associations, and indigenous communities, and can directly facilitate the participation of these stakeholders in IRBM or represent their needs.

Finally, NGOs in the region play an important role in defining and promoting the research priorities for landscape- or basin-scale management. Independent scientists in academic institutions often lack the access to and leverage with large donors, including foundations, bilaterals (e.g., US Agency for International Development), and multilaterals (e.g., World Bank) that NGOs have. Nongovernmental organizations in the Amazon, in partnership with scientists in academia, are effectively engaging these large donors and developing mechanisms to support applied research for IRBM. Thus far NGOs such as WCS and The Nature Conservancy have played an important role in capturing significant resources from large philanthropic foundations and are actively engaged in conversations with potential donors.

There are three key challenges facing successful implementation of large-landscape conservation in the Amazon: (1) establishment of a scientific and publicly available body of knowledge as well as a long-term monitoring database about the basin's ecology, economy, and ongoing and potential impacts of human activities that will support landscape (aquatic and terrestrial) management decisions; (2) creation of a core network of government, civil society, community, and business partners in each major subbasin and for the Amazon Basin as whole who agree that IRBM is necessary to accomplish shared goals and are ready to collaborate toward that end; and (3) public policies at the national, state, and international levels that support IRBM management and monitoring, prevent irreversible effects, and help advance practical climate change adaptation strategies. Addressing these challenges at the scale of the entire Amazon Basin will be important to the long-term conservation of biodiversity and ecosystem services.

Case Study 3: Lake Tahoe Basin and Western Great Basin Desert, Nevada

The Great Basin is the largest area (~541,000 km²) of contiguous endorheic watersheds in North America. It is a region known for its arid climate and the basin and range topography that varies from the North American low point to the highest point of the contiguous United States in the adjacent Sierra Nevada Mountains. The Great Basin also has one of the lowest human footprints in the contiguous United States (Sanderson et al. 2002).

Black bears (*Ursus americanus*) and grizzly bears were fairly common in the Great Basin of Nevada until logging to supply gold and silver mines with timber support structures, combined with unregulated hunting and removals arising from conflict with humans and their livestock, led to their extirpation in the 1930s. In many regions across the globe, recovery of extirpated populations of large carnivores is extremely difficult and rarely accomplished owing to a variety of factors, one of which is the large-scale space that carnivores must have to live. This is particularly true for apex predators, such as bears, that have large home ranges and occur at low densities especially in arid landscapes. Thus being able to successfully recover large carnivore populations, a rare feat, requires identifying threats to their existence across the landscape at large scales, mitigating those threats, and monitoring population responses over large scales of space and time in response to conservation efforts.

The Great Basin provides an excellent example of natural recolonizing processes by large carnivores owing to conservation efforts. The system is home to a population of black bears that is currently expanding in numbers and geographical extent into historic range along a colonizing front because of the long-term cooperative effort of a state agency (Nevada Department of Wildlife [NDOW]), a university (University of Nevada, Reno), and an NGO (WCS).

Prior to the 1980s in Nevada, black bear sightings, management issues, and bear deaths from vehicles were considered rare events (Goodrich 1993, Goodrich and Berger 1994). The director of NDOW at the time stated at the first Western Black Bear Workshop: "Nevada has no bear, except for an occasional one that strays in along the Sierras adjacent to Lake Tahoe in California. Therefore, we have no management responsibilities" (LeCount 1979:63). But by the 1980s, a population of bears was known to be present in western Nevada (Goodrich 1993). By the mid-1990s, conflicts between humans and black bears began to rise sharply in the Lake Tahoe Basin and the western portion of the Great Basin Desert in Nevada (Fig. 19.5). A 10-fold increase in the annual number of complaints and a 17-fold increase in bear mortalities due to collisions with vehicles were reported between the early 1990s and early 2000s (Beckmann and Berger 2003a, 2003b). Motivated by these increasing bear-human conflicts, but without knowing the relative importance of various potential catalysts influencing the increase, a new effort began to understand black bear ecology and conservation in the region (Lackey et al. 2013).

Understanding whether the conflicts were influenced by a growing black bear population, a redistribution of the animals owing to human food sources such as garbage, or a combination of factors, would provide the context in which NDOW could make decisions regarding management options. Given the lack of recent history of bears in the state, however, NDOW had no funding for research and had only a

Figure 19.5. American black bear (*Ursus americanus*) cubs in a garbage dumpster in the Great Basin, Nevada, USA. Photo by J. Beckmann.

small amount of funds and a relatively loose set of protocols for dealing with nuisance bears.

Black bear populations are recovering in many parts of their natural range and increasing in areas with substantial anthropogenic influence (Garshelis and Noyce 2011, Lackey et al. 2013). Understanding the factors that influence variability in habitat use can guide management toward protecting core habitat and can also identify linkage zones for movement between core habitats to allow for adequate space as populations expand (Frary et al. 2011, Beckmann et al. 2015). Landscape-level data, to inform management and conservation at this scale, are almost always important for low-density and wide-ranging large carnivores. It is equally important to have the full complement of partners (state, university, federal, and NGO) to maximize the effectiveness of those data in determining management practices and policies that allow large carnivores a place in human-altered landscapes.

Thus, in 1997, biologists at the University of Nevada, Reno, and NDOW began a long-term study of Nevada's black bears that continues today (Beckmann and Lackey 2008). After the initial phase of the research was funded by university grants, one of the principal investigators transitioned to WCS, and the NGO began to help fund and conduct the effort, and moved it from research to conservation and management action at a landscape scale. In many years where NDOW lacked funding for the research aspects of the project, NGO and university funding was pivotal in starting and keeping the long-term research and monitoring going, particularly in the first decade of the project.

The NDOW and WCS have collaborated through the entire process of collecting landscape-level field data and then using those data to inform and guide management decisions and conservation efforts (Beckmann and Lackey 2008, Lackey et al. 2013). Landscape-level data have been extremely important in guiding management decisions for the recovery of this large carnivore across the Great Basin. Global Positioning System location data from bears have been used to develop resource selection function models across the Great Basin, identifying core bear habitat in areas where bears currently occur and key areas of habitat in the historic range where bears have yet to recolonize. These data have allowed policymakers and decision makers the opportunity to work with various communities and land management agencies to proactively prepare for bear expansion into historic ranges across the Great Basin. WCS and NDOW have also worked together to understand and model habitat connectivity between these core areas of habitat at a landscape level to mitigate threats at large scales. As one example, road mitigation measures are currently being examined to enhance connectivity for bears and other species of large mammals in the region. The Wildlife Conservation Society and NDOW have also worked with university partners to model the genetics of this population that experienced extirpation followed by recolonization to understand the genetic consequences of carnivore recovery as a result of conservation at a landscape level (Malaney et al. 2018).

The Wildlife Conservation Society and NDOW have engaged in a variety of conservation efforts across the Great Basin for more than 20 years. These joint efforts include the abovementioned research and landscape-scale analyses identifying core habitats and key areas for connectivity that are being used by decision makers and planners at local and landscape levels. They also include putting in place over $3 million (USA) in bear-resistant trash cans and dumpsters throughout the western Great Basin study site; new bear-resistant garbage container ordinances in the Lake Tahoe Basin; the first ever long-term study investigating the effects of the wildland-urban interface on American black bear behavior and demography at a landscape scale (Beckmann and Berger 2003a, 2003b, Beckmann and Lackey 2008); and a nonlethal deterrent techniques program that uses dogs to alter behavior of nuisance bears (Beckmann and Lackey 2004, Beckmann et al. 2004). As an NGO, WCS partnered on the research and used the data to advocate for many of these conservation actions to a much greater extent than the other partners could. Owing to these landscape-level data and joint conservation efforts between a state wildlife management agency and an NGO, coupled with recovering habitat in the region, the bear population has increased by an average lambda (λ) of 1.16 (i.e., an average annual growth rate of 16% of the population) over the past decade. There are now 400–600 individuals, and bears are now present in Great Basin mountain ranges, some of which had an absence of bears for >80 years (Fig. 19.6; Lackey et al. 2013).

The current understanding of the effects of human-altered landscapes on bears is only possible through the long-term, joint program by NDOW and WCS with data collected across the entire landscape comparing urban and wildland bears. Without this partnership it is likely that the data would have never been collected, or the usefulness of the data in guiding management and policy decisions would have been much more limited, including the contentious decision by NDOW to initiate the state's first ever bear hunt in 2011, a decision that some NGOs oppose. Continued expansion of the black bear population in the Great Basin will require conservation and management planning, including a detailed analysis of how changes in patterns of human

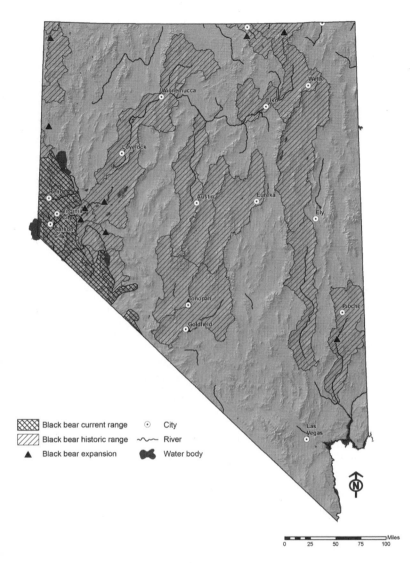

Figure 19.6. American black bear (*Ursus americanus*) historical and current distribution in Nevada, USA, and selected examples of recent sightings (1988–2015) of bears (triangles) indicating possible expansion into historical range that has been unoccupied for >80 years. Modified from Lackey et al. (2013:fig. 2).

Legend:

- ▨ Black bear current range
- ▨ Black bear historic range
- ▲ Black bear expansion
- ⊙ City
- ∿ River
- ● Water body

activity throughout the landscape may influence overall habitat suitability and connectivity for bears. Ultimately, the protection of habitat is important to continued recovery throughout this vast region (Nielsen et al. 2006). The partnerships between science-based NGOs that work at a landscape scale and state agencies, such as the one described in this case study, will become increasingly important as funding for wildlife management agencies is potentially reduced further and the Great Basin continues to experience stresses from an increasing human footprint, coupled with recovering large carnivore populations (e.g., wolves are likely to also expand back into the Great Basin in the future).

Continued expansion of the black bear population in the Great Basin will require conservation and management planning that considers how changes in human activity patterns throughout the landscape influence overall habitat suitability and connectivity for black bears. At the time of this writing, NDOW and WCS are working together with university partners to develop additional resource selection probability function models, connectivity models including genetic structure, and mortality risk models to identify current and potential core habitat areas and stay ahead (e.g., educate and work with various communities, decision makers, and constituents to adequately prepare for living with bears) of the re-colonizing front (Wynn-Grant et al. 2018). Given that Nevada is one of the fastest-growing states in the United States (World Population Review 2016), the rapidly increasing human population will make continued expansion of the bear population challenging. Only through the state-NGO-university partnerships developed in Nevada will it be possible to address all the emerging threats to the continued recovery of this species across the Great Basin landscape.

A final challenge will be that the Great Basin is one of the most water-limited systems in North America, one that already has experienced serious drought conditions in recent years (Beckmann and Berger 2003a). Projections are for the severity and frequency of such droughts to increase over time with climate change (Coats et al. 2006, Dolanc et al. 2014). Water availability will likely continue to influence many aspects of black bear dispersal and habitat use (Obbard et al. 2010, Atwood et al. 2011). A complete understanding of how male and female bears use various sources of water and how climate change, interacting with an increasing human population in the region, will affect water availability and continue to be a threat to the continued recovery of this population is needed. Research-based conservation NGOs will be key partners in making sure these data are collected and enter into the policy and management decision process to continue recovery of black bears in the Great Basin.

Summary

One of the most significant innovations in conservation science in the 20th century was the understanding that parks and protected areas alone are insufficient to ensure the long-term survival of healthy populations of wild species, primarily because many species migrate over large expanses far beyond individual parks and because human effects extend across park boundaries. The field of large-landscape conservation emerged, providing a new perspective on the scale required to achieve conservation, far beyond the bounds of most individual jurisdictions.

Nongovernmental organizations have absorbed the need to achieve conservation at a landscape scale and are playing a critical role in pushing for the realization of this science to guide conservation on the ground. Because of their ability to act independently, advocate publicly, access resources unavailable to other sectors, litigate, and work across boundaries and a spectrum of networks and partnerships, NGOs are uniquely positioned to contribute to landscape management objectives.

This chapter summarized the unique attributes of NGOs and offered examples of how they are helping to achieve significant conservation gains in many regions. As the world enters the uncertain future of a changed climate, cooperating at large scales to help sensitive species adapt will challenge existing institutions and management regimes. Nongovernmental organizations will continue to be leaders in efforts to innovate and advocate in response to these challenges.

LITERATURE CITED

Apps, C. 2013. Assessing cumulative impacts to wide-ranging species across the Peace Break Region of northeast British Columbia. Yellowstone to Yukon Conservation Initiative, Canmore, Alberta, Canada.

Asner, G. P., W. Llactayo, R. Tupayadhi, and E. R. Luna. 2013. Elevated rates of gold mining in the Amazon revealed through high-resolution monitoring. Proceedings of the National Academy of Sciences 110:18,454–18,459.

Atwood, T. C., J. K. Young, J. P. Beckmann, S. W. Breck, J. Fike, O. E. Rhodes, and K. D. Bristow. 2011. Modeling connectivity of black bears in a desert sky island archipelago. Biological Conservation 144:2851–2862.

Barthem, R. B., M. Goulding, R. G. Leite, C. Cañas, B. Forsberg, E. Venticinque, P. Petry, M. L. de B. Ribeiro, J. Chuctaya, and A. Mercado. 2017. Goliath catfish spawning in the far western Amazon confirmed by the distribution of mature adults, drifting larvae and migrating juveniles. Scientific Reports 7:41784.

Batista, J. S., and J. A. Alves-Gomes. 2006. Phylogeography of *Brachyplatystoma rousseauxii* (Siluriformes-Pimelodidae) in the Amazon Basin offers preliminary evidence for the first case of "homing" for an Amazonian migratory catfish. Genetics and Molecular Research 5:723–740.

Batista, J., K. F. Aquino, I. P. Farias, and J. A. Gomes. 2005. Variabilidade genética da dourada e da piramutaba na bacia Amazônica. Pages 15–19 in N. N. Fabré and R. B. Barthem, editors. O manejo da pesca dos grandes bagres migradores. Instituto Brasileiro do Meio Ambiente e dos Recursos Naturais Renováveis, ProVárzea, Manaus, Brazil.

Beckmann, J. P., and J. Berger. 2003a. Rapid ecological and behavioural changes in carnivores: the responses of black bears (Ursus americanus) to altered food. Journal of Zoology 261:207–212.

Beckmann, J. P., and J. Berger. 2003b. Using black bears to test ideal-free distribution models experimentally. Journal of Mammalogy 84:594–606.

Beckmann, J. P., and C. W. Lackey. 2004. Are desert basins effective barriers to movements of relocated black bears (Ursus americanus)? Western North American Naturalist 64:269–272.

Beckmann, J. P., and C. W. Lackey. 2008. Carnivores, urban landscapes, and longitudinal studies: a case history of black bears. Human-Wildlife Conflicts 2:168–174.

Beckmann, J. P., and C. W. Lackey. 2018. Lessons learned from a 20-year collaborative study on American black bears. Human-Wildlife Interactions 12(3):396–404.

Beckmann, J. P., C. W. Lackey, and J. Berger. 2004. Evaluation of deterrent techniques and dogs to alter behavior of "nuisance" black bears (Ursus americanus). Wildlife Society Bulletin 32:1141–1146.

Beckmann, J. P., L. P. Waits, A. Hurt, A. Whitelaw, and S. Bergen. 2015. Using detection dogs and RSPF models to assess habitat suitability for bears in Greater Yellowstone. Western North American Naturalist 75:396–405.

Berger, J., and S. L. Cain. 2014. Moving beyond science to protect a mammalian migration corridor. Conservation Biology 28(5):1142–1150.

Blumstein, D. T. 2012. How important is long-term ecological research? Huffington Post. 4 Dec 2012. https://www.huffpost.com/entry/how-important-is-long-term-ecological-research_b_2232076.

Castello, L., D. G. McGrath, L. L. Hess, M. T. Coe, P. A. Lefebvre, P. Petry, M. N. Macedo, V. F. Renó, and C. C. Arantes. 2013. The vulnerability of Amazon freshwater ecosystems. Conservation Letters 6:217–229.

Chester, C. C. 2006. Conservation across borders: biodiversity in an independent world. Island Press, Washington, DC, USA.

Clark, W. R. 2010. Principles of landscape ecology. Nature Education Knowledge 3(10):34.

Clevenger, A. P., and N. Waltho. 2005. Performance indices to identify attributes of highway crossing structures facilitating movement of large mammals. Biological Conservation 121:453–464.

Coats, R., J. Perez-Losada, G. Schladow, R. Richards, and C. Goldman. 2006. The warming of Lake Tahoe. Climatic Change 76:121–148.

Coe, M. T., M. H. Costa, and B. S. Soares-Filho. 2009. The influence of historical and potential future deforestation on the stream flow of the Amazon river-land surface processes and atmospheric feedbacks. Journal of Hydrology 369:165–174.

Coe, M. T., E. M. Latrubesse, M. E. Ferreira, and M. L. Amsler. 2011. The effects of deforestation and climate variability on the streamflow of the Araguaia River, Brazil. Biogeochemistry 105:119–131.

Dean, C. 2006. Wandering wolf inspires project. New York Times. 23 May 2006. http://www.nytimes.com/2006/05/23/science/earth/23wolf.html?_r=0.

Dolanc, C. R., H. D. Safford, J. H. Thorne, and S. Z. Dobrowski. 2014. Changing forest structure across the landscape of the Sierra Nevada, CA, USA, since the 1930s. Ecosphere 5:101.

Erb, L. P., C. Ray, and R. Guralnick. 2011. On the generality of a climate-mediated shift in the distribution of the American Pika (Ochotona princeps). Ecology 92:1730–1735.

Farhadinia, M. S., H. Akbari, S. Mousavi, M. Eslami, M. Azizi, J. Shokouhi, N. Ghoikhani, and F. Hosseini-Zavarei. 2013. Exceptionally long movements of the Asiatic cheetah Acinonyx jubatus venaticus across multiple arid reserves in central Iran. Oryx 47:427–430.

Finer, M., and C. N. Jenkins. 2012. Proliferation of hydroelectric dams in the Andean Amazon and implications for Andes-Amazon connectivity. PLOS One 7:e35126.

Forsberg, B. R., J. M. Melack, T. Dunne, R. B. Barthem, M. Goulding, R. C. D. Paiva, M. V. Sorribas, U. L. Silva Jr., and S. Weisser. 2017. The potential impact of new Andean dams on Amazonian fluvial ecosystems. PLOS One 12:e0182254.

Frary, V. J., J. Duchamp, D. S. Maehr, and J. L. Larkin. 2011. Density and distribution of a colonizing front of the American black bear (Ursus americanus). Wildlife Biology 17:404–416.

Garshelis, D., and K. Noyce. 2011. Status of Minnesota black bears, 2010. Final Report to Bear Committee, Minnesota Department of Natural Resources, St. Paul, Minnesota, USA.

Goodrich, J. M. 1993. Nevada black bears: ecology, management, and conservation. Nevada Department of Wildlife Biological Bulletin 11, Reno, Nevada, USA.

Goodrich, J. M., and J. Berger. 1994. Winter recreation and

hibernating black bears *Ursus americanus*. Biological Conservation 67:105–110.

Goulding, M., R. Barthem, and E. Ferreira. 2010. The Smithsonian atlas of the Amazon. Smithsonian Books, Washington, DC, USA.

Graumlich, L., and W. L. Francis, editors. 2010. Moving toward climate change adaptation: the promise of the Yellowstone to Yukon Conservation Initiative for addressing the region's vulnerability to climate disruption. Yellowstone to Yukon Conservation Initiative, Canmore, Alberta, Canada.

Heller, N. E., and E. S. Zavaleta. 2009. Biodiversity management in the face of climate change: a review of 22 years of recommendations. Biological Conservation 142(1):14–32.

Hilty, J., G. Worboys, A. Keeley, S. Woodley, B. Lausche, H. Locke, M. Carr, I. Pulsford, J. Pittock, W. White, et al. 2019*b*. (Draft) guidance on safeguarding ecological corridors in the context of ecological networks for conservation. International Union for Conservation of Nature, Gland, Switzerland. https://www.iucn.org/sites /dev/files/content/documents/2019_6-28_consultation _draft_safeguardingecologicalcorridorsinthecontext ._.pdf.

Hilty, J. A., A. T. Keeley, W. Z. Lidicker Jr., and A. M. Merenlender. 2019*a*. Corridor ecology: linking landscapes for biodiversity conservation and climate adaptation. Island Press, Washington, DC, USA.

Huijser, M., W. Camel-Means, E. R. Fairbank, J. P. Purdum, T. D. Allen, A. Hardy, J. Graham, J. S. Begley, P. Basting, and D. Becker. 2016. US 93 North post-construction wildlife-vehicle collision and wildlife crossing monitoring and research on the Flathead Indian Reservation between Evaro and Polson. FHWA/MT-16-009/8208, prepared for the State of Montana Department of Transportation, Helena, Montana, USA. https://pdfs .semanticscholar.org/5662/5eb656ed8f640cded03013 d13f57fac898d9.pdf.

Humphreys, D. 1996. Regime theory and non-governmental organisations: the case of forest conservation. Journal of Commonwealth and Comparative Politics 34:90–115.

Inman, R. M., K. H. Inman, M. L. Packila, and A. J. McCue. 2012. Spatial ecology of wolverines at the southern periphery of distribution. Journal of Wildlife Management 76:778–792.

Jacobson, C., J. F. Organ, D. J. Decker, G. R. Batcheller, and L. Carpenter. 2010. A conservation institution for the 21st century: implications for state wildlife agencies. Journal of Wildlife Management. 74:203–209.

Kendall, K. C., A. C. Macleod, K. L. Boyd, J. Boulanger, J. A. Royle, W. F. Kasworm, E. Paetkau, M. Proctor, K. Annis, and T. A. Graves. 2015. Density, distribution, and ge-

netic structure of grizzly bears in the Cabinet-Yaak Ecosystem. Journal of Wildlife Management 80:314–331.

Lackey, C. W., J. P. Beckmann, and J. Sedinger. 2013. Bear historical ranges revisited: documenting the increase of a once-extirpated population in Nevada. Journal of Wildlife Management 77:812–820.

LeCount, A. 1979. Proceedings of the first western black bear workshop. Arizona Game and Fish Department, Phoenix, Arizona, USA.

Lees, A. C., C. A. Peres, P. M. Fearnside, M. Schneider, J. A. S. Zuanon. 2016. Hydropower and the future of Amazonian biodiversity. Biological Conservation 25:451–466.

Malaney, J. L., C. W. Lackey, J. P. Beckmann, and M. D. Matocq. 2018. Natural rewilding of the Great Basin: genetic consequences of recolonization by black bears (*Ursus americanus*). Diversity and Distributions 24:168–178.

Mazza, E. 2016. North Dakota's first wolverine in 150 years is immediately shot and killed by rancher. Huffington Post. 13 May 2016. https://www.huffpost.com/entry /wolverine-killed-north-dakota_n_573545ffe4b060aa7 819ef8a.

Mikle, N. L., T. A. Graves, and E. M. Olexa. 2019. To forage or flee: lessons from an elk migration near a protected area. Ecosphere 10(4):e02693.

Nielsen, S. E., G. B. Stenhouse, and M. S. Boyce. 2006. A habitat-based framework for grizzly bear conservation in Alberta. Biological Conservation 130:217–229.

Nuwer, R. 2012. Safe passage for pronghorn. New York Times. 17 Oct 2012. https://green.blogs.nytimes.com /2012/10/17/safe-passage-for-pronghorns/.

Obbard, M. E., M. B. Coady, B. A. Pond, J. A. Schaefer, and F. G. Burrows. 2010. A distance-based analysis of habitat selection by American black bears (*Ursus americanus*) on the Bruce Peninsula, Ontario, Canada. Canadian Journal of Zoology 88:1063–1076.

Parks Canada. 2014. Banff National Park wildlife crossings research and monitoring. https://www.pc.gc.ca/en/pn -np/ab/banff/info/gestion-management/enviro/transport /tch-rtc/passages-crossings/recherche-research. Accessed 28 June 2016.

Proctor, M. F., W. F. Kasworm, K. M. Annis, A. G. MacHutchon, J. E. Teisberg, T. G. Radandt, and S. Servheen. 2018. Conservation of threatened Canada-USA trans-border grizzly bears linked to comprehensive conflict reduction. Human–Wildlife Interactions 12(3):348–372.

Proctor, M. F., D. Paetkau, B. N. McLellan, G. B. Stenhouse, K. C. Kendall, R. D. Mace, W. F. Kasworm, C. Servheen, C. L. Lausen, M. L. Gibeau, et al. 2012. Population fragmentation and inter-ecosystem movements of grizzly bears in western Canada and the northern United States. Wildlife Monographs 180:1–46.

Pulsford, I., J. Fitzsimons, and G. Wescott, editors. 2013. Linking Australia's landscapes: lessons and opportunities large-scale conservation networks. CSIRO, Collingwood, Victoria, Australia.

Ray, J. C., D. B. Cichowski, M.-H. St-Laurent, C. J. Johnson, S. D. Petersen, and I. D. Thompson. 2014. Conservation status of caribou in the western mountains of Canada: protections under the species at Risk Act, 2002–2014. Rangifer 35:49–80.

Sanderson, E. W., M. Jaiteh, M. A. Levy, K. A. Redford, A. W. Wannebo, and G. Woolmer. 2002. The human footprint and the last of the wild. Bioscience 52:891–904.

Saintato, M. 2015. Protected areas in the United States too small, disconnected to preserve biodiversity, studies find. Earth Island Journal. 4 June 2015. https://www .earthisland.org/journal/index.php/articles/entry /protected_areas_in_usa_too_small_disconnected _to_preserve_biodiversity/.

Shepard, B. B., B. B. May, and W. Urie. 2005. Status and conservation of westslope cutthroat trout within the western United States. North American Journal of Fisheries Management 25:1426–1440.

Shockman, E. 2016. Climate change is a huge threat to our national parks. Public Radio International. 19 June 2016. http://www.pri.org/stories/2016-06-19/climate-change-huge-threat-our-national-parks.

Sorribas, M. V., R. C. D. Piava, J. M. Melack, J. M. Bravo, C. Jones, L. Carvalho, E. Beighley, B. Forsberg, and M. H.

Costa. 2016. Projections of climate change effects on discharge and inundation in the Amazon Basin. Climate Change 136:555–570.

Stewart, F. E., N. A. Heim, A. P. Clevenger, J. Paczkowski, J. P. Volpe, and J. T. Fisher. 2016. Wolverine behavior varies spatially with anthropogenic footprint: implications for conservation and inferences about declines. Ecology and Evolution 6:1493–1503.

Theobald, D. M. 2001. Land-use dynamics beyond the American urban fringes. Geographical Review 91:544–564.

Watson, N. W. 2004. Integrated river basin management: a case for collaboration. International Journal of River Basin Management 2:243–257.

Weber, B. 2004. The arrogance of America's designer ark. Conservation Biology 18:1–3.

Wildlife Conservation Society. 2017. Amazon waters. http://amazonwaters.org/the-initiative/the-strategy/. Accessed 8 Sep 2017.

World Population Review. 2016. Nevada population. http:// worldpopulationreview.com/states/nevada-population/. Accessed 28 June 2016.

Wynn-Grant, R. J., J. Ginsberg, C. W. Lackey, E. Sterling, and J. P. Beckmann. 2018. Risky business: modeling mortality risk near the urban-wildland interface for a large carnivore. Global Ecology and Conservation 16:e00443.

Yellowstone to Yukon. 2019. Cabinet-Purcell Mountain Corridor. https://y2y.net/work/where-by-region/cabinet -purcell-mountain-corridor. Accessed 8 Aug 2019.

——20—— Part IV Synthesis
Translating Landscape Ecology to Management

DAVID M. WILLIAMS

Having established the foundational content and considerations essential for wildlife managers and landscape ecologists to work more effectively together to accomplish landscape-scale conservation (Parts II and III), Part IV transitions to explore how those concepts and strategies are applied. Chapters 15 and 18 explore novel ways that landscape analyses can be coupled with decision processes to facilitate strategic management of broad landscapes. Chapters 16, 17, and 19 explore how existing entities (joint ventures, Landscape Conservation Cooperatives [LCCs], and nongovernmental organizations [NGOs]) are uniquely positioned to facilitate wildlife management across broad spatial extents. Those entities leverage partnerships and the unique roles organizations can play to address many of the challenges outlined in Part III.

In Chapter 15, Keller proposes a way of assessing and managing landscapes based on four attributes that influence species occurrence: age, size, configuration, and context. Those attributes can be used to identify threshold conditions for species presence (e.g., too old, too small, poorly shaped, too isolated). These landscape attributes must be quantified to be useful in an assessment tool and thus are subject to many of the previously discussed issues for scale (see

Chap. 8). Keller reminds us that all of these landscape attributes should be evaluated relative to the species and ecological process of interest, and habitat management is most effective when it matches the grain and scale (see Chap. 4) of the landscape to which the population or organism is responding. Changes in scale or classification system, however, can affect the predicted relationships between occurrence and these landscape attributes (e.g., fine-scale features like individual trees not captured by landcover data). Thus Keller introduces a term to describe the range of scale and grain over which the pattern–process relationships are consistent: *domain*. Domain, then, represents the range of resolution and level of classification most appropriate for habitat analyses because it is less sensitive to changes in scale.

Keller provides a set of questions to guide strategic management and that demonstrate how, when mapped, thresholds in species occurrence related to these landscape attributes can be used to identify land parcels for protection or habitat improvement, what management is needed, and whether seemingly good habitat is isolated or represented in adjacent areas. Because age is an important covariate in explaining these species distributions, managers can also evaluate how long habitat manipulations will last and how the system will change over time. Finally, in addition to identifying the potential for gains in habitat, Keller stresses how such spatial models can be used to prioritize opportunities to partner with adjacent landowners by identifying the potential gain from adding new parcels or manipulating existing lands.

In Chapter 18, Lott et al. discuss a similar mapping approach identifying Priority Areas for Conservation for golden-winged warblers (*Vermivora chrysoptera*) in the Appalachian Mountains, USA. Their approach involved acquiring presence data for the golden-winged warbler, developing spatial models with habitat covariates related to probability of occurrence, and predicting habitat use across a large landscape. Of particular interest is their discussion on the implications of the management decision context for how errors (false positives or negatives) in classification of probability of occurrence data should be treated. If the management objective is to identify or create a reserve network, false positives should be avoided because managers do not want a model that predicts areas to have a species when it does not. Similarly, if the objective is to increase important habitat, managers want to avoid false negatives so that important habitat associations are not missed.

Lott et al. extended their model in two additional ways that demonstrate how these processes can be important means for identifying needs or areas with potential for success. First, similar to Keller, they identified collaborative opportunities using their model. Lott et al. identified that much of the surrounding landscape contained suitable nesting habitat. Because that land was predominantly privately owned, they concluded that federal assistance programs for private landowners could play a major role in achieving management goals for golden-winged warblers. Second, they reduced the amount of priority habitat predicted by the model to areas where suitable management might occur. For example,

areas where forestry was not a management option or areas where restoration is impossible (e.g., open water) were removed. Removing areas predicted by the model represents an important practical accommodation that is illustrative of the approaches needed for wildlife managers and landscape ecologists to overcome barriers to landscape-level management.

Application of species occurrence models is only one component of translating landscape ecology to management. While habitat management for birds often occurs at local scales, migratory birds traverse and use habitat that can span continents. Effective management of those birds requires strategic planning, broad collaboration, and a diverse expertise of population, landscape, political, and social processes. Bird habitat joint ventures were first established for management of North American waterfowl (Soulliere and Al-Saffar, Chap. 16); however, the program has expanded to include other bird taxa. Joint ventures consist of a partnership of agencies, organizations, corporations, tribes, and individuals who are tasked with implementing bird conservation plans that link population processes with landscape attributes. Soulliere and Al-Saffar suggest that early in the program those efforts were based on limited scientific knowledge and lacked a landscape ecology perspective, and decisions were opportunistic rather than evidence based. An important shift has occurred since then, with landscape ecology principles driving the very structure of the organization. Beyond an organizational structure, joint venture organizations became leaders in the development of repeatable decision support tools that are grounded in population ecology, landscape ecology and change, and social science. Those tools include relevant objectives in a decision matrix, spatially explicit maps for each objective, parameter weighting, and an amalgamated map to direct habitat management. Tools like these can engage decision makers in bridging the gap between landscape ecology and wildlife management.

One of the challenges wildlife managers face when trying to implement these decision tools are jurisdictional constraints. Landscape Conservation Cooperatives serve to help wildlife managers cooperatively manage landscapes beyond jurisdictional boundaries. In Chapter 17, Jacobson et al. review the vision of the LCC program and the landscape conservation design framework for conservation planning. The 22 LCCs in the United States differ in conservation priorities, but all are working to maintain landscapes that will sustain both natural and cultural resources into the future. Landscape conservation design is a five-part framework for conservation planning that can serve as a model for others trying to bring a diverse group together to accomplish wildlife management across a broad landscape. The design consists of visioning; setting goals, objectives, and priorities; conducting landscape assessments and evaluating scenarios; implementing conservation actions; and monitoring and evaluating the effects of those actions (Table 17.1). Landscape ecology expertise is essential to that process.

Jacobson et al. highlight three LCC case studies focused on how landscape ecology principles translate to those conservation plans. In the Northwest Boreal LCC the role of the intervening landscape, or matrix, is highlighted as critical for multiple ecological processes of interest. Pattern, process, and scale explicitly shaped the criteria used to evaluate options and success. The Southern Rockies LCC is an example of how a diverse partnership working together was essential for climate change adaptation efforts. The final case study relates a project involving six LCCs working together to identify priority areas for wildlife and habitat conservation for the entire southeastern United States. The authors stress that the diversity of such a large group can make it difficult to align conservation goals. An overarching initiative can provide the structure to keep objectives aligned. All three case studies highlight that often the challenges to implementing landscape-scale management are social and not scientific. Intentionally integrating managers, scientists, and or-ganizations often and early can keep the products applicable.

Nongovernmental organizations have been integral to promoting and supporting shifts in management toward large-landscape conservation. In Chapter 19, Hilty et al. use case studies to illustrate some of the unique aspects of NGOs that enable them to facilitate wildlife management across broad landscapes. While NGOs may have limitations of staff and income sources (compared to agencies) and cannot make policy, they are able to engage an array of approaches to landscape management. They can join with managers and landscape ecologists for collaborative research, act independently, advocate with stakeholders, fundraise, and litigate. As part of the integrated river basin management for the Amazon Basin, NGOs were able to work across significant jurisdictional boundaries (countries) because they had staff positioned in multiple countries that shared the same objectives, had good communication, and shared financial systems. The NGOs were also able to successfully engage large donors. Finally, NGO involvement in Nevada helped fund an initial project on black bears that ultimately moved from research to landscape-scale conservation and management. The NGO used the data to advocate for the project and for conservation actions in ways that other partners cannot. Hilty et al. point out that many NGOs play an important role in holding governments to their policies and best management practices.

Translating landscape ecology practices to management is a daunting and essential task. While there are some barriers to effective wildlife management across broad landscapes, many hurdles are challenges that can be overcome. The wildlife manager will benefit from shaping their thinking about the ecological processes of wildlife from a landscape ecology perspective, by taking the time to learn the theory and vocabulary, and by taking advantage of the many advances in analyses that allow us to relate landscape patterns to ecological processes. The landscape ecologist will benefit from communicating with the wildlife manager early and often to ensure evaluated relationships translate into practicable management. Finally, neither group can overcome many hurdles simply by learning from the other. They must work together and learn from examples that demonstrate how landscape ecology analyses can be used in the decision-making processes of managers. They should participate in and model new efforts after existing collaborations that are successfully accomplishing wildlife management and conservation across broad landscapes by collaborating with agencies, organizations, tribes, corporations, and individuals.

Index